269
Topics in Current Chemistry

Editorial Board:
V. Balzani · A. de Meijere · K. N. Houk · H. Kessler · J.-M. Lehn
S. V. Ley · S. L. Schreiber · J. Thiem · B. M. Trost · F. Vögtle
H. Yamamoto

Topics in Current Chemistry
Recently Published and Forthcoming Volumes

Sulfur-Mediated Rearrangement II
Volume Editor: Schaumann, E.
Vol. 275, 2007

Sulfur-Mediated Rearrangement I
Volume Editor: Schaumann, E.
Vol. 274, 2007

Bioactive Conformation II
Volume Editor: Peters, T.
Vol. 273, 2007

Bioactive Conformation I
Volume Editor: Peters, T.
Vol. 272, 2007

Biomineralization II
Mineralization Using Synthetic Polymers and Templates
Volume Editor: Naka, K.
Vol. 271, 2007

Biomineralization I
Crystallization and Self-Organization Process
Volume Editor: Naka, K.
Vol. 270, 2007

Novel Optical Resolution Technologies
Volume Editors:
Sakai, K., Hirayama, N., Tamura, R.
Vol. 269, 2007

Atomistic Approaches in Modern Biology
From Quantum Chemistry
to Molecular Simulations
Volume Editor: Reiher, M.
Vol. 268, 2006

Glycopeptides and Glycoproteins
Synthesis, Structure, and Application
Volume Editor: Wittmann, V.
Vol. 267, 2006

Microwave Methods in Organic Synthesis
Volume Editors: Larhed, M., Olofsson, K.
Vol. 266, 2006

Supramolecular Chirality
Volume Editors: Crego-Calama, M., Reinhoudt, D. N.
Vol. 265, 2006

Radicals in Synthesis II
Complex Molecules
Volume Editor: Gansäuer, A.
Vol. 264, 2006

Radicals in Synthesis I
Methods and Mechanisms
Volume Editor: Gansäuer, A.
Vol. 263, 2006

Molecular Machines
Volume Editor: Kelly, T. R.
Vol. 262, 2006

Immobilisation of DNA on Chips II
Volume Editor: Wittmann, C.
Vol. 261, 2005

Immobilisation of DNA on Chips I
Volume Editor: Wittmann, C.
Vol. 260, 2005

Prebiotic Chemistry
From Simple Amphiphiles to Protocell Models
Volume Editor: Walde, P.
Vol. 259, 2005

Supramolecular Dye Chemistry
Volume Editor: Würthner, F.
Vol. 258, 2005

Molecular Wires
From Design to Properties
Volume Editor: De Cola, L.
Vol. 257, 2005

Novel Optical Resolution Technologies

Volume Editors: Kenichi Sakai · Noriaki Hirayama · Rui Tamura

With contributions by
G. Coquerel · F. Faigl · E. Fogassy · D. Fujimoto · N. Hirayama
B. Kaptein · R. M. Kellogg · H. Murakami · H. Nohira · K. Sakai
R. Sakurai · J. Schindler · H. Takahashi · R. Tamura · T. Ushio
T. R. Vries · R. Yoshioka

 Springer

The series *Topics in Current Chemistry* presents critical reviews of the present and future trends in modern chemical research. The scope of coverage includes all areas of chemical science including the interfaces with related disciplines such as biology, medicine and materials science. The goal of each thematic volume is to give the nonspecialist reader, whether at the university or in industry, a comprehensive overview of an area where new insights are emerging that are of interest to a larger scientific audience.

As a rule, contributions are specially commissioned. The editors and publishers will, however, always be pleased to receive suggestions and supplementary information. Papers are accepted for *Topics in Current Chemistry* in English.

In references *Topics in Current Chemistry* is abbreviated Top Curr Chem and is cited as a journal.

Visit the TCC content at springerlink.com

ISSN 0340-1022
ISBN 978-3-540-46317-7 Springer Berlin Heidelberg New York
DOI 10.1007/978-3-540-46320-7

This work is subject to copyright. All rights are reserved, whether the whole or part of the material is concerned, specifically the rights of translation, reprinting, reuse of illustrations, recitation, broadcasting, reproduction on microfilm or in any other way, and storage in data banks. Duplication of this publication or parts thereof is permitted only under the provisions of the German Copyright Law of September 9, 1965, in its current version, and permission for use must always be obtained from Springer. Violations are liable for prosecution under the German Copyright Law.

Springer is a part of Springer Science+Business Media

springer.com

© Springer-Verlag Berlin Heidelberg 2007

The use of registered names, trademarks, etc. in this publication does not imply, even in the absence of a specific statement, that such names are exempt from the relevant protective laws and regulations and therefore free for general use.

Cover design: WMXDesign GmbH, Heidelberg
Typesetting and Production: LE-T$_E$X Jelonek, Schmidt & Vöckler GbR, Leipzig

Printed on acid-free paper   02/3100 YL – 5 4 3 2 1 0

Volume Editors

Dr. Kenichi Sakai

Toray Fine Chemical Co. Ltd.
Sepciality Chemical Research Laboratory
Ooe-cho 9-1,
Nagoya 455-8502, Japan
Kenichi_sakai@tfc.toray.co.jp

Prof. Rui Tamura

Kyoto University
Graduate School
of Human & Environmental Studies,
Sakyo-ku
Kyoto 606-8501, Japan
Tamura-r@mbox.kudpc.kyoto-u.ac.jp

Prof. Noriaki Hirayama

Tokai University
School of Medicine
Basic Medical Science
and Molecular Medicine
Bohseidai, Isehara,
Kanagawa 259-1193, Japan
hirayama@is.icc.u-tokai.ac.jp

Editorial Board

Prof. Vincenzo Balzani

Dipartimento di Chimica „G. Ciamician"
University of Bologna
via Selmi 2
40126 Bologna, Italy
vincenzo.balzani@unibo.it

Prof. Dr. Armin de Meijere

Institut für Organische Chemie
der Georg-August-Universität
Tammanstr. 2
37077 Göttingen, Germany
ameijer1@uni-goettingen.de

Prof. Dr. Kendall N. Houk

Department of Chemistry and
Biochemistry
University of California
405 Hilgard Avenue
Los Angeles, CA 90024-1589
USA
houk@chem.ucla.edu

Prof. Dr. Horst Kessler

Institut für Organische Chemie
TU München
Lichtenbergstraße 4
86747 Garching, Germany
kessler@ch.tum.de

Prof. Jean-Marie Lehn

ISIS
8, allée Gaspard Monge
BP 70028
67083 Strasbourg Cedex, France
lehn@isis.u-strasbg.fr

Prof. Steven V. Ley

University Chemical Laboratory
Lensfield Road
Cambridge CB2 1EW
Great Britain
Svl1000@cus.cam.ac.uk

Prof. Stuart L. Schreiber

Chemical Laboratories
Harvard University
12 Oxford Street
Cambridge, MA 02138-2902
USA
sls@slsiris.harvard.edu

Prof. Dr. Joachim Thiem

Institut für Organische Chemie
Universität Hamburg
Martin-Luther-King-Platz 6
20146 Hamburg, Germany
thiem@chemie.uni-hamburg.de

Prof. Barry M. Trost

Department of Chemistry
Stanford University
Stanford, CA 94305-5080
USA
bmtrost@leland.stanford.edu

Prof. Dr. F. Vögtle

Kekulé-Institut für Organische Chemie
und Biochemie
der Universität Bonn
Gerhard-Domagk-Str. 1
53121 Bonn, Germany
voegtle@uni-bonn.de

Prof. Dr. Hisashi Yamamoto

Department of Chemistry
The University of Chicago
5735 South Ellis Avenue
Chicago, IL 60637
USA
yamamoto@uchicago.edu

Topics in Current Chemistry
Also Available Electronically

For all customers who have a standing order to Topics in Current Chemistry, we offer the electronic version via SpringerLink free of charge. Please contact your librarian who can receive a password or free access to the full articles by registering at:

springerlink.com

If you do not have a subscription, you can still view the tables of contents of the volumes and the abstract of each article by going to the SpringerLink Homepage, clicking on "Browse by Online Libraries", then "Chemical Sciences", and finally choose Topics in Current Chemistry.

You will find information about the

- Editorial Board
- Aims and Scope
- Instructions for Authors
- Sample Contribution

at springer.com using the search function.

Foreword

After the end of the 20th century, the science of crystallization reached a truly exciting stage where new opportunities emerged in both theory and experiment. Various physical methods are capable of resolving the surface as well as the inside structure of crystals at the atomic level while new high-performance computing resources afford the capability of modeling the complex large-scale alignments necessary to simulate crystallization in real systems. As a result, the science of crystallization has shifted gradually from static to dynamic science and considerable progress now underlies the complex but beautiful crystallization process. I believe that if a definitive history of 21st century science is ever written, one of the highlights will be the science of crystallization.

This science has the following characteristics: infinite advances in sophistication, unlimited opportunities not only for intellectual excitement but also for industrial merit, strong collaboration with biology and material science, as well as with all areas of chemistry. The vast potential of crystallization as an important field of science is far beyond the simple technology of pharmaceutical industries during the 20th century.

Optical resolution was one small area of chemistry in the last century. This was more a technology than a science, largely because trial and error was the only method to obtain good results. However, the situation is now changing. There are so many appealing, hidden findings in the process of crystallization. Historically, crystallization began in an old laboratory in academia and then gradually shifted to industry. Now, it is making its comeback in academia due to several new research branches trying to discover what is going on during the crystallization process. I believe this field of science is now growing as a result of the wonderful coupling between industry and academia.

I read a prepublication draft of *Novel Optical Resolution Technologies*, and found that each one of these general characteristics of science had a reality and sharpness that I had not expected. While it was a sheer delight to revisit each of these triumphs guided by the wise insights and analyses found throughout the book. There is a good balance between the underlying historical material and the design and execution aspects of each topic.

I would like to congratulate Dr. Sakai on this fine edition. I wish him well in his further work as a scientist and as an author.

The University of Chicago   Hisashi Yamamoto

Preface

The prototype of two important optical resolution methods via crystallization, preferential crystallization and the diastereomeric salt formation method, was discovered by Louis Pasteur et al. in the middle of the 19th century. After gradual progress, in the latter half of the 20th century, these methods grew as one of the most important technologies in the chemical industry for obtaining various enantiopure chemical compounds useful to humankind, especially in the fields of the pharmaceuticals and other functional materials. Despite such remarkable progress, however, the molecular mechanism of the resolution phenomena was not sufficiently elucidated until recently. For the last decade, in accord with the substantial development of analytical tools, particularly in crystal structure solution, resolution chemistry has been revisited by many scientists and the situation has dramatically changed.

With this situation in mind, this volume is devoted to current topics concerning novel optical resolution technologies and is composed of four sections: (1) a theoretical approach to preferential crystallization, (2) the mechanism of preferential enrichment, (3) the practical use of preferential crystallization and various applications of the diastereomeric salt formation method, and (4) a statistical overview of the synthetic method of chiral compounds in pharmaceutical industry. In the first two sections, you will find an in-depth interpretation of preferential crystallization and the mechanism of a newly found resolution phenomenon called preferential enrichment. In the third section, you will see practical examples of the preferential crystallization and the diastereomeric salt formation method, which help one understand these mechanisms. In the final section, you will find the current situation of the resolution method in the pharmaceutical industry. Due to the limited space, optical resolution using the inclusion complexation method was excluded.

The editors would be delighted if this volume helps you better understand contemporary resolution chemistry and can be used to reinforce your chemistry methodology.

The editors would like to thank all authors in this volume and are obliged to Professor Hisashi Yamamoto of The University of Chicago for his encouragement to complete this volume.

Toray Fine Chemicals Co. Ltd., Nagoya, September 2006   Kenichi Sakai
Tokai University School of Medicine, Isehara   Noriaki Hirayama
Kyoto University, Kyoto   Rui Tamura

Contents

Preferential Crystallization
G. Coquerel 1

Mechanism and Scope of Preferential Enrichment,
a Symmetry-Breaking Enantiomeric Resolution Phenomenon
R. Tamura · H. Takahashi · D. Fujimoto · T. Ushio 53

Racemization, Optical Resolution
and Crystallization-Induced Asymmetric Transformation
of Amino Acids and Pharmaceutical Intermediates
R. Yoshioka 83

Advantages of Structural Similarities of the Reactants
in Optical Resolution Processes
F. Faigl · J. Schindler · E. Fogassy 133

Dutch Resolution of Racemates
and the Roles of Solid Solution Formation and Nucleation Inhibition
R. M. Kellogg · B. Kaptein · T. R. Vries 159

New Resolution Technologies Controlled
by Chiral Discrimination Mechanisms
K. Sakai · R. Sakurai · H. Nohira 199

Molecular Mechanisms of Dielectrically Controlled Resolution (DCR)
K. Sakai · R. Sakurai · N. Hirayama 233

From Racemates to Single Enantiomers –
Chiral Synthetic Drugs over the last 20 Years
H. Murakami 273

Author Index Volumes 251–269 301

Subject Index 311

Contents of Volume 265

Supramolecular Chirality

Volume Editors: Mercedes Crego-Calama · David N. Reinhoudt
ISBN: 3-540-32151-9

Chiral Spaces in Supramolecular Assemblies
A. Scarso · J. Rebek, Jr.

Dynamic Helical Structures: Detection and Amplification of Chirality
K. Maeda · E. Yashima

Supramolecular Chirogenesis in Host–Guest Systems Containing Porphyrinoids
V. V. Borovkov · Y. Inoue

Supramolecular Chirality in Coordination Chemistry
G. Seeber · B. E. F. Tiedemann · K. N. Raymond

Dynamic Chirality: Molecular Shuttles and Motors
D. A. Leigh · E. M. Pérez

Supramolecular Surface Chirality
K.-H. Ernst

Supramolecular Chiral Functional Materials
D. B. Amabilino · J. Veciana

Preferential Crystallization

Gérard Coquerel

UC2M2, UPRES EA 3233, Université de Rouen—IRCOF,
76821 Mont Saint Aignan Cedex, France
gerard.coquerel@univ-rouen.fr

1	Foreword	3
1.1	Historical Background	3
1.2	Structure of this Contribution	4
2	**Nature of the System**	4
3	**Definition and Detection of a Conglomerate**	5
3.1	Definition	5
3.2	How to Detect a Conglomerate?	9
3.3	Competition Between Stable, Metastable and Unstable States, Impact of Polymorphism and Formation of Solvates	15
3.4	Statistics About Conglomerate Formation	15
4	**Benefits of Conglomerates**	16
4.1	Optimal Enantiomeric Purification	16
4.2	Resolution by Shape Recognition	17
4.3	Preferential Nucleation	18
4.4	Resolution by Using Replacing Crystallization	18
5	**Preferential Crystallization (PC)**	19
5.1	PC in Binary Systems	19
5.2	Heterogenous Stable and Metastable Equilibria which Govern the Entrainment Effect in a Ternary System R–S–A (A = Solvent or Homogeneous Mixture of Solvents)	24
5.3	Entrainment Effect in Solution	28
5.3.1	Ideal Seeded Isothermal Preferential Crystallization (SIPC)	28
5.3.2	Ideal Seeded Polythermic Preferential Crystallization (S3PC)	29
5.3.3	Ideal Auto-Seeded Polythermic Programmed Preferential Crystallization (AS3PC)	30
5.4	Cyclicity of the Operations	30
5.5	Preferential Crystallization of Solvates in a Ternary System	32
5.6	Preferential Crystallization in a Reciprocal Ternary System or System of Order Greater Than 3	33
5.7	Comparison Between the Different Processes (SIPC, S3PC, and AS3PC) Under Real Operative Conditions	34
5.8	The Seeding Problems	34
5.9	Cooling Program	36
5.10	Stirring Mode and Stirring Rate	37
5.11	Metastable Zones (for the Crystallizing Enantiomer and the Noncrystallizing Enantiomer)	37

6	Yield and Practical Limitations of PC ("Internal" and "External" Parameters)	42
6.1	PC in a System with a Detectable Metastable Racemic Compound	43
6.2	R–S Epitaxy	45
6.3	Irreversible Adsorption on a Given (hkl) Orientation	45
6.4	Solid Solution Between Enantiomers	46
6.5	Surfaces of the Seeds	46
6.6	PC and Polymorphism	46
7	Simultaneous Crystallization with a Controlled Crystal Size (Preferential in CSD)	48
8	Preferential Crystallization and In Situ Racemization	48
9	Continuous Process	48
10	Conclusion and Perspectives	49
References		50

Abstract This contribution focuses primarily on chiral discrimination in the solid state, i.e. the formation of conglomerates, how to detect them and the various benefits that can be retrieved from chiral recognition in crystal lattices.

The core of the work is devoted to phase diagrams (mainly binary and ternary). The stable and metastable heterogeneous equilibria are depicted and used to understand different variants of preferential crystallization and how the entrainment effect proceeds. Seeded and auto-seeded processes are analyzed minutely. The seeding procedure has a significant impact on the control of stereoselective secondary nucleation and crystal growth.

The review ends with a systematic analysis of the reasons that can put, sometimes, severe limitations on the entrainment effect.

Keywords Conglomerate · Entrainment · Phase diagrams · Seeding

Abbreviations

PC	preferential crystallization
P	pressure
T	temperature
T_B	highest temperature of the AS3PC process (a suspension is obtained)
T_F	lowest temperature of the preferential crystallization process
T_L	temperature of dissolution of the racemic fraction of a mixture of enantiomers
T_c	critical temperature
T_{homo}	temperature of homogenization of the system
$T_e, T_\pi, T_\varepsilon$	temperatures of the binary eutectic, peritectoid and eutectoid, respectively
$\langle S \rangle; \langle R \rangle$	crystallized enantiomers S and R, respectively
$\langle RS \rangle$	crystallized racemic compound RS
V	achiral solvent
a and b	acid and base, respectively
R_a and S_a	couple of enantiomers with an acidic character
R_b and S_b	couple of enantiomers with a basic character

(R)$_{solv.}$ and (S)$_{solv.}$	solvated R and solvated S molecules, respectively
e	composition of the binary eutectic liquid between enantiomers (e.e. = 0)
ε	ternary eutectic point
TfR = TfR	melting temperature of the enantiomers
SHG	second harmonic generation
e.e.	enantiomeric excess = (R−S)/(R+S)
X_F	final composition at the end of the entrainment
SIPC	seeded isothermal preferential crystallization
S3PC	seeded polythermic programmed preferential crystallization
AS3PC	auto-seeded polythermic programmed preferential crystallization
L	point on the phase diagram corresponding to the complete fusion (binary system) or dissolution (ternary system) of the racemic mixture only (i.e. the whole enantiomeric excess remains as a solid phase)
Z	$d(T_{homo})/d(e.e.)$ at $(+/-)$ = constant: parameter quantifying the temperature versus solubility variation of the enantiomer in excess for a constant concentration in racemic solute
e.e.$_{Bmax}$	the maximum attainable enantiomeric excess in the mother liquor at the beginning of the entrainment fulfilling the condition of cyclicity
e.e.$_{Fmax}$	the maximum attainable enantiomeric excess in the mother liquor at the end of the entrainment. e.e.$_{Fmax}$ > e.e.$_{Bmax}$
H (and H′)	composition of the solution at T_{homo}
B (and B′)	composition of the solution at T_B
U	$(T_L - T_F)/(T_{homo} - T_L)$ parameter revealing both: how steep is the solubility curve and the supersaturation of the medium attainable at the end of the entrainment respecting the condition of cyclicity
α_{mol}	molar solubility of the racemic mixture/molar solubility of the enantiomer = $S(\pm)/S(+) = S(\pm)/S(-)$
Um.eq	$(T_L - T_Fm.eq)/(T_{homo} - T_L)$ ideal thermodynamic parameter which accounts for the ideal cooling along metastable equilibria and the condition of cyclicity of the operations
M_{Total}	total mass of the system
C (and C′)	point representative of the overall synthetic mixture at the end of the entrainment
SMB	simulated moving bed
Fm.eq	metastable solubility of an enantiomer that allows the condition of cyclicity to be fulfilled for an ideal entrainment

1
Foreword

1.1
Historical Background

In 1866, Gernez [1], one of Pasteur's students, was the first to detect the entrainment effect. This discovery did not arouse much attention before A. Werner (Nobel Price 1913) carried out his famous work on chiral metal complexes. Duchinsky (1936) and later (in the 1950s) Amiard and Velluz revitalized interest in the resolution by preferential crystallization and its application for large-scale production (several thousands of tons per year). The first rational approach to the preferential crystallization process has to be credited to Secor [2]. Later, Nohira's group introduced new ideas on partial salt formation which provide clear benefits in terms of productivity and robustness of the process. Jacques and his co-workers issued in 1981 the first edition of the famous book "Enantiomers, Racemates, and Resolution" [3] which helped once again to renew interest in resolution by entrainment. In 1994, the research group led by G. Coquerel in Rouen performed a thorough analysis of preferential crystallization by using polythermic stable and metastable heterogeneous equilibria. On the basis of this analysis the so-called "auto-seeding" was proposed. In the author's opinion, the story is not finished here; several improvements will probably be made in the future making preferential crystallization even more attractive at the laboratory as well as the industrial scale ...

1.2
Structure of this Contribution

Before discussing preferential crystallization (PC) in detail, various prerequisites are examined.

The nature of the binary heterogeneous systems in which the PC takes place are outlined.

In sequence, the definition, detection methods and benefits of the paramount concept of conglomerate will be examined.

The PC is then presented in three different variants; firstly in its simplest context, i.e., in an ideal binary system. Necessary fundamentals in ternary heterogeneous equilibria are also presented; then three ideal PC applications in solution are detailed by using the three variants. Comparisons between the variants and the rationale behind PC limitations end the overview on this preparative resolution process.

Note that chirality resulting from assemblage of nonasymmetric chemical entities such as metal–organic complexes (e.g. Werner's complexes) are not considered throughout this contribution but the majority of paragraphs also apply to this class of compounds.

2
Nature of the System

In addition to nonchiral solvents, it is supposed hereafter that the system is composed of mirror image components which, in the domain investigated cannot inter-convert. In other words, a left-handed (right-handed) molecule will remain left-handed (right-handed) in the P, T, solvent(s) domain investigated.

It will also be postulated that the systems investigated contain every couple of chiral molecule. The composition spaces are then symmetrical with reference to the racemic composition [4, 5]. For instance, we can differentiate the following systems:

N° 1 R–S binary system of enantiomers.
N° 2 R–S–V ternary system composed of the couple of enantiomers and an achiral solvent V.
N° 3 R_ab–S_ab–V ternary system composed of: an achiral solvent V and the couple of salts formed by the acidic chiral molecules and an achiral base.
N° 4 R_ba–S_ba–V ternary system composed of: an achiral solvent V and the couple of salts formed by the basic chiral molecules and an achiral acid.
N° 5 R_a–S_a–b–V quaternary system composed of: an achiral solvent V and the couple of enantiomers (acidic chiral molecules) and an achiral base.
N° 6 R_b–S_b–a–V quaternary system composed of: an achiral solvent V and the couple of enantiomers (basic chiral molecules) and an achiral acid.
N° 7 $R_aS_aR_bS_b$ reciprocal ternary system composed of two couples of enantiomers (one with an acidic character, the other with a basic character).
N° 8 $R_aS_aR_bS_b$–V reciprocal quaternary system composed of two couples of enantiomers (one with an acidic character the other with a basic character) and an achiral solvent V.

It is worth noting that the number of solvents can be increased, adding up in the order of the heterogeneous system; a great variety of heterosolvates can therefore be contemplated. Furthermore, extra achiral components can participate in the system so that co-crystals of chiral species can be obtained [6]. As the couple of enantiomers can also be added there is no conceptual limit in the order of the system! ...

Obviously, system N° 1 is a sub-set of all the other systems (N° 2–8). It is worth noting that system N° 5 differs from system N° 3 by the possibility of departure from stoichiometry; indeed, in system N° 5 any base/acid ratio can be considered. This is an important difference since abnormal salt [7] or non-congruent salt (solvated or not) can be obtained as crystallized phases [8]. In a similar way, system N° 6 enlarges the possibilities encompassed in system N° 4, i.e. system N° 4 is a particular section of system N° 6.

3
Definition and Detection of a Conglomerate

3.1
Definition

In binary systems a conglomerate is simply a eutectic mixture. This definition remains correct for some systems of higher degrees, but this definition cannot hold when considering solvates, adducts, noncongruent solubility compounds, etc. A more general definition is therefore needed:

A conglomerate is a mixture of mirror-image crystallized phases exhibiting symmetrical enantiomeric excesses.

Figures 1a–m illustrate examples or counter examples of this definition.

Figures 1a–h correspond to system N° 1. Figure 1a represents a stable conglomerate and refers to a eutectic mixture. The two pure enantiomers co-exist as separated crystallized phases for any temperature below T_e. Figure 1b,c shows that the competition between the free enthalpy of the racemic compound and that of the mixture of enantiomers is temperature dependent. As illustrated in Fig. 1b, above T_π (temperature of the peritectoid invariant) the conglomerate is stable. Below T_π the racemic compound is stable, the conglomerate is therefore metastable. By contrast, in Fig. 1c the conglomerate is stable below T_ε (temperature of the eutectoid invariant) and the racemic compound is stable above T_ε. These invariants, corresponding to reversible heterogeneous reactions, should not be confused with polymorphism since three phases are in equilibrium at T_π and T_ε instead of two only for polymorphism.

The crystallization of a conglomerate without solid solution represents a nice (and up to now unpredictable) phenomenon of complete chiral discrimination during a self-assembling process. There are some cases in which the chiral discrimination is less selective: in Fig. 1d,e various cases of conglomerates associated with different domains of partial solid solutions are illustrated. The solid state chiral discrimination is poorer than in case 1a; these data have been used to quantify the chiral character of an asymmetric molecule [9]. When a complete solid solution exists there is no longer any conglomerate since a single phase only exists between the two chiral components. Nevertheless, at low temperature a chiral discrimination can appear. This phenomenon results from a solid–solid miscibility gap as illustrated on Fig. 1f. At a temperature above T_c there is no conglomerate since a complete solid solution spans over the complete range of composition between the pure enantiomers. Below T_c, there is a stable conglomerate since two mirror-image solid solutions co-exist. Their symmetrical compositions are temperature dependent and read on the solvus curve. This case should not be confused with Fig. 1g and Fig. 1h where a racemic compound exists inside the miscibility gap at the solid state (Fig. 1g) or as an ordered superstructure of the complete solid solution (Fig. 1h).

Figure 1i–m illustrate situations which are encountered in systems N° 2, 3, and 4. Figure 1i represents a classical stable conglomerate of asolvated pure components whatever the temperature below the binary eutectic. Figure 1k illustrates a ternary system corresponding to the binary system represented in Fig. 1d at temperature T_2. Figure 1j shows a stable nonsolvated racemic compound whatever the temperature. Nevertheless, in solvent V and below the temperature of the reversible decomposition of the mirror image solvates, a conglomerate is stable. It is worth noting that these solvates might only be

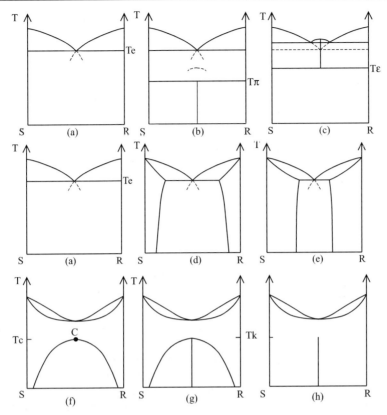

Fig. 1 a Stable conglomerate whatever the temperature. **b** Stable conglomerate for $T > T\pi$. **c** Stable conglomerate for $T < T\varepsilon$. **d,e** Stable conglomerate with partial miscibility at the solid state. **f** Complete solid solution at $T_c < T < T$ melting and a conglomerate with partial solid solution at $T < T_c$. **g** Complete solid solution at $T_c < T < T$ melting and a racemic compound inside the miscibility gap for $T < Tk$ **h** Complete solid solution at $T_c < T < T$ melting. For $x = 0.5$ and $T < Tk$ a superstructure exists

stable in equilibrium with their saturated solutions. Indeed, they can undergo a swift desolvation when isolated from their mother liquor [10].

Figure 1l depicts a system of type N° 5 in which a couple of enantiomers R and S of weak acidic character can be converted into solvated salts only if an excess of strong base is introduced in the medium. Note that for the sake of clarity of this isothermal quaternary system, components b (the strong base) and V (the solvent) have been expanded to form two parallel segments. The corresponding inserts show the geometrical transformation applied to the original tetrahedron; the two symmetrical S_a-b-V and R_a-b-V ternary systems are thus well separated. At this temperature, C_R and C_S are two enantiomerically pure solvate salts which form a stable conglomerate in this basic medium.

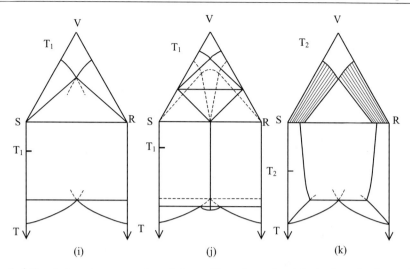

Fig. 1 i Fig. 1a in a ternary system. **j** A specific solvent V might lead to the formation of a conglomerate of solvates whereas the binary system corresponds to a stable racemic compound whatever the temperature. **k** Fig. 1d in a ternary system at $T = T_2$

These phases are called noncongruent solubility compounds [8], because C_R and C_S placed in the pure solvent V undergo a partial decomposition if a small amount of V is added, or a complete solvolysis if a great deal of solvent is added. In terms of phase diagrams, this means that the segment VC_R joining V (representative point of the nonchiral solvent) to the representative point of the solvated chiral phase C_R does not intersect the stable solubility curve r_1–r_2 of the compound $R_a bV$ but instead, it crosses the stable solubility curve of the R enantiomer.

Figure 1m represents a reciprocal quaternary system (system of type N° 8) composed of two enantiomers of a chiral acid, two enantiomers of a chiral base and an achiral solvent V. If, no solid solution nor solvate are postulated here, the system leads to five different possible solids: $\langle R_a R_b \rangle$, $\langle S_a S_b \rangle$, $\langle R_a S_b \rangle$, $\langle S_a R_b \rangle$ and the racemic compound $\langle R_a R_b S_a S_b \rangle$. This list corresponds to two couples of diastereomers and the double salt. If a couple of enantiomers, for example $\langle R_a R_b \rangle$, $\langle S_a S_b \rangle$ are thermally more stable than $\langle R_a S_b \rangle$, $\langle S_a R_b \rangle$ then the mixture of the four species can lead to the crystallization of a conglomerate of diastereomeric salts [11].

By changing the temperature, the other couple of enantiomers can become more stable (i.e. crux reversible decomposition [8]). By varying the nature of the solvent V', another couple of enantiomers can become less soluble and therefore give another conglomerate $\langle R_a S_b - V'_n \rangle$, $\langle S_a R_b - V'_n \rangle$.

The list of systems depicted here is by no means limitative, for instance one can enlarge the conglomerate screening domain by investigating the possible formation of co-crystals. An amine hydrochloride can form various (1–1), (1–

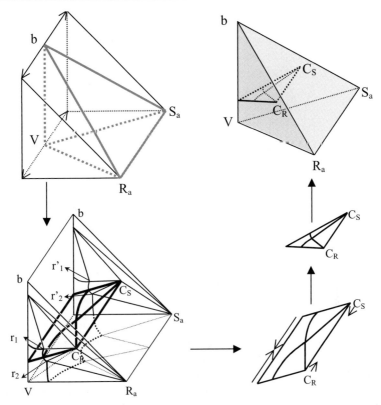

Fig. 1 | Presence of a stable conglomerate of noncongruent soluble compounds. VC_R and VC_S segments do not intersect the corresponding solubility curves (r_1r_2) of C_R and C_S ($r'_1r'_2$) resulting in the noncongruence of the compounds. The geometric transformation detailed by the successive arrows, is designed to facilitate the visualization of the pair of noncongruent salts

2), ... crystallized adducts or co-crystals with simple carboxylic acids such as benzoic acid, succinic acid, fumaric acid, etc. [6]

3.2
How to Detect a Conglomerate?

According to the definition of a conglomerate the two mirror-image phases should have the same intensive scalar properties (temperature of fusion, enthalpy of fusion, density, lattice parameters, solubility in a nonchiral solvent, in the case of solvate the same vapor pressure of decomposition, etc.) and when considering vectorial properties the same modulus but of opposite direction. Among organic compounds, some of these characteristics may not be accessible or difficult to measure. For instance, chemical decomposition can happen prior to melting, a solvate might only be stable in its mother liquor

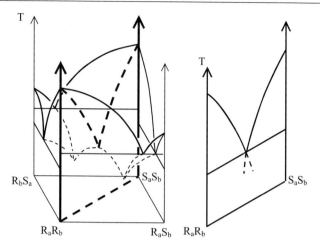

Fig. 1 m Isothermal reciprocal quaternary system (solvent V is rejected to the infinity) $R_aR_bS_aS_bV$. At this temperature the racemic mixture crystallizes as a R_aR_b–S_aS_b stable conglomerate.

(efflorescence), a salt can decompose under normal conditions of storage (CO_2 and/or H_2O sensitive). In practice, racemic compounds, conglomerates, and complete solid solutions are, by far, the most populated instances. Other alternatives such as "exotic" stoichiometries (2–1) (1–2) [12, 13] and partial solid solutions [14, 15] are rare. Here are listed various methods applicable to conglomerate detection:

(1) Because of hemihedrism [16, 17], shape recognition (Fig. 2) is a subtle way to detect conglomerates that need a minute amount of racemic mixture. Pasteur spotted the first conglomerate by using shape recognition of the tetrahydrated sodium ammonium tartrate—a significant proportion of particles exhibit the {111} hemihedral faces (space group $P2_12_12$) which makes single crystals of opposite chirality macroscopically differentiable. (±) Threitol (1,2,3,4-butanetetrol) obtained in ethanol under weak supersaturation is another chiral component that displays hemihedrism allowing a hand sorting of the enantiomers (G. Coquerel, unpublished results).

(2) Investigations on the binary phase diagrams of enantiomers allow us to spot conglomerates close to their melting point only. In the case of diagram 1b if T_ε is too far from the temperature of the molten state the poor diffusion rate at the solid state is likely to hinder the conglomerate ↔ racemic compound transition. Moreover, this method is basically restricted to systems whose components can stand fusion without chemical decomposition.

(3) Investigations in a ternary system (R and S enantiomers plus solvent) are of special interest since: (i) there is less restriction on the type of systems submitted to investigation (types N° 2, 3 and 4 see Sect. 2 above), (ii) the system is maintained in (or close to) thermodynamic equilibrium,

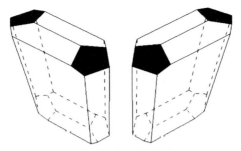

Fig. 2 Morphologies of Pasteur salts obtained at 15 °C in a slightly supersaturated aqueous solution with an excess of ammonia

(iii) it corresponds to data readily applicable for enantio-purification and PC. This method offers robust data yet it is time consuming for a preliminary conglomerate screening. When one enantiomer and the racemic mixture are available in sufficient quantities simple tests based on dissolution of the enantiomer deliver a fast answer on the nature of the solid(s). Indeed, a conglomerate of nonsolvated species without any partial solid solution implies that no dissolution nor any evolution in composition of the crystals occurs when particles of pure enantiomer are added to a saturated racemic solution. Obviously, only systems in which the pure enantiomer is less soluble than the 50 – 50 composition have to be considered. Figure 3a,c clearly show that starting from a stable racemic saturated solution, addition of a suspension of crystals of enantiomer in equilibrium with their saturated solution does not lead to any dissolution in the case of a conglomerate. By contrast, on Fig. 3b, the same experiment leads to a dissolution (total or partial depending on the quantity added) of the crystals of the pure enantiomer.

(4) It is possible to identify a conglomerate by dissolving a single particle isolated from the racemic mixture in a nematic phase [3]. When the particle contains an enantiomeric excess, the local composition leads to the formation of a cholesteric phase. This mesophase induces several optical effects (colors, striations, etc.) rendering the method very sensitive. Nevertheless, this technique needs some degree of training and is of interest for covalent compounds soluble in the nematic phase.

(5) Chromatographic (chiral GC or HPLC) or polarimetric measurements on single particles isolated from the racemic mixture can deliver evidence of formation of conglomerates. The former method being in general more sensitive than the latter. One problem is to make sure a single particle only is extracted and not a bundle of fibers which might be of random chirality.

(6) As a conglomerate is a physical mixture of single crystals, spectroscopic features of the solid particles are of prime importance in establishing the nature of the crystallized racemic mixture. In the case of a conglomerate and in the absence of partial solid solution, the IR, XRPD, ss NMR, raman patterns of the pure enantiomer and the racemic mixture should be superimposable.

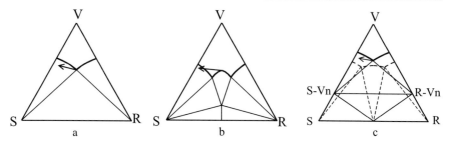

Fig. 3 Detection of a conglomerate by using the dissolution test. Addition of a suspension of ⟨S⟩ and ⟨S–Vn⟩ enantiomer to a saturated racemic solution leads to no dissolution in **a** and **c**. By contrast, in case **b** a partial or total dissolution of the crystal occurs

X-ray diffraction (Fig. 4a,b) is, among others, a method that can be recommended:

- It allows one to check that the phases are crystallized;
- It is very sensitive to even minor modifications of the crystal lattice such as those observed with partial solid solutions. These minor differences are significant if an internal standard is mixed with the solid phases to be compared;
- Temperature-resolved XRPD allows one to check if any peritectoid invariant (Fig. 1b) or eutectoid invariant (Fig. 1c) or desolvation occurs on heating and if the phenomena are reversible on cooling. These phenomena limit the stability domain of the conglomerate and the application of the preferential crystallization;
- XRPD analysis (as well as single crystal analysis-see below) can be carried out in the exact environment of the existing phases: closed chamber with a saturated vapor pressure in the solvent with "humid" solids (solid + its saturated solution for efflorescent solvate, CO_2 and H_2O sensitive salts, etc.).

(7) When a single crystal of sufficient size and quality for X-ray diffraction studies can lead to the resolution of the structure and when the representativeness of the single crystal from the bulk is checked (by comparing calculated and experimental XRPD patterns), a nonequivocal identification of the conglomerate is performed.

The crystallized phases which form conglomerates cannot possess symmetry operators of the second kind. That is to say: no center of symmetry, mirror or glide mirror and inverted axis (e.g. – 3; – 4; – 6) can exist in the crystal lattice. Only 65 out of 230 space groups are thus possible. Among them 11 pairs have mirror-related screw axes (e.g. $P4_1$ and $P4_3$ or $P6_122$ and $P6_522$); if for instance, one enantiomer crystallizes in the space group $P4_1$, its antipode crystallizes in the space group $P4_3$. The occurrence of the 65 space groups is by far not evenly distributed: the famous $P2_12_12_1$ represents ca. 58% of the

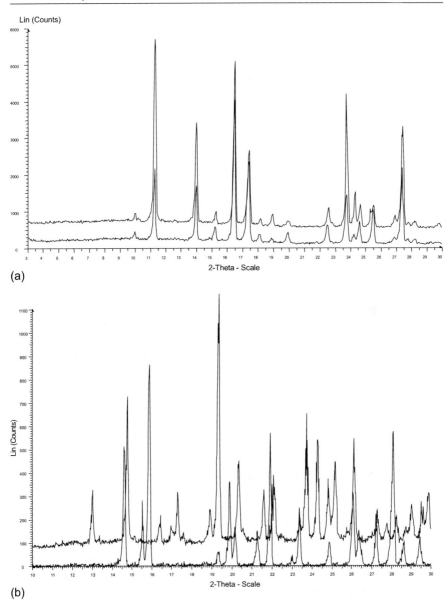

Fig. 4 a Superimposable XRPD patterns of the racemic mixture and the enantiomer means that the racemic mixture is a conglomerate. See [27] **b** Nonsuperimposable XRPD patterns of the racemic mixture and the enantiomer means that there is no conglomerate (see text for possible difficulties)

observed space groups for "small" asymmetric organic molecules (labeled by Zorky as the "whale" [18]). $P2_1$ ranks number two and accounts for almost 30%. If C2 and P1 space groups are added a 90 to 95% chance to encounter one of these four space groups is observed. The usual trend of low symmetry space groups among organic molecules is thus observed in the sub class of asymmetric molecules [19].

By contrast, a very small fraction of racemic compounds crystallizes in a chiral space group. A survey in the CCSD [20] shows the clear trend of centrosymmetric pairs of molecules at the solid state.

(8) The conglomerate being a physical mixture of crystals corresponding to a chiral space group makes possible the screening of solid-state discrimination by using Second Harmonic Generation (SHG). We will, first, briefly recall the principle of the technique. Under irradiation with a light beam of wavelength λ, materials with noncentrosymmetric crystal space groups generate a beam of wavelength $\lambda/2$ whose intensity depends on the $\chi^{(2)}$ tensor [21–24].

The main interest of the technique lies in the fact that in centrosymmetric structures all the elements of the $\chi^{(2)}$ tensor are zero. This is the case for the overwhelming majority of racemic compounds which are thus expected to be nonSHG materials. Conversely, crystallized samples exhibiting SHG must be composed of noncentrosymmetric crystals. Therefore, crystallized derivatives of a racemic mixture exhibiting a SHG positive effect are likely to crystallize as a conglomerate.

Diagnosis of the nature of a racemic mixture, conglomerate or not by using SHG, requires a small quantity only of the racemic mixture (a few mg typically) and there is a priori no need for comparison between results of SHG tests on the racemic mixture and the pure enantiomer. Therefore, the conglomerate screening can be undertaken even when the pure enantiomer is not available and particles are too small for direct tests (single crystal, chromatography, dissolution in nematic crystals). The search can thus be carried out at an early stage of the development of the molecule. The answer delivered is instantaneous and the method is nondestructive if the intensity of the laser beam is maintained within reasonable limits.

Besides the interest of the SHG method (including its compatibility for a high throughput screening), its reliability has appeared somewhat questionable. Two drawbacks affect the SHG detection of conglomerates. These are mainly related to the properties of the $\chi^{(2)}$ tensor: (i) nondetected SHG positive racemic mixtures and (ii) false-positive.

(i) Nondetected SHG positive materials can be attributed to:

- A $\chi^{(2)}$ parameter that is too small (i.e. a low hyperpolarizability of the molecule);
- An absorption at the wavelength of irradiation (e.g. 1064 nm) or at the wavelength of the re-emitted radiation (e.g. 532 nm);

- A nonactive SHG space group. Indeed, some chiral space groups are SHG inactive (those related to point groups 422, 622 and 432 [25]. However, these highly symmetrical groups are poorly represented among organic materials.

(ii) False-positive response can occur:

- For racemic compounds crystallizing in a noncentrosymmetric space group, e.g. $Pca2_1$ (not frequent);
- For racemic compounds crystallizing in a chiral space group, e.g. $P2_1$ (rare);
- If the sample is composed of too fine particles.

3.3
Competition Between Stable, Metastable and Unstable States, Impact of Polymorphism and Formation of Solvates

Any extensive detection of conglomerates should proceed by a systematic search. A simple comparison of spectroscopic data of the solids isolated from their mother liquors resulting from crystallizations without precaution can lead to an erroneous answer, i.e. missing a good opportunity. For instance:

- A racemic compound can be less stable than the corresponding conglomerate but according to the Oswald law of stages [26] it might appear first. A seeding by the enantiomer of the racemic mixture can reveal the stable chiral discrimination at the solid state;
- Polymorphism of the enantiomer can lead to the crystallization of one form when pure and to another form when mixed with its antipode;
- Efflorescence of solvates can jeopardize the detection of conglomerates with interesting properties.

In conclusion, despite numerous attempts, it must be stressed that prediction of the conglomerate is beyond the current possibilities. The physico-chemist must then rely on trial and error methods just as for polymorph screening.

3.4
Statistics About Conglomerate Formation

It is usually accepted that between 5 to 10% of the racemic mixtures crystallize as a conglomerate [3]. It is important to note that:

- This figure corresponds to stable conglomerates which are, most of the time, stable under ambient conditions. Only a few works deal with research of solvates at low temperature which could increase the number of conglomerates [27]. Even less investigations are published in systems described in Figs. 1i or 1j. Probable additional opportunities are then unfortunately missed ...

- These statistics show noticeable heterogeneities. For some components, an extensive screening is necessary to find a derivative crystallizing without solid solution nor racemic compound. Conversely, Collet [28] has pointed out that mandelic acid derivatives have a propensity to form conglomerates. The research group in Rouen [29] has also pointed out a series of 5-alkyl-5-aryl hydantoin derivatives which form eutectic mixtures with a high frequency.

Even if some racemic compounds appear much more thermally stable than the corresponding enantiomers (anti-conglomerate [9]), the difference in energy between the racemic compound and the enantiomers is usually small. The structural similarities between the crystal lattices of the racemic compounds and the corresponding enantiomers have long ago be recognized [30] and account, at least partially, for the close competition in terms of energy. Some controversies concerning Wallach's rule [31] have shown that the densities of the crystal lattices are not solely responsible for the overwhelming proportion of racemic compounds.

For a given chiral amine (or acid) dozens of salts can be tested, each of them in different solvents or mixture of solvents (the latter aimed at crystallizing heterosolvates) and at different temperatures. These multi-parameter screenings often result in one or several detections of stable conglomerates. Covalent chiral components with no possibility of salt formation in extreme media offer less chance (systems type N° 6 and N° 7). These components neither possess a sufficient acid or basic character nor undergo a chemical decomposition in extreme pH media. Nevertheless, a screening for co-crystals and solvates can also result in the detection of stable conglomerates.

4
Benefits of Conglomerates

Beside the fundamental scientific interest in understanding the reason(s) for spontaneous resolution, and application of the PC (also called entrainment), Sects. 4.1 to 4.4 are devoted to the additional practical interests of conglomerates.

4.1
Optimal Enantiomeric Purification

Inasmuch as no partial solid solution exists between the enantiomers in equilibrium, any mixture containing an enantiomeric excess can be quantitatively enantio-purified by selective crystallization. For conglomerates obtained in systems N° 2 to 8 listed in Section 2, the choice of an appropriate solvent and a careful management of the crystallization usually lead to an e.e. greater

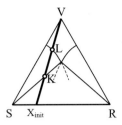

Fig. 5 The best enantiopurification of the mixture of composition X_{init} by recrystallization in the solvent V at temperature T1 consists of adding the exact amount of solvent V so that the overall synthetic mixture composition corresponds to point K. Along KL segment any solid-liquid mixture in equilibrium is composed of a saturated solution plus ⟨S⟩ crystals

than 99% or even above 99.5% with a quantitative yield. This is achievable at laboratory scale as well as at industrial scale without difficulties because the method relies on stable thermodynamic equilibria. Figure 5 illustrates the best conditions for purification purposes. The solvent is chosen so that the slurry is not too viscous and a "good" filterability is ensured (in practice the solvent is chosen according to the following rule: the higher the purity of the starting mixture to be purified, the poorer the solubility of the solute). The management of the crystallization is also a key step when there is a need to reach a high e.e. Seeding and an appropriate cooling rate are the usual general recommendations which apply. But more sophisticated means are sometime necessary (interfacial crystallization with two nonmiscible solvents, specific filtration procedure, etc.) when problems of crystallinity or retention of mother liquor in the filtration cake arise. Conglomerates obtained in systems N° 5 or 6 can provide more difficulties in tuning the enantiopurification step. Nevertheless, time spent on finding the optimal conditions (both thermodynamics and kinetics need adjustment) is also rewarded by a quantitative and easy-to-scale-up process.

4.2
Resolution by Shape Recognition

As mentioned above, Louis Pasteur performed in 1848 the first ever resolution by hand sorting of the tetrahydrated sodium ammonium tartaric salt at a temperature below 27 °C based on the shape recognition of hemihedrism [32].

Whereas it is possible to identify a conglomerate by shape recognition of hemihedral morphology, it might also help in some absolute configuration assignments in a series of similar components having isomorphous crystal structures [16]. Preparative resolution by shape recognition has not yet been exploited at the large scale; this method remains more one of historical interest than being actually useful in practice. Nevertheless, appropriate conditions of nucleation and crystal growth (including the nature of the solvent and tailor-made impurities to obtain large single crystals exhibiting hemihedrism) associated with a robot programmed for shape recognition and sorting could lead to a revival of this method.

4.3
Preferential Nucleation

When a racemic mixture crystallizes as a conglomerate, a chiral molecule that possesses extensive similarities with the ($\pm$) mixture can interfere, even when added in small amounts ($\ll 1\%$), with the nucleation rate of the two symmetrical chiral solutes. The method is schematized on Fig. 6. The "Rule of reversal" [33] states that the solute with the same absolute configuration as that of the additive undergoes the strongest disrupting effect on its nucleation rate.

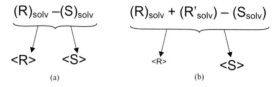

Fig. 6 Principle of the preferential nucleation. **a** Without chiral tailor-made additive, R and S nucleation rate and crystal growth rate are equal. **b** After addition of a chiral tailor-made additive such as R′, the crystallization of R is more strongly inhibited than that of ⟨S⟩

This method, based on kinetic effects, is of general applicability [34]. The first hundred milligrams of an enantiomer can easily be obtained by using this method when it crystallizes as a conglomerate and a similar derivative is available as a pure chiral co-solute.

Though it is not clear what exactly is a "similar molecule", the rule of reversal is an additional tool for absolute configuration assignment. "Similar molecules" often means species which are able to participate in the same strong bond network in the crystal lattice (ionic bonds if they exist, H-bonds) but differ by organic substituents involved in the so-called van der Waals interactions.

4.4
Resolution by Using Replacing Crystallization

Optical resolution by using replacing crystallization is another procedure designed to obtain a single enantiomer from a racemic mixture crystallizing as a conglomerate. This process is based on interactions between each of the enantiomers and an optically active co-solute whose structure is "similar" to that of the racemic mixture to resolve [35]. Moreover, the additive should be in similar concentration to that of the solute and be more soluble than the racemic mixture [36]. For instance DL-threonine crystallizing as a conglomerate in water shows stereoselective interactions in the presence of a significant amount of L-proline (approximately the same amount as that of threonine) so that D-threonine and L-threonine solubilities differ. In addition to the break in symmetry of the heterogeneous equilibria (the solubilities in that case) it seems that some agonist effects exist in the nucleation kinetics of the two enantiomers forming the conglomerate. This is consistent with the results obtained by preferential nucleation (Sect. 4.3) and justifies the efficiency of this method in the series of natural amino-acids.

5
Preferential Crystallization (PC)

5.1
PC in Binary Systems

So far, only a very few examples have been published concerning PC without solvent [37]. Two reasons explain this scarce applicability: (i) most of the organic materials have a poor stability, if any, at the molten state, (ii) the viscosity of the liquid phase makes the separation of the solid from the slurry difficult [38].

Notwithstanding the almost anecdotic character of PC as a stereoselective melt crystallization, we will start describing the stereoselective crystallization in the S-R binary systems. This preliminary analysis will be extended to ternary systems, quaternary systems, etc.

Figure 7 shows a binary system between two enantiomers which crystallize as a stable conglomerate without any partial solid solution. Starting from an

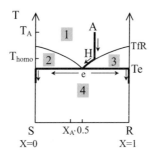

Fig. 7 System in thermodynamic equilibrium. Behavior on cooling for a 20% e.e. mixture in a binary system which always remains in equilibrium. 1: liquid, 2: ⟨S⟩ + liquid, 3: ⟨R⟩ + liquid, 4: ⟨S⟩ + ⟨R⟩

overall composition X_A at T_A (point A with a 25% enantiomeric excess of R — i.e. 65% R and 35% S corresponding to M mass units of <R> as enantiomeric excess), the system will be submitted to a slow cooling. From T_A to T_{homo} the system remains monophasic: only the liquid phase exists. If we suppose that the system remains in equilibrium, at T_{homo} the first crystal of ⟨R⟩ appears. From T_{homo} to T_e the solid phase ⟨R⟩ nucleates and grows; simultaneously the liquid phase composition evolves from 25% e.e. (point H) to the eutectic point (e) whose composition is 50% S—50% R (i.e. 0% e.e.). The temperature of the eutectic (e) corresponds to the lowest temperature for the liquid phase—when the system remains in equilibrium. At this constant temperature T_e, ⟨S⟩ and ⟨R⟩ crystallize simultaneously in equal proportion leading to a complete solidification of the medium. When the eutectic liquid is exhausted, i.e. transformed into a eutectic mixture of ⟨S⟩ and ⟨R⟩, the system resumes its cooling.

Figure 8a shows the same starting situation. Now the cooling is supposed to be swift so that the system skips the spontaneous crystallization of ⟨R⟩ from T_{homo} to T_e; the system is therefore out of equilibrium as soon as: $T < T_{homo}$. T_F is defined so that in the actual conditions of the experiment no crystallization takes place during the cooling step from T_e to T_F. Moreover, we suppose that for a given period of time (e.g. 2 hours) no spontaneous crystallization occurs. At T_F (point C), if the system is seeded by a small amount of pure crystals of ⟨R⟩, (the more supersaturated phase) it will induce the stereoselective secondary nucleation and crystal growth of ⟨R⟩. The composition of the liquid phase evolves from X_A (point C) to 0.5 and continues towards X_F (point F) as the stereoselective crystallization proceeds. The ideal final composition X_F is nothing else than the composition of the metastable liquidus of the R enantiomer at T_F. At point F, the mother liquor and the solid phase <R>

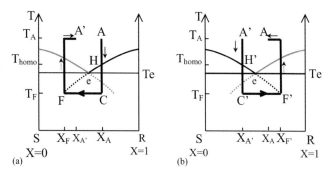

Fig. 8 Ideal Seeded and Isothermal Preferential Crystallization (out of equilibrium process). **a** Crystallization of ⟨R⟩. From A to C: fast cooling without spontaneous nucleation. At C: inoculation of the seeds. From C to F: stereoselective crystallization of the R enantiomer. At point F, the ⟨R⟩ crystals are separated from their liquid phase. **b** Crystallization of ⟨S⟩. Idem. S substitutes for R, A' for A, C' for C, F' for F

are separated. The liquid is re-heated up to T_A and then added with the same mass of racemic mixture as that harvested after this first entrainment, i.e. 2 M mass units. This ideal process leads to point A′ symmetrical to point A with reference to the vertical line $X = 0.5$. Figure 8b is symmetrical to Fig. 8a; starting from point A′ with an excess of M mass units of S enantiomer the rapid cooling down to T_F and the inoculation of very pure ⟨S⟩ seeds induce a specific isothermal crystallization of ⟨S⟩ which ideally yields 2 M mass units of the pure enantiomer S. Another addition of 2 M mass units of racemic mixture puts the system back to square one. The following operations just consist of repeating the cycle described above and is schematized in Fig. 8a,b. This cyclic process leads also to:

$$2\left(0.5 - X_{A'}\right)/0.5 = \left(X_{F'} - X_{A'}\right)/X_{F'} \tag{1}$$

$$2\left(0.5 - X_{A'}\right)/0.5 = \left(X_{F'} - X_A\right)/\left(X_{F'} - 0.5\right) \tag{2}$$

$$= \left[X_{F'} - \left(1 - X_{A'}\right)\right]/\left(X_{F'} - 0.5\right)$$

A continuous recycling of the liquid phase and the addition of 2 M mass units of racemic mixture after every crystallization allow us to resolve any quantity of racemic mixture without the aid of any resolving agent or the use of solvent. The ideal yield will depend on the supercooled state attainable for a sufficient period of time and the shape of the liquidus curves (stable and metastable). Usually, the composition versus temperature relation can fairly well be estimated by using the Schröder–Van Laar equation [39, 40]. In practice, the crystal growth rate decreases steadily from point C (respectively C′) to point F (respectively F′) where it reaches zero, so that waiting for the very maximum can be nonproductive and even dangerous because the counter enantiomer can start crystallizing fast at any time.

The purification of the crude crops can be carried out by the sweating method [38]; zone refining [41] can even be used to access a very high e.e.

Figure 9 illustrates a first variant of the so-called SIPC method (Seeded Isothermal Preferential Crystallization) detailed above. The S3PC process (Seeded Polythermic Programmed Preferential Crystallization) differs from

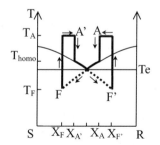

Fig. 9 Ideal Seeded Polythermic Programmed Preferential Crystallization (S3PC). Same procedures as those described in Fig. 8a,b except that the seeding is performed at T_{homo}

SIPC by the seeding temperature at T_{homo} rather than T_F. The inoculation is performed at the highest possible temperature where the crystals of the pure enantiomer ($\langle S \rangle$ or $\langle R \rangle$) can be in equilibrium with the saturated liquid. The crystallization spans over the $T_{homo} - T_F$ interval in temperature; the cooling rate is adapted to the crystal growth rate so that a smooth enlargement of crystals is favored at the expense of a noncontrolled high secondary nucleation rate.

Figure 10 illustrates a second variant of the SIPC method: the Auto-Seeded Polythermic Programmed Preferential Crystallization (AS3PC [14, 42, 43]). The main difference is that the highest temperature T_B is lower than T_{homo} but above T_e. The cyclic process will be explained by starting from point F' that is to say the liquid phase obtained at the end of the previous run. The liquid phase is heated to T_B so that: $T_L < T_B < T_{homo}$ and added with the same amount of racemic mixture as the mass of solid harvested after the previous crystallization. As the point representative of the overall synthetic mixture (E_B) is situated in the biphasic domain $\langle R \rangle$ + saturated liquid, this implies that a preferential fusion takes place up to a complete melting of $\langle S \rangle$ when the system reaches equilibrium. Note that at the beginning of the AS3PC process the overall compositions of the system X_{EB} and $X_{EB'}$ and the corresponding compositions of the liquid phase X_B and $X_{B'}$ are different and thus there is no need for seeding. By using an accurate temperature control, T_B is preferably adjusted so that a maximum of $\langle R \rangle$ crystals remains present with the molten phase i.e. T_B is adjusted as close as possible to T_e. During the entrainment, the cooling rate, stirring rate and stirring mode are adjusted to favor mainly the crystal growth of the pure $\langle R \rangle$. The mass transfer between the liquid and the enantiomerically pure solid phase is ensured by an efficient agitation which renews the molten liquid phase around the crystals with less possible damag-

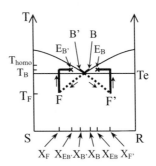

Fig. 10 Ideal Auto Seeded and Polythermic Programmed Preferential Crystallization (AS3PC). The highest temperature (T_B) of the process, corresponds to the selective melting of the enantiomer in default. The cooling is adapted so that a quasi ideal composition of the liquid phase is maintained throughout the process, down to T_F. B (B' respectively) is the point representative of the liquid phase at the beginning of the entrainment which delivers R (S respectively)

ing effects on the growing faces. Indeed, any creation of high energy surfaces on the solid particles is likely to promote the heteronucleation of the counter enantiomer.

Everything that can add to a smooth process should be favored. AS3PC starts with a thermodynamic equilibrium (hence very reproducible) and the entrainment can thus be conducted according to optimized kinetic parameters.

Like SIPC and S3PC processes, AS3PC is presented here as an ideal process which could be operated up to the extreme metastable limits regardless of kinetic and other practical limitations. Figure 11 schematizes the PC cycle of operations as well as the inputs and outputs before and after each operation, without a temperature versus time profile. It has been demonstrated that a single run of preferential crystallization is an asymmetric process during which: one enantiomer remains supersaturated, i.e. out of equilibrium, while the other enantiomer crystallizes towards its metastable equilibrium. In the three variants detailed above the differences lie in the way the irreversible process (from a thermodynamic point of view) of crystallization is conducted. From SIPC via S3PC to AS3PC the evolution of the system is conducted in a smoother way; the benefits of this will be detailed below.

The next section will treat the PC in its usual context: with at least one solvent.

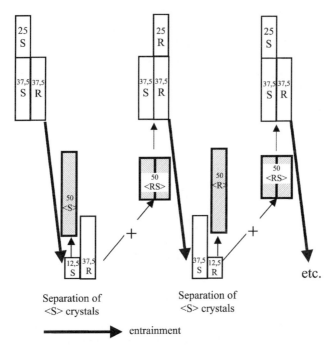

Fig. 11 Temperatureless scheme which illustrates the successive operations which deliver alternately S enantiomer (every odd operation) and R enantiomer (every even operation)

5.2
Heterogenous Stable and Metastable Equilibria which Govern the Entrainment Effect in a Ternary System R–S–A (A = Solvent or Homogeneous Mixture of Solvents)

Before going into detail on PC operated within a ternary system, the preliminary part of this section will shed light on the background of heterogeneous stable and metastable equilibria which is necessary to understand the physico-chemistry behind the entrainment effect. This "return" to fundamentals is also necessary to understand some of the difficulties that can be encountered while trying to apply this elegant resolving method.

It is important to note that the $(S_{0.5}R_{0.5})$–V–T is a plane of symmetry in every R–S–V ternary system. In order to simplify the representations the following assumptions will be made: (i) there is no solid solution (complete or partial) between the enantiomers which therefore crystallize as pure substances; (ii) there is no stable or metastable solvate in this system; (iii) there is no unstable or metastable racemic compound. Figure 12 gives, in accordance with these three hypotheses, a schematic representation of the stable and metastable equilibria of an isothermal section. The phase diagram is divided into four regions:

Region 1: a monophasic domain constituted by an undersaturated solution;

Regions 2 and 3: two symmetrical biphasic domains: ⟨S⟩ + saturated solution in S; ⟨R⟩ + saturated solution in R;

Region 4: a triphasic domain: ⟨S⟩ + ⟨R⟩ + doubly saturated solution with respect to S and R (represented by point I) and limited by lines SI and RI called tie-lines. Because of the symmetry between the chiral components S and R, this solution is perfectly racemic (e.e. = 0). In this triphasic domain, the dashed lines represent the metastable solubility of S and R. They pro-

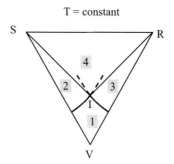

Fig. 12 Isothermal section of a ternary system R-S-V, R and S crystallizing as a stable conglomerate. 1: Undersaturated solution. 2: ⟨S⟩ + saturated solution in S. 3: ⟨R⟩ + saturated solution in R. 4: ⟨S⟩ + ⟨R⟩ + doubly saturated solution I. *Dashed lines* stand for metastable solubilities

long the stable solubility curves which limit the biphasic domains and the monophasic domain. These lines, representing metastable equilibria, are of great importance in the PC procedure and will be treated in detail in the paragraph on the performance of the PC.

Starting from the collection of ternary isotherms represented on Fig. 13 one can easily deduce the polythermic representation displayed on Fig. 14. Three surfaces delineate the boundaries in the composition—temperature space between the undersaturated solution and the polyphasic domains; two symmetrical surfaces for the two enantiomers plus their saturated solutions and one surface, at low temperature, for the solvent plus the saturated solution. The three surfaces intersect two by two forming three eutectic curves

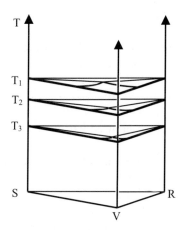

Fig. 13 Three different isotherms in the polythermic ternary system R–S–V (for clarity, metastable equilibria are omitted)

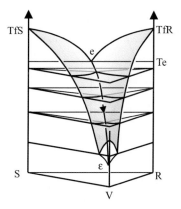

Fig. 14 Symmetrical surfaces of solubility (and crystallization) of the R–S–V ternary system (metastable solubilities omitted for clarity)

(also called monovariant lines) gathering at the ternary eutectic point ε. The monovariant line eε is also of great importance in PC procedure; it corresponds to the solubility curve of the racemic mixture versus temperature. The S and R solubility surfaces (i.e. the crystallization surfaces) interpenetrate to account for the metastable solubility of each enantiomer (not represented for the sake of clarity).

The three-dimensional representation on Fig. 15 defines the so-called conoids. These surfaces are the locus of the tie-lines separating the three phase domains from the two contiguous biphasic domains. Two out of six conoids are represented; the four conoids at low temperature are omitted.

On Fig. 16a,b the vertical R–Y–T plane is an isoplethal section (the S/V ratio is constant throughout this section) which is uppermost in the under-

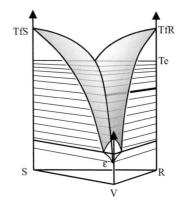

Fig. 15 Representation of the conoids related to the two enantiomers. The segment in bold illustrates the tie line brought to light in Fig. 16a,b

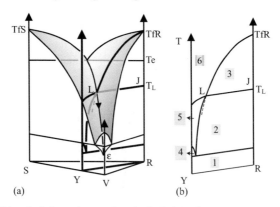

Fig. 16 a T–R–Y isoplethal section. Points included in this section correspond to a fixed ratio S/V. **b** Nature of the different domains in the T–R–Y isoplethal section: *1* ⟨R⟩ + ⟨S⟩ + ⟨V⟩, *2* ⟨R⟩ + ⟨S⟩ + doubly saturated solution, *3* ⟨R⟩ + saturated solution, *4* ⟨S⟩ + ⟨V⟩ + saturated solution, *5* ⟨S⟩ + saturated solution, *6* undersaturated solution

standing of the diversity of crystallization behaviors in the S–R–V system. A fundamental property of this section is that during a crystallization in a medium of overall composition situated in this vertical plane, as long as the solution point remains in the RYT section, only ⟨R⟩ crystallizes. The R–Y–T section has been extracted to clearly label the different domains. One of the important features in this section is the tie-line LJ which is the boundary between the biphasic domain: ⟨R⟩ + saturated solution and the triphasic domain ⟨R⟩ + ⟨S⟩ + doubly saturated solution. The metastable solubility of the R enantiomer prolongs the stable solubility curve T_fR–L into the three phase domain (in this case, but we will see that the metastable solubility curve of ⟨R⟩ can also enter in the ⟨S⟩ + saturated solution domain).

Figure 17 shows the behavior of a system on cooling when it constantly remains in thermodynamic equilibrium.

Starting from an undersaturated solution containing a slight enantiomeric excess in R and represented in A (the solution exhibits the YV/SY fixed ratio), the following steps should be observed on cooling:

- From T_A to T_{homo}, the system consists of a single phase which remains an undersaturated solution.
- From T_{homo} to T_L, the crystallization of pure ⟨R⟩ takes place. Simultaneously, the liquid phase evolves from point H to point L in the isoplethal section.
- At T_L the mother liquor exhibits no enantiomeric excess, the second enantiomer ⟨S⟩ starts crystallizing in equal quantity per time unit as ⟨R⟩ does.
- From T_L to T_ε, the trajectory of the liquid phase has made a sharp turn at T_L, evolving now along the eutectic valley$(S_{0.5}R_{0.5})$–V–T down to the ternary eutectic point ε. The liquid phase which is doubly saturated in S and in R remains perfectly racemic (i.e. no enantiomeric excess).

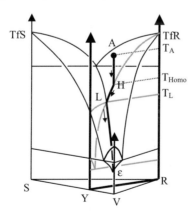

Fig. 17 Trajectory of the solution point when the system initially homogenized (A) is slowly cooled down from T_A to $T\varepsilon$. Points A, H, and L belong to the T–R–Y isoplethal section. The system is supposed to remain in thermodynamic equilibrium throughout the process

- At T_ε, the three components of the system S, R, and V crystallize simultaneously until no liquid remains.
- Below T_ε, the three solid phases ⟨S⟩, ⟨R⟩, and ⟨V⟩ co-exist as pure components.

5.3
Entrainment Effect in Solution

The polythermic heterogeneous equilibria detailed above will now be used to give a complete description of the entrainment effect carried out with three different modes SIPC, S3PC, and AS3PC. In every case the starting situation will consist of a mixture A slightly deviated from the racemic composition (e.g. containing 10% enantiomeric excess of R enantiomer). These descriptions are idealized for reasons of simplification.

5.3.1
Ideal Seeded Isothermal Preferential Crystallization (SIPC)

Figure 18 shows an out of equilibrium evolution of the system corresponding to an isothermal entrainment effect. Starting from the same point A at T_A as that used in the previous section (thermodynamic equilibrium) the system is maintained at T_A long enough to ensure a complete dissolution of every particle of the chiral solute. When this preliminary step is completed, a fast cooling of the system is applied down to T_F so that no spontaneous nucleation of any solid occurs even if, in that domain, both enantiomers should crystallize simultaneously. At T_F, seeds of very pure ⟨R⟩ are added, initiating a stereoselective crystallization of ⟨R⟩. As the crystallization pro-

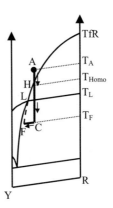

Fig. 18 Ideal Seeded Isothermal Preferential Crystallization (SIPC). The system is homogenized at T_A and swiftly cooled from T_A to T_F, seeded at T_F by very pure and fine ⟨R⟩ crystals. As the isothermal preferential crystallization proceeds, the representation point of the solution moves from C to F (F is the metastable solubility of R at T_F)

gresses, the points representative of the solution and the overall composition of the system are no longer merged. Neglecting the exothermic effect of the crystallization, the solution point evolves horizontally from C to F. F is the metastable solubility of R at T_F. If the phase diagram is expressed in mass fraction, the lever rule can be applied and thus the amount m of pure enantiomer R harvested at the end of this run is:

$$m = 2M = M_T \cdot FC/FT_F \quad (M_T \text{ equal to the total mass of the system}). \quad (3)$$

5.3.2
Ideal Seeded Polythermic Preferential Crystallization (S3PC)

Figure 19 illustrates the pathway of the mother liquor during the S3PC process.

From T_A (point A) to T_{homo} (point H) the solution remains clear of any solid particle.

At T_{homo} the solution is saturated and seeded with very pure ⟨R⟩ crystals. A slow cooling program is applied to the system so that the thermodynamic equilibrium is assumed and the point representative of the liquid phase runs along the stable solubility curve from H to L. When in L (at T_L) there is no longer any enantiomeric excess in the mother liquor.

The adapted cooling program keeps on being applied to the system from T_L to T_F. The point representative of the liquid phase moves along the curve LF inside the isoplethal section R–Y–T; thus only the R enantiomer crystallizes.

S3PC is run in thermodynamic equilibrium from T_A to T_L, and in accordance to a dual status, from T_L to T_F, out of equilibrium for the S enantiomer and in metastable equilibrium for the R enantiomer.

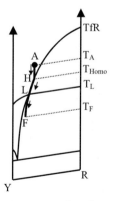

Fig. 19 Ideal Seeding Polythermic Programmed Preferential Crystallization (S3PC). Same process as SIPC (Fig. 18) except that the seeding is performed at T_{homo} and the cooling rate is smooth. During the entrainment, the solution point moves along the HLF curve

5.3.3
Ideal Auto-Seeded Polythermic Programmed Preferential Crystallization (AS3PC)

Figure 20 shows the framework in which the AS3PC process is operated. The temperature of the system is initially set at T_B so that the point representative of the system is located in the biphasic domain: ⟨R⟩ + saturated solution. The starting point E_B of the AS3PC process corresponds to a suspension (and not a solution) in thermodynamic equilibrium. A specific temperature ramp is then applied to the system so that the system can be considered to be conducted along stable equilibrium down to T_L; the point representative of the mother liquor moves on the solubility curve of the R enantiomer from B down to point L. In perfect continuity with this first part of the cooling, the point representative of the mother liquor moves along the metastable solubility curve of the R enantiomer down to temperature T_F. During the cooling, the solution point remains in the isoplethal section R–Y–T indicating that only pure ⟨R⟩ crystallizes. The solid phase ⟨R⟩ already present at T_B acts as seeds hence the adjective: "auto-seeded" associated to this particular stereoselective crystallization.

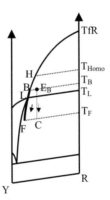

Fig. 20 Ideal Auto Seeded Polythermic Programmed Preferential Crystallization (AS3PC). At T_B the biphasic system is in thermodynamic equilibrium and composed of a saturated solution (B point) plus a Mtotal x (BE$_B$/BT$_B$) mass of ⟨R⟩ crystals. The auto-seeded suspension is smoothly cooled from T_B to T_F. The point representative of the solution point moves from B to L and to F; a filtration will yield a Mtotal x (FC/FT$_F$) mass of ⟨R⟩ crystals

5.4
Cyclicity of the Operations

As each crystallization results in a fairly low mass of chiral product, it is (at least economically) necessary to recycle the mother liquor and to implement another PC and so on. By using these successive recyclings it is thus theoretically possible (see next section: limitations of the PC) to resolve any amount of racemic mixture from fractions of milligrams to thousands of tons or more.

At the end of the entrainment, the mother liquor (supposedly separated from the solid by a fast and efficient separation procedure), contains an excess of the counter enantiomer (namely S in the three above ideal examples). After addition of an equal mass of racemic mixture as that collected by filtration, a similar procedure as the one used for the first crystallization is applied. The only difference concerns the nature of the seeds for SIPC and S3PC processes which are now ⟨S⟩. Figures 21–23, detail the projections and perspective views of the successive operations. These illustrations show that for every odd crystallization, 2 M mass units of pure ⟨R⟩ are harvested and every even crystallization, an equal mass of 2 M mass units of pure ⟨S⟩ is collected.

Whatever the amount of the first investment the system will gradually evolve batch after batch towards two steady states in which A and A′ ini-

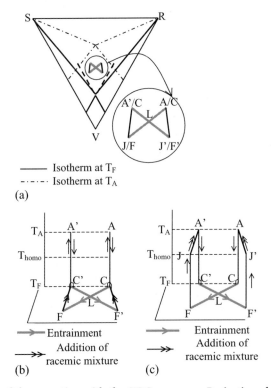

Fig. 21 Cyclicity of the operation with the SIPC process. **a** Projection along the temperature axis. **b** and **c** differ only by the temperature at which the 2 M mass units of racemic mixture are added to the previous mother liquors. The sequence of operations is: seeding at T_F by ⟨R⟩, entrainment of R along CF, addition of (±) along FC′ (or JA′), seeding at T_F by ⟨S⟩, entrainment of S along C′F′, addition of (±) along F′C (or J′A) and so on

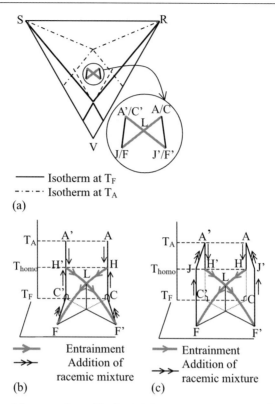

Fig. 22 Cyclicity of the operation with the S3PC process. **a** Projection along the temperature axis. **b** and **c** differ only by the temperature at which the 2 M mass units of racemic mixture are added to the previous mother liquors. The sequence of operations is: seeding at T_{homo} by ⟨R⟩, entrainment of R along HLF, addition of (±) along FC′ (or JA′), seeding at T_{homo} by ⟨S⟩, entrainment of S along H′L′F′, addition of (±) along F′C (or J′A)

tial compositions are simply symmetrical with regards to the racemic section ($S_{0.5}R_{0.5}$)–A–T (i.e. they have the same absolute value of the initial enantiomer excess). This also means that if a technical failure affects a given run, whether the operation can be started again or progressively, batch after batch, the system will evolves back to its initial performances.

5.5
Preferential Crystallization of Solvates in a Ternary System

In principle, there is no fundamental difference between the entrainment effect of a solvate and a nonsolvate. Nevertheless, the possible efflorescent character of the solvate, if it exists, must be addressed. In SIPC and S3PC processes, it might be necessary to store the seeds under specific conditions to avoid partial or total desolvation of the phase which is supposed to give the

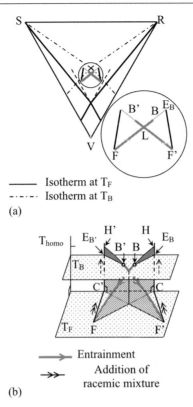

Fig. 23 Cyclicity of the operation with the AS3PC process. **a** Projection along the temperature axis **b** at T_B suspension of ⟨R⟩ crystals in equilibrium with their saturated solution, entrainment of R along BF from T_B to T_F, addition of (±) along FC' at T_B; suspension of ⟨S⟩ crystals in equilibrium with their saturated solution, entrainment of S along B'F' from T_B to T_F, addition of (±) along F'C

stereoselective nucleation and crystal growth. In the AS3PC process, the whole racemic mixture needs to be composed of the right solvate unless it is possible to reload the reactor by the nonsolvated racemic mixture (whatever its nature: racemic compound, solid solution, or another conglomerate) and the right solvated phase is allowed to crystallize swiftly and reproducibly at T_B.

5.6
Preferential Crystallization in a Reciprocal Ternary System or System of Order Greater Than 3

The number of constitutive components of the conglomerate does not bring about more limitations in the principle and the application of the PC (e.g. for a couple of diastereomers forming heterosolvates, four different species constitute the building blocks of the crystal lattice). Some difficulties come when noncongruent solubility phases have to be handled; here the seeds cannot be separated from their mother liquor (SIPC and S3PC mode). In AS3PC mode it is often more convenient to crystallize in situ the conglomerate, e.g. the weak chiral organic acid to be resolved is added together with the stoichiometric quantity of base in the filtrate (containing already an excess of base) of the previous batch.

5.7
Comparison Between the Different Processes (SIPC, S3PC, and AS3PC) Under Real Operative Conditions

All the variants of PC presented so far were performed in an ideal way; the entrainment effect could be obtained without time limit, no spontaneous nucleation of the counter enantiomer was occurring, the seeds were 100% pure, the mother liquor could evolve up to the metastable solubility (SIPC) or could continuously evolve along the stable and the metastable solubility equilibria (S3PC and AS3PC), the filtrations were ideal. In practice almost all these simplifications do not hold. In laboratories as well as in industry, crystallizations are operated under a certain supersaturation which, in other words, means that the equilibrium is the remote state towards which the system evolves. The closer to the metastable solubility of the crystallizing enantiomer the lower the crystal growth rate. Moreover, PC has a dual character: crystallization towards a metastable equilibrium for one solute and simultaneously maintaining an out of equilibrium state (no crystallization of the antipode). Time is thus a twofold inescapable parameter which puts stringent limits on the process.

5.8
The Seeding Problems

SIPC and S3PC differ by the temperature of seed inoculations. SIPC implies a seeding in the three-phase domain thus there is no enantio-purifying effect on the solid particle introduced, S3PC allows a slight enantio-purification of the seeds because the solid phase is introduced in the biphasic domain. This effect will depend on the cooling program: the resident time between T_{homo} and T_L.

Beside the enantio-purity of the seeds, the homogeneous dispersion of the seeds can also be problematic. Typically, 1% of the future crops is introduced into the medium. These crystals must be small and rapidly dispersed in the mother liquor. It may occur that the fine particles at the top of the stirred solution flock together into large aggregates dropping the secondary nucleation rate and introducing lack of reproducibility in the successive batches. In the case of solvate, the operator has to make sure that the solid introduced really corresponds to the solvated phase (especially in the case of the efflorescent character of the phase). With a noncongruent solubility phase, inoculation of solid only is impossible. In these two latter cases particularly but preferably for all the others, small crystals of the pure enantiomer should be dispersed in their saturated solution (of the same medium) prior to seeding in the reactor.

The AS3PC process skips the major part of the seeding problem; usually between 25 to 45% of the future crop exists as a pure solid in equilibrium with the liquid phase. This proportion is adjusted by the temperature T_B; the closer to T_L the better. Factor Z defined by Eq. 4 is a useful parameter in this fine temperature adjustment.

$$Z = d(T_{homo})/d(e.e.) \quad \text{at} \quad [+/-] = \text{constant} \tag{4}$$

Z is thus the slope of the solubility curve of the pure enantiomer from point L (e.e. = 0) to the melting point in the isoplethal section R–Y–T or S–Y'–T. In the context of application of AS3PC we are just interested in the part of the solubility curve close to point L from e.e. = 0% to e.e.$_{Bmax}$ (usually < 20%). e.e.$_{Bmax}$ stands for the maximum initial enantiomeric excess in the mother liquor at the beginning of the entrainment, so that the condition of cyclicity is fulfilled. For most instances, it is a good approximation to take Z constant in the interval: e.e. = 0% to e.e.$_{Bmax}$ < 12%. When considering a concentration close to point L, Z is commonly in the range: 0.3 °C/e.e.% – 1.2 °C/e.e.%. As a rough rule of thumb, the lower values are for ionic derivatives in a polar medium and conversely the upper values are for covalent components. Taking into account the hypothesis of linearity of the law $Z = f(e.e.)_{(\pm)=constant}$ in the 0 to 12%e.e. region a simple relation can be derived:

$$T_{homo} - T_L \approx Z \cdot e.e._{Bmax} . \tag{5}$$

This relation establishes how easy it is to adjust the temperature of the system at T_B so that it is in the biphasic domain.

When $T_B = (T_{homo} + T_L)/2$, ca. 25% of the solid which will be collected by filtration is already present in the initial suspension (Z being kept constant in the range 0–e.e.$_{Bmax}$). If for example e.e.$_{Bmax}$ = 6%, Eq. 5 shows that for Z = 0.5 °C/e.e.%, about 3 °C separate T_{homo} from T_L. Twice this value is obtained, if Z = 1 °C/e.e.%, that is 6 °C. In the former case T_B must be adjusted 1.5 °C above T_L in the latter case 3 °C.

These simple simulations show that the greater e.e.$_{Bmax}$, and higher Z, the wider the $T_{homo} - T_L$ interval and the easier it is to regulate the temperature of the system in the biphasic domain.

Under favorable conditions (e.g. e.e.$_{Bmax} \approx 10\%$ and $Z \approx 1$ °C/e.e.%) T_B is set closer to T_L; the percentage of pure solid already present at T_B can be increased up to 45% with a sufficient margin of security (100% being the mass of the crop after the optimal entrainment). That is to say, $T_B - T_L$ remains large enough in comparison to the temperature regulation capability of the system. Obviously 50% of the future crop as initial solid remains the limit. In comparison to the SIPC and S3PC processes where 1% only of the future crop is present at the beginning of the process, AS3PC appears in essence more as a stereo-selective crystal growth than the usual crystallization with an important secondary nucleation step.

Therefore, in the AS3PC process, the "annealing" at T_B corresponds to a selective dissolution of the enantiomer in default. The selective dissolution step brings about a shift in the crystal size distribution of the particle of the enantiomer in excess. An in situ grinding can restore a fine population of particles leading to a high surface ready for crystal growth as soon as the temperature is decreased.

5.9
Cooling Program

SIPC: from T_B to T_F the faster the better; at laboratory scale this is easily achieved but on up-scaling technical difficulties arise. When T_F is reached the system remains isothermal; there is no control of the nucleation and the crystal growth. The nucleation rate is often high after seeding [44]. The crystal size distribution shifts towards bigger particles in the second half of the entrainment.

S3PC: from T_{homo} to T_L, the temperature ramp has little influence on the results. Some enantio-purifying effect can be expected and some in situ grinding can be operated. From T_L to T_F, the temperature program must be adjusted on a case-by-case basis [45]. As far as possible one should try to promote soft secondary nucleation and crystal growth.

AS3PC: the temperature vs. time law also includes the heating and duration of the preliminary step at T_B ($T_{homo} > T_B > T_L$). The preferential dissolution step at T_B needs to last long enough to ensure a complete dissolution of the crystal of the enantiomer in default. A simple polarimetric monitoring of the mother liquor is usually sufficient to make sure that the system at T_B is indeed situated in the biphasic domain (⟨R⟩ + saturated solution or ⟨S⟩ + saturated solution). It also helps in adjusting T_B as close as possible from T_L. Usually less than 30 minutes are sufficient to complete this step when the crystal size distribution of the racemic mixture is purposely fine. Concomitant ultra-sound pulses or in situ smashing can also be beneficial: (i) to speed up this dissolution, (ii) to yield a fine population of crystals offering a large surface, (iii) to standardize the crystal size distribution at T_B when using different batches of racemic mixtures, and (iv) to lead to particles with fresh surfaces when possible treatment of the racemic mixture might have poisoned or coated by impurities the surface of the particles (see Sect. 6.5). In order to optimize every parameter a short relaxation period should follow these harsh conditions so that damaged surfaces are healed and less chance is left for hetero-nucleation of the counter enantiomer on high energy surfaces as soon as the temperature of the system drops below T_L.

From T_B to T_L, for a well-tuned process, the gap is narrow.

From T_L to T_F the program is adapted to the specificity of each system. As the number of particles is great the aim is to allow the large number of crystals to grow under smooth conditions. Starting from the thermodynamic equilibrium at T_B the cooling program must be conceived as a major way of keeping the system under control. It is worth noting that after a certain number of crystallizations some impurities contained in the racemic mixture can accumulate in the mother liquor. These additional components can introduce a shift in the temperatures (usually a lowering of the temperatures): T_{homo}, T_B, T_L, T_F are then adjusted accordingly.

In the case of low dT/d (solubility), i.e. a large variation of solubility versus temperature, filtration must also be thermostated at T_F. High vapor pressure of the organic solvent(s); large and well-shaped crystals ease the filtration. This decisive step must be conducted as fast as possible without initiating the nucleation of the opposite enantiomer.

5.10
Stirring Mode and Stirring Rate

Two parameters are important and balance between opposite effects have to be found.

- The higher the stirring rate the better the renewal of the mother liquor around the crystals, allowing faster crystal growth and elimination of a concentration gradient around the growing particles [3, 11, 16].
- The higher the stirring rate the more likely the crystals will be broken and scratched favoring the secondary nucleation of the crystallizing antipode.
- The higher the stirring rate the more likely the crystals will be damaged creating high-energy surfaces which possibly prompt a hetero-nucleation of the counter enantiomer.

In order to speed up the selective dissolution step, high-stirring speed can be applied for T inside the interval $T_B - T_L$. As the entrainment proceeds, the viscosity of the suspension increases (this effect is more pronounced for SIPC, intermediate with S3PC, and less pronounced for AS3PC) therefore, it is sometime necessary to increase the stirring rate when approaching the end of the entrainment.

5.11
Metastable Zones (for the Crystallizing Enantiomer and the Noncrystallizing Enantiomer)

Application of any PC procedure depends on the width of the metastable zones and the corresponding crystallization rates inside these domains. That is to say this paragraph deals with kinetics. It is well known that every set up will lead to different operative zones and that every batch of racemic mixture will contain a different impurity profile. Nevertheless, the discussion here will focus on the general aspect of the supersaturation zones and their comparisons according to variations of process, solvent, temperature etc.

In 1897 W. Ostwald [26] defined the first concept of the metastable zone. Let us suppose a binary system (Fig. 24), there is a ribbon roughly parallel (when a small ΔT is considered) limited on the one side by the solubility curve and on the other side by the limit of the metastable zone below which a spontaneous crystallization takes place. If a solution of pure solute is progressively evaporated (pathway N° 1 on Fig. 24), the solution is in sequence: undersaturated, saturated, supersaturated without spontaneous nucleation, then beyond the supersaturated zone a spontaneous nucleation is very likely to occur. The Ostwald limit is, in fact, the limit of primary nucleation. The location of this limit depends on the rate of evaporation (how fast the supersaturation is created), the stirring mode and the stirring rate, the nature of the inner wall, etc. When the supersaturation is created by cooling the same concept applies (pathway n° 2 on Fig. 24). When an antisolvent is added the system becomes a ternary system but the concept is also applicable.

From experimental studies [46], it has been demonstrated that a subdivision of the metastable zone can be defined. The additional frontier displayed on Fig. 25 corresponds to a change in the crystal growth regimes. Below a certain supersaturation the crystal growth rate is poor, beyond the limit a rapid crystal growth is observed as well as a significant secondary nucleation.

As PC in solution is usually operated in at least a ternary system, it is necessary to investigate how the presence of another component modifies the metastable zones. We will assume hereafter that the domain investigated is not too close to the binary system; i.e. the temperature of interest for PC is significantly lower than the temperature of fusion of the racemic mixture. Figures 12 and 20 show the typical situations which can be encountered with

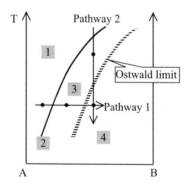

Fig. 24 *1*: undersaturated solution, *2*: Solubility curve vs. *T*, *3*: zone of concentration without spontaneous nucleation (metastable zone), *4*: zone of concentration with spontaneous nucleation (labile zone)

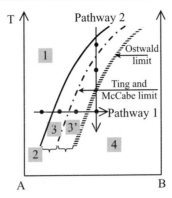

Fig. 25 *1*: undersaturated solution, *2*: Solubility curve vs. *T*, *3*: zone of concentration without spontaneous nucleation (metastable zone) and slow growing rate, *3'*, zone of concentration without spontaneous primary nucleation but with a high secondary nucleation and crystal growth rate, *4*: zone of concentration with spontaneous nucleation (labile zone)

covalent chiral components; the metastable solubility of the R enantiomer enters in the three-phase domain: ⟨S⟩ + ⟨R⟩ + doubly saturated solution. Ideally, Meyerhoffer's rule is observed: $\alpha_{mol} = S(\pm)/S(+) = S(\pm)/S(-) = 2$ [47]. The solubility of each enantiomer is not affected by the presence of its counterpart, it is also somewhat likely that the metastable zone is not affected too much.

By contrast, Fig. 26a,b illustrate a case where the metastable solubility of the R enantiomer enters in the biphasic domain: ⟨S⟩ + saturated solution. This behavior comes from the effect of a common ion between S and R. Ideally when (1-1) stoichiometric salts are considered $\alpha_{mol} = S(\pm)/S(+) = S(\pm)/S(-) = \sqrt{2}$, the solubility of one enantiomer decreases upon addition of the other. For systems such as N° 5 and N° 6, where large excesses of achiral counter ion are introduced, the alpha ratio decreases further [48]. The location of the metastable solubility curve of the crystallizing enantiomer results from the interactions between the enantiomers in solution.

Whatever the PC considered, the conditions of cyclicity cause the entrainment effect to persist down to a precise temperature T_Fm.eq corresponding to a metastable solubility of the crystallizing enantiomer so that after filtration addition of the same mass of racemic mixture as the mass of ⟨R⟩ collected, the e.e. of the next initial state is identical to the e.e. of the corresponding initial state of the previous entrainment (this initial e.e. should be preferably close, if not equal, to ee$_{Bmax}$). Because of the curvatures of the stable and metastable solubility curves, the ideal thermodynamic parameter Um.eq = $(T_L - T_F\text{m.eq})/(T_{homo} - T_L)$ needs to be greater than 1. In practice and keeping in mind the condition of cyclicity, $U = (T_L - T_F)/(T_{homo} - T_L) >$ Um.eq is usually situated in the interval (1.5 – 3), which implies that the experimenter has to increase the driving force (i.e; the supersaturation) for

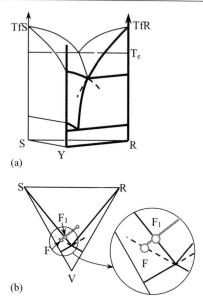

Fig. 26 a Stable and metastable equilibria applicable to enantiomer showing a common achiral counter ion (this counter ion can be in excess in the medium) or submitted to a dissociative solvent. **b** Isotherm at T_F

a successful PC with optimum: yield, duration, and robustness of the process. Two interconnected reasons can justify the difference between Um.eq and U.

1) It is necessary to keep in mind that PC is a stereoselective crystallization where ca. half of the solvated chiral molecules act as impurities (i.e. the antipode of the crystallizing enantiomer). At the end of the entrainment even more than half of the chiral species in solution are considered to be impurities in the crystal growth process. Therefore, the growth inhibiting effect of the counter enantiomer, already strong at the beginning of the process, becomes even stronger as the entrainment proceeds. Thus, there is a need to adjust T_F (sometimes quite far) below the strict application of the metastable equilibrium: T_Fm.eq. U can be twice or more Um.eq.

2) Even without considering the antipode effect, in order to speed up the crystal growth (and the secondary nucleation) of the crystallizing enantiomer, there is a need to enter in the second sub domain of the metastable zone. Figure 27a–c depict the trajectories of the solution points inside the isoplethal section R–Y–T. Of course these trajectories depend on the cooling program applied and to a lesser extent on the stirring mode and stirring rate.

For $\alpha_{mol} + S(\pm) < 2$ the metastable solubility curve of ⟨R⟩ enters in the biphasic domain: ⟨S⟩ + saturated solution. One may wonder if the stereoselective crystallization (of ⟨R⟩) can persist in a domain where the same crystals should

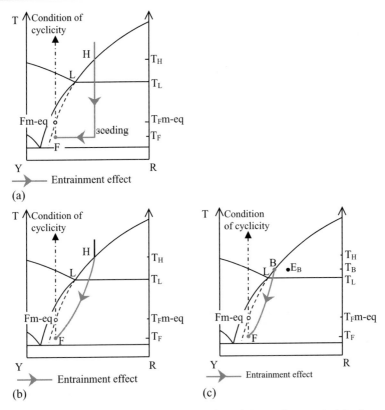

Fig. 27 Preferential crystallization under practical conditions. The practical final temperature T_F has to be adjusted below T_Fm-eq if the condition of cyclicity (within these given initial conditions) has to be fulfilled. **a** SIPC, **b** S3PC, **c** AS3PC

in fact dissolve—i.e. can the PC still continue along the F_1F segment (see Fig. 26b). As such, the question appears confusing, in fact metastable and stable equilibria cannot be mixed as they are equilibria of different stabilities which cannot co-exist. When considering the exclusive crystallization of R, only the metastable solubility of R should be taken into account. This implies that there is no crystal of ⟨S⟩ in the medium. When an evolution towards equilibrium is considered (for instance after the spontaneous nucleation of ⟨S⟩), a partial dissolution of ⟨R⟩ should occur and the system should end its evolution when the mother liquor is perfectly racemic in equilibrium with ⟨S⟩ and ⟨R⟩. A conclusive experimental demonstration has recently been given by using two consecutive AS3PC processes without adding racemic mixture after the first crystallization [49].

On the basis of the shape of the metastable solubility curve only, the ideal PC of salt derivatives should be more productive than that of the covalent species [50–53]. Nevertheless, experiments show that productivity is a multisided question which does not depend on extension of metastable equilibria exclusively.

Hongo, Yamada, and Chibata [54] have quantitatively defined two adjacent zones in a concentration–time diagram which could be correlated with the sub metastable zones of Ting and McCabe. The metastable region of the unseeded enantiomer could be divided into two regions in a concentration versus time diagram. In the first metastable region (the less supersaturated) no spontaneous crystallization was observed for the counter enantiomers for more than a given time limit (e.g. 4 hours). In the second metastable region no spontaneous crystallization takes place until the integrated area of supersaturation degree $(C-C_{sat.})$ as a function of the cooling time reaches a fixed value.

6
Yield and Practical Limitations of PC ("Internal" and "External" Parameters)

A successful application of PC requires an optimization of: (i) the yield of each batch, i.e. the productivity, (ii) the duration of each batch, (iii) the robustness of the whole process.

Figures 21–23, expressed in mass fraction show that the yield is just the application of the lever rule.

$$Y = M_{Total} \cdot FC/FR = M_{Total} \cdot F'C'/F'S. \tag{6}$$

The duration of each batch includes the dissolution of the racemic mixture (complete dissolution for SIPC and S3PC, partial and stereoselective dissolution for AS3PC), cooling bringing about the entrainment, filtration, and addition of the racemic mixture. This sequence of operations lasts typically ca. 2 hours. The robustness of the process usually includes a limitation of the entrainment effect as a security margin.

Because of a far better control and the benefit of the auto-seeding with a large initial population of crystals, the AS3PC process gives better results than S3PC and SIPC. The benefits are even enhanced on up-scaling because the initial step of the crystallization starts from the thermodynamic equilibrium, the supersaturation can be adapted to the necessary driving force required by the PC. When well-conducted, the crystals are bigger with smooth surfaces; the filtration step is thus facilitated. In good cases [42, 43], the mother liquor can reach ca. 15–20% e.e. at the end of the entrainment, so that about 1/3 of the total mass of racemic mixture dissolved in the solution can be retrieved as pure enantiomer per run. It is worth noting that the better the entrainment effect the easier the application of the AS3PC variant and the more beneficial in terms of yield, cost, and robustness.

Nevertheless, despite intense efforts to find the optimum conditions (solvent, temperature domain, different impurity profiles of the racemic mixture, tests of the three PC variants, etc.) the results of the preferential crystallization remain sometimes poor. The next paragraph will detail the rationale behind these observations.

6.1
PC in a System with a Detectable Metastable Racemic Compound

The dilemma: conglomerate–racemic compound is not as clear cut as it seems at first sight. In the binary system S–R (considered without any solid solution, this possibility will be envisaged later), $G(\langle S \rangle) + G(\langle R \rangle)$ is in competition with $G(\langle RS \rangle)$ and the difference:

$$\Delta G = G(\langle RS \rangle) - \left(G(\langle S \rangle) + G(\langle R \rangle) \right) \tag{7}$$

is a function of temperature (pressure will not be considered here [55]). On Fig. 1b, ΔG is successively negative for $T < T_\pi$; naught for $T = T_\pi$ and positive for $Tf(\langle RS \rangle) > T > T_\pi$. The reverse situation is depicted on Fig. 1c where ΔG is successively negative for $T > T_\varepsilon$; naught for $T = T_\varepsilon$ and positive for $T < T_\varepsilon$. Even if a conglomerate appears to be stable whatever the temperature, various metastable or unstable racemic compounds can be considered. Molecular modeling tools [56–58] can be used to generate such crystallized racemic compounds. As a matter of simplification, we will consider here only the one with the lowest free enthalpy, supposedly the same whatever the temperature. Figure 28a–c shows, for a given conglomerate, various relative thermal stabilities of the metastable racemic compound. Figure 28a corres-

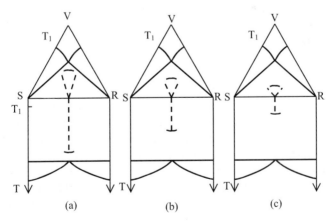

Fig. 28 Various metastable racemic compounds and their metastable solubility curves in the ternary systems. At T_1, from **a** to **c** $\Delta G = G(\langle RS \rangle) - [G(\langle S \rangle) + G(\langle R \rangle)]$ increases, i.e. the relative stability of the racemic compound versus the conglomerate diminishes. A good correlation is found between the PC performances and the ΔG obtained by Eq. 7. One can foresee better PC results for **c** than **b** and **b** than **a**

ponds to a racemic compound with a ΔG Eq. 7 slightly positive only. It is likely that a quench of the molten state will deliver, for a given period of time, the transient ⟨RS⟩ phase. Indeed, this metastable racemic compound has both: its nucleation and its rapid growth favored by kinetic factors [59]. The free migration paths of S and R chiral molecules to dock on the ⟨RS⟩ surface are several orders of magnitude lower than that on the physical mixture of ⟨S⟩ and ⟨R⟩. Indeed for the latter, every molecule (e.g. R) needs to dock on the surface of the right particle (e.g. ⟨R⟩) whereas for the former both enantiomers can be incorporated on the surface of the racemic compound. If the molten state is not accessible, in a ternary system it is sometimes possible to reach this metastable phase by imposing a swift supersaturation. From the reasoning above and in accordance with the Ostwald law of stages, a single crystal of the metastable racemic can be obtained. Consequently, the crystal structure can be known [59, 60]. Figure 28b shows a metastable racemic compound which is much less stable than the corresponding conglomerate. On Fig. 28c, the metastable intermediate compound (1 – 1) is even more metastable and it will be difficult to isolate because of the poor stability of this phase and the drastic conditions of crystallization which can be beyond any possibility of nucleation and growth (i.e. formation of a glass).

In order to apply any PC process, the necessity of the stability of the conglomerate has already been emphasized [61] but what is the impact of a detection of a metastable phase ⟨RS⟩?

There is a good correlation between poor performances of an optimized PC and the detection of a metastable racemic compound [59, 62]. Usually e.e.$_{Fmax}$ does not exceed 5–6% when a metastable racemic compound is detected. This does not mean that a racemic compound crystallizes during the entrainment; the detection of this phase simply shows that the conglomerate is just a little bit more stable than the corresponding intermediate compound ⟨RS⟩. In other words, some heterochiral interactions at given (hkl) solid–solution interfaces are very competitive in terms of energy compared to the homochiral interactions at the same interfaces. At the beginning of a PC the level of enantiomeric impurity is already as great as $\approx 45\%$; but at the end of the entrainment the counter enantiomer outnumbers the crystallizing component; it can peak at 55–58% of the whole solute. Therefore, statistically the frequency of wrong docking increases as the PC proceeds and can deeply hinder the homochiral self-assembling process. The two-way correlation has been put forward to estimate the relative stabilities of the metastable racemic compound and the conglomerate. Some examples have shown that e.e.$_{Fmax}$ can be as low as 1% in case of an easy isolable ⟨RS⟩ compound [62]. This good reliability has led us to propose some routine tests at the early stage of development of a PC to spot a possible lack of productivity [59, 60].

6.2
R–S Epitaxy

This phenomenon has been wrongly described as twinning. Indeed, this appellation is confusing because twinning supposes that a given component has specific crystallographic relations with itself. Moreover, this is not a priori an anathema for the success of the PC. By contrast, epitaxy is a phenomenon which binds according to a specific orientation of the mirror-imaged crystal lattices of the two components (here the two opposite enantiomers). When a R–S epitaxy occurs the PC will usually lead to poor results. This happens when on a single (hkl) interface the heterochiral interactions are competitive with the homochiral interactions. A two-dimensional racemic compound is then formed but this R–S association cannot expand in the third dimension due to a too high cost in energy.

In terms of crystal growth in a quasi-racemic medium, when a single particle of a given chirality grows, the mother liquor in the vicinity of the solid–liquid interface is depleted of the crystallizing enantiomer and along the way enriched in the counter enantiomer. It might happen that the threshold of hetero-secondary nucleation is overcome leading to the spontaneous crystallization of the opposite enantiomer [63–67]. Moreover, the phenomenon can be repeated several times leading to some sort of oscillatory crystallization. The final particle is thus a stack of parallel homochiral layers that alternate in chirality along crystallographically equivalent orientations. Some dedicated experiments [65] have shown that a vigorous stirring helps in diminishing the number of heterogeneous domains. This is consistent with the idea that by increasing the turbulences around the crystal, it is possible to drop the gradient of counter enantiomer concentration at the solid–liquid interface. When considering the agitation at its best, the concentrations of R and S enantiomers around the growing particles will tend to equal those in the bulk.

6.3
Irreversible Adsorption on a Given (hkl) Orientation

R–S interactions along a specific direction can be so strong that the crystal growth is almost inhibited under a reasonable supersaturation. This strong poisoning effect has been pointed out on a salt derivative [68] starting from large single crystals obtained in a solution of the pure enantiomers. The single particles immersed in a racemic solution show no crystal growth on the face with the largest morphological index. The crystal ends its growth when it reaches the shape of a lens. No metastable racemic compound has been detected in this case.

6.4
Solid Solution Between Enantiomers

Up to now we have considered that the solid phases are pure ⟨S⟩, ⟨R⟩, and ⟨RS⟩. In some instances, conglomerates with partial solid solutions have been pointed out [69]. The solid solutions considered here are not those obtained under harsh conditions of precipitation [70] but correspond to phases in thermodynamic equilibrium.

The best composition of the solid obtained under ideal conditions of PC will be equal to the metastable limit of the solid solution. As exemplified in Fig. 1k, the tie-lines connecting the metastable solubility to the composition of the solid phase will determine the best possible composition of the crystallized material retrieved from a PC. An efficient entrainment in terms of mass of the crops would be limited by the e.e. of the solid phase. The orientations of the tie-lines in the three-phase domain have therefore an important role in limiting the purity of the solid phases isolable by means of PC.

6.5
Surfaces of the Seeds

Figure 29a,b illustrates different shapes of the same chemical [71]. When using seeds depicted in Fig. 29a the performance of the entrainment is half that obtained by using the seeds represented on Fig. 29b. The overall purities of the two samples are the same (even slightly better for the seeds shown in Fig. 29a); the poor entrainment effect results from some sort of coating by impurities at some hundredth ppm levels resulting from excessive heating during drying.

Among the four types of limitation described above, heterochiral interactions competitive with homochiral interactions revealed by a metastable racemic compound and epitaxy are the more populated. They account for the large majority of the "difficult" cases which render, in practice, the PC not so attractive if e.e.$_{Fmax}$ < ca. 5%.

6.6
PC and Polymorphism

A study [44] has shown the impact of polymorphism on the optical resolution by means of PC. Two very similar conglomerates of the same solute have been characterized and used in PC. In contrast to their structural and energetic analogies, the difference in the entrainments increased as the supersaturation increased. Under smooth conditions (AS3PC variant) both varieties lead to stereoselective secondary nucleation and crystal growth with comparable results. With SIPC mode when the entrainment was performed under high initial supersaturation, the metastable form led to an uncontrolled offspring

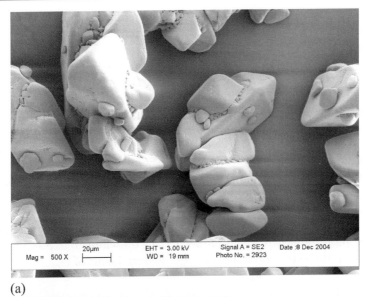

Fig. 29 Although both populations of particles exhibit the same overall chemical purities, the solid displayed in **a** — due to overheating—leads to significantly poorer performances during PC than that displayed in **b** (G. Coquerel, unpublished results)

of small crystals and a nonfilterable solid phase even at two-liter scale (no industrial application can therefore be contemplated); the stable form does not accommodate this high supersaturation and coats itself with nuclei of the metastable form. In this latter case, the solid is filterable but is constituted by 99% of the metastable form (the initial 1% of seeds have almost remained unchanged).

7
Simultaneous Crystallization with a Controlled Crystal Size (Preferential in CSD)

The literature reveals some possibilities of concomitant resolution of both enantiomers by a fine management of the crystal size distribution [72]. For example, large crystals of ⟨S⟩ are inoculated with small crystals of ⟨R⟩ under smooth conditions of crystal growth. A combined sieving and filtration procedure could be used to isolate the components. The efficiency of the process depends on the ability to avoid the secondary nucleation of the large crystals and to limit the entrainment to crystal growth only.

8
Preferential Crystallization and In Situ Racemization

Compared to asymmetric synthesis, any resolution method suffers from the 50% limitation in each enantiomer. When only a single chiral component is desired (the eutomer) the second enantiomer (the distomer) can undergo a racemization process under which it is transformed into a racemic mixture and re-injected in the PC loop. A combined PC and racemization procedure can ultimately yield 100% of pure enantiomer. Some very elegant processes have been discovered which allow in one pot the crystallization of the eutomer and the in situ racemization [73].

9
Continuous Process

For laboratory and industrial medium size production, PC operated by batch is efficient and easy to handle. When large-scale productions are necessary the duration of the filtration becomes a critical parameter. This duration must be kept within strict limits before the spontaneous nucleation of the counter enantiomers starts and jeopardizes the whole resolution. In conventional crystallization reactors of several hundred liters, there is then a need to start the filtration well before reaching the optimum e.e.$_{Fmax}$ value attainable with pilot reactors. In order to cure the problem, a swap to semi-continuous or even continuous processes have been proposed [74, 75].

10
Conclusion and Perspectives

lthough the applicability of PC suffers from two kinds of limitation:
- The necessary presence of a stable conglomerate with a limited domain of solid solution between the enantiomer if any;
- The homochiral interactions are stronger than heterochiral interactions at the interface crystal–mother liquor.

It is a simple, robust, and cost-effective resolution method which deserves careful consideration every time a chiral resolution is at stake. It is very likely that nice and productive opportunities have been (are still) missed because chemists have simply not enlarged enough the conglomerate screening or have been (are still!) reluctant to use a process based on metastable equilibria. This review shows that a conglomerate screening can be vast. This is why emerging high-throughput techniques are a great help with this issue. After simple tests assessing the relative hetero versus homochiral interactions, the favorable cases can be spotted rather early in the development.

Beside its applications at the laboratory and industrial scales, the chiral discrimination at the solid state is an interesting scientific area where: chemists, crystallographers, solid-state physico-chemists etc. flock together. The answers to queries on the structural, energetic, and dynamic aspects of the stereoselective self-assembling process could help in the understanding of spontaneous break in symmetry phenomena.

Fundamental and applied researches will probably continue to be active. During the last few years the combination: PC + SMB (Simulated Moving Bed) has emerged as an efficient dual technique [76]. The large-scale chromatographic facilities are used at their best in terms of productivity, the second step being the selective crystallization of the enriched mixture. The second step can be more productive when the PC can be used.

Without being exhaustive several areas need to be investigated or re-investigated: (i) a continuous process in order to gain access to a high productivity, (ii) molecular modeling and prediction (or at least insights on the first "orientations" of the screening) of conglomerates, (iii) the role of the impurities and the design of tailor-made additives which could enhance the entrainment performances.

Acknowledgements Dr. M.-N. Petit must be thanked for help with the illustrations and the following companies: Cephalon Inc. (USA), Boeringer-Ingelheim Pharma GmbH (D), Merck Darmstadt KGaA (D), DSM (NL), Sanofi-Aventis ex HMR (D) for collaborations in the field of preferential crystallization.

References

1. Gernez D (1866) C R Acad Sci 63:843
2. Secor RM (1963) Chem Rev 63:297
3. Jacques J, Collet A, Wilen SH (1994) Enantiomers, Racemates and Resolutions. Krieger Publishing Company, Malabar (Florida)
4. Scott RL (1977) J Chem Soc Faraday Trans II 3:356
5. Coquerel G (2000) Enantiomer 5:481
6. Childs SL, Chyall LJ, Dunlap JT, Smolenskaya VN, Stahly BC, Stahly GP (2004) J Am Chem Soc 126:13335
7. Gicquel-Mayer C, Coquerel G, Petit M-N, Bouaziz R (1993) Anal Sci 9:25
8. Ricci JE (1966) The Phase Rule and Heterogeneous Equilibrium. Dover Publications Inc., New York
9. Coquerel G (2004) J Phys IV France 113:11
10. Dowd MK (2003) Chirality 15:486
11. Dufour F, Perez G, Coquerel G (2004) Bull Chem Soc Jpn 77:79
12. Gallis HE, Van Ekeren PJ, Cornelis Van Miltenburg J, Oonk HAJ (1999) Thermochim Acta 326:83
13. Toke L, Acs M, Fogassy E, Faigl F, Gàl S, Sztaticz J (1979) Acta Chim Acad Sci Hung 102:59
14. Aubin E, Petit M-N, Coquerel G (2004) J Phys IV France 122:157
15. Lajzerowicz-Bonneteau J, Lajzerowicz J, Bordeaux D (1986) Phys Rev B 34:6453
16. Coquerel G, Petit S (1993) J Cryst Growth 130:173
17. Tobe Y (2003) Mendeleev Commun 3:93
18. Belsky VK, Zorkaya ON, Zorkii PM (1995) Acta Cryst A 51:473
19. Pratt Brock C, Dunitz JD (1994) Chem Mater 6:1118
20. Steiner T (2000) Acta Crystallogr B 56:673
21. Andrews DL, Allcock P, Demidov AA (1995) Chem Phys 190:1
22. Strachan CJ, Rades T, Lee CJ (2005) Opt Lasers Eng 43:209
23. Timofeeva TV, Nestorov VN, Dolgushin FM, Zubavichus YV, Goldshtein JT, Sammeth DM, Clarck RD, Penn B, Antipin MY (2000) Crystal Eng 3:263
24. Viedma C (2004) J Cryst Growth 261:118
25. Dougherty JP, Kurtz SK (1976) J Appl Crystallogr 9:145
26. Ostwald W (1897) Z Phys Chem 119:227
27. Kostyanovsky RG, Lakhvich FA, Philipchenko PM, Lenev DA, Torbeev VY, Lyssenko KA (2002) Mendeleev Commun 12:147
28. Collet A (1990) Problems and wonders of chiral molecules: the homochiral versus heterochiral packing dilemma. Akadémiai Kiado, Budapest
29. Coquerel G, Petit M-N, Robert F (1993) Acta Crystallogr C 49:824
30. Benedetti E, Pedone C, Sirigu A (1973) Acta Crystallogr B 29:730
31. Pratt Brock C, Schweizer WB, Dunitz JD (1991) J Am Chem Soc 113:9811
32. Tobe Y (2003) Mendeleev Commun 3:93
33. Addadi L, Van mil J, Lahav M (1981) J Am Chem Soc 103:1249
34. Ndzié E, Cardinaël P, Petit M-N, Coquerel G (1999) Enantiomer 4:97
35. Amaya K (1961) Bull Chem Soc Jpn 34:1803
36. Shiraiwa T, Yamauchi M, Yamamoto Y, Kurokawa H (1990) Bull Chem Soc Jpn 63:3296
37. Jensen H (1970) German Patent 1 807 495
38. Ulrich J, Glade H (2003) Melt Crystallization. Shaker, Aachen
39. Schroeder I (1893) Z Phys Chem 11:449

40. van Laar JJ (1903) Arch Neerl 11:264
41. Wilcox WR, Friedenberg R, Back N (1966) Chem Rev 64:187
42. Courvoisier L, Ndzié E, Petit M-N, Hedtmann U, Sprengard U, Coquerel G (2001) Chem Lett 4:364
43. Ndzié E, Cardinaël P, Schoofs A-R, Coquerel G (1997) Tetrahedron: Asymmetry 8:2913
44. Courvoisier L, Mignot L, Petit M-N, Coquerel G (2003) Org Process Res Dev 7:1007
45. Wang XJ, Ching CB (2006) Chem Eng Sci 61:2406
46. Ting HH, Mc Cabe WL (1934) Ind Eng Chem 26:1201
47. Meyerhoffer W (1904) Ber 37:2604
48. Asai S, Ikegami S (1982) Ind Eng Chem Fundam 21:181
49. Lefèbvre L, Coquerel G (2006) CGOM7, Rouen, France
50. Nohira H (1992) J Synth Org Chem Jpn 50:14
51. Nohira H, Ehara K, Miyashita A (1970) Bull Chem Soc Jpn 43:2230
52. Nohira H, Terunuma D, Kobe S, Asakura I, Miyashita A, Ito I (1986) Agric Biol Chem 50:675
53. Toda F (2004) Enantiomer Separation (Fundamentals and Practical Methods). Kluwer Academic Publishers, Rotterdam
54. Hongo C, Yamada S, Chibata I (1981) Bull Chem Soc Jpn 54:1905
55. Collet A (1995) New J Chem 19:877
56. Gervais C, Coquerel G (2002) Acta Crystallogr B 58:662
57. Pauchet M, Coquerel G (2004) J Phys IV France 122:177
58. Pauchet M, Gervais C, Courvoisier L, Coquerel G (2004) Cryst Growth Des 4:1143
59. Dufour F, Gervais C, Petit M-N, Perez G, Coquerel G (2001) J Chem Soc, Perkin Trans II, p 2022
60. Houllemare-Druot S, Coquerel G (1998) J Chem Soc, Perkin Trans II, p 2211
61. Jacques J, Gabard J (1972) Bull Soc Chim Fr 342
62. Coquerel G, Catroux L, Combret Y (1997) PCT FR97/02158
63. Green BS, Knossow M (1981) Science 214:795
64. Torbeev VY, Lyssenko KA, Kharybin ON, Antipin MY, Kostyanovsky RG (2003) J Phys Chem B 107:13523
65. Beilles S, Cardinaël P, Ndzié E, Petit S, Coquerel G (2001) Chem Eng Sci 56:2281
66. Davey RJ, Black SN, Williams LJ, McEwan D, Sadler DE (1990) J Cryst Growth 102:97
67. Potter GA, Garcia C, Raymond MC, Brian A, André C (1996) Angew Chem Int Ed Engl 35:1666
68. Coquerel G, Perez G, Hartman P (1988) J Cryst Growth 88:511
69. Wermester N, Aubin E, Coste S, Coquerel G (2005) ISIC 16, Dresden, Germany, p 223
70. Gallis HE, van der Miltenburg JC, Oonk HAJ (2000) Phys Chem Chem Phys 2:5619
71. Broquaire M, Courvoisier L, Mallet F, Coquerel G, Frydman A (2004) WO02004014846, France
72. Doki N, Yokota M, Kido K, Sasaki S, Kubota N (2004) Crystal Growth Des 4:1359
73. Sheldon RA (1993) Chirotechnology, Industrial Synthesis of Optically Active Compounds. Marcel Dekker, New York
74. Nogushi-Institute (1970), GB Patent 1 197 809
75. Grabowski EJJ (2005) Chirality 17:S249
76. Lorenz H, Sheehan P, Seidel-Morgenstern A (2001) J Chromatogr A 908:201

Mechanism and Scope of Preferential Enrichment, a Symmetry-Breaking Enantiomeric Resolution Phenomenon

Rui Tamura[1] (✉) · Hiroki Takahashi[1] · Daisuke Fujimoto[2] · Takanori Ushio[3]

[1]Graduate School of Human and Environmental Studies, Kyoto University, Sakyo-ku, 606-8501 Kyoto, Japan
tamura-r@mbox.kudpc.kyoto-u.ac.jp

[2]Department of Chemical Science and Engineering,
Ariake National College of Technology, 150 Higashihagio-machi, Omuta, 836-8585 Fukuoka, Japan

[3]Taiho Pharmaceutical Co. Ltd., 2-4-3 Hitotsubashi, Chiyoda-ku, 101-0003 Tokyo, Japan

1	Introduction	54
1.1	Origin of Symmetry Breaking to Biomolecular Homochirality	54
1.2	Crystallization as an Event of a Complexity System	55
1.3	Enantiomeric Resolution by Crystallization	55
1.4	Discovery of Preferential Enrichment	56
2	Features of Preferential Enrichment	58
3	Mechanism of Polymorphic Transition Causing Preferential Enrichment	61
3.1	Assembly Mode of Enantiomers in Solution	61
3.2	Crystal Structure	62
3.2.1	General Crystalline Nature of a Racemic Mixed Crystal Showing Preferential Enrichment	62
3.2.2	Metastable γ-Form	63
3.2.3	Stable δ-Form	65
3.3	Mode of Polymorphic Transition	67
4	Relationship Between Molecular Structures and Preferential Enrichment	71
5	Origin of Preferential Enrichment	72
6	Induction and Inhibition of Preferential Enrichment	73
6.1	Significant Influence of Minor Molecular Modification on the Mode of Polymorphic Transition	73
6.2	Induction and Inhibition of Preferential Enrichment by Controlling the Mode of Polymorphic Transition with Seed Crystals	76
7	Conclusions and Scope	79
	References	80

Abstract The mechanism of "preferential enrichment", an unusually symmetry-breaking enantiomeric resolution phenomenon that is observed upon simple recrystallization of a certain kind of organic racemic crystal from the usual organic solvents without any external chiral element, has been rationalized in terms of a complexity system involving multistage processes that affect each other. These processes comprise: (1) preferential homochiral molecular association to give one-dimensional (1D) R and S chains in solution; (2) formation of γ-form prenucleation aggregates consisting of the same homochiral 1D chains; (3) nucleation and crystal growth of the metastable γ-form crystal composed of irregular alignment of the homochiral 1D R and S chains; (4) the solvent-assisted solid-to-solid type of polymorphic transition of the incipient γ-form crystal into the more stable polymorphic form, such as the α-, δ-, or ε-form; and (5) partial crystal disintegration in the transformed crystal to release excess R or S molecules into solution. Based on this mechanism, both induction and inhibition of preferential enrichment have been achieved by controlling the mode of polymorphic transition during crystallization with appropriate seed crystals. We call this forced polymorphic transition on the surface of a seed crystal "epitaxial transition".

Keywords Enantiomeric resolution · Epitaxial transition · Polymorphic transition · Preferential enrichment · Symmetry-breaking complexity system

1
Introduction

1.1
Origin of Symmetry Breaking to Biomolecular Homochirality

Since Pasteur's discovery of dissymmetric (chiral) crystals [1], scientists have been fascinated with trying to rationalize the origin of biomolecular homochirality on Earth [2–7]. Thus far, there have been two potent explanations for the source of symmetry breaking in cosmic chirality: (1) the parity-violating energy difference (in the order of 10^{-14} J mol$^{-1} \approx 10^{-15}$% ee) between more stable L-amino acids and less stable D-amino acids, based on parity violation in the weak interactions observed upon the radioactive decay of polarized ^{60}Co nuclei [2–9]; and (2) the predominant degradation of one enantiomer caused by chiral cosmic forces such as circularly polarized light (CPL), resulting in enrichment (0.1% ee) of the L-amino acids over the D-amino acids [2–7, 10]. To prove these hypotheses, however, a reliable and reproducible, significant enantiomeric amplification mechanism has to be demonstrated.

Apart from the cosmic origin, it seems much more realistic to ascribe such symmetry breaking to an event of a "complexity" system. For the last two decades, a concept of complexity has been extensively developed and recognized as a universal nonlinear theory that governs various dynamic behaviors observed in natural science, computer science, economics, social science, philosophy, and so on [11–13]. It is noteworthy that with increasing complexity, symmetry is locally and globally broken by phase transitions of chaos in dynamical systems to generate a variety of new order or chaos.

1.2
Crystallization as an Event of a Complexity System

Crystallization can be regarded as an event of a complexity system in terms of the kinetic behavior, far from thermal equilibrium, involving formation of metastable prenucleation molecular aggregates, nucleation and crystal growth, and polymorphic transition. Furthermore, these processes affect each other in feedback and feed-forward manners. Accordingly, crystallization is strongly affected by the surrounding conditions such as additives (seed crystals and impurities), solvent, temperature, concentration, and so on.

Therefore, crystallization of the racemates of chiral compounds is expected to afford a chance of observing a symmetry breaking in chirality [14]. However, until our discovery of preferential enrichment, an unusually symmetry-breaking enantiomeric resolution phenomenon observed upon recrystallization of organic racemic crystals [15, 16], this fascinating subject was not realized.

1.3
Enantiomeric Resolution by Crystallization

The methods for enantiomeric resolution of racemates comprising non-racemizable enantiomers by crystallization are classified into two categories [17]: one is an indirect method using an external chiral element, such as diastereomeric salt formation followed by fractional crystallization [18–25] or a diastereoselective host–guest inclusion complexation [26, 27]; the other is a straightforward method for separating enantiomers by crystallization in the absence of an external chiral element. As a typical example of this latter category, the "preferential crystallization" method is well known to resolve *a racemic conglomerate* composed of a mixture of homochiral *R* and *S* crystals in which, by repeated crystallization from the supersaturated solution with the aid of its enantiopure seed crystals, the enantiomerically enriched crystals are efficiently deposited in the alternating chirality sense [28–31]. However, the racemates existing as a racemic conglomerate only occupy less than 10% of the characterized crystalline racemates, and more than 90% of them are supposed to belong to racemic crystals [32], which are further classified into either *a racemic compound* consisting of a regular packing of a pair of *R* and *S* enantiomers or *a racemic mixed crystal* (in other words, a pseudoracemate or a solid solution) composed of a random alignment of the two enantiomers in the defined positions (Fig. 1) [17]. It has been believed for over a century that there is no way for the resolution of these racemic crystals by simple crystallization in the absence of an external chiral element. Accordingly, if one could find a spontaneous enantiomeric resolution phenomenon for these racemic crystals, it would have a great impact on industrial and academic communities.

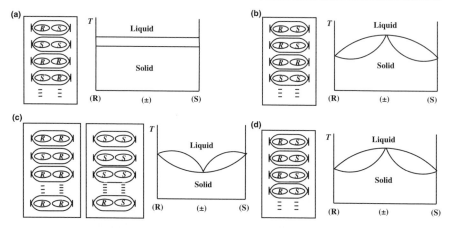

Fig. 1 General packing modes of two enantiomers in a racemic mixed crystal with $Z = 2$ and a plausible binary melting-point phase diagram of **a** disordered type, **b** fairly disordered type that resembles a racemic compound, **c** fairly disordered type that resembles a conglomerate, and **d** ordered type

Contrary to this common recognition, we have found the first instance where such an ideal enantiomeric resolution by simple recrystallization of a series of racemic crystals is feasible.

1.4
Discovery of Preferential Enrichment

In 1996, we reported an unprecedented, symmetry-breaking, spontaneous enantiomeric resolution phenomenon that can be applied to racemic crystals [15, 16]. This phenomenon is completely opposed to preferential crystallization; in preferential enrichment, it is in the mother liquor that substantial enantiomeric enrichment occurs by recrystallization, and at the same time slight enrichment of the opposite enantiomer always occurs in the deposited crystals (Fig. 2). In 1998, this phenomenon was referred to as preferential enrichment [33]. In 2002, we proposed the mechanism of preferential enrichment in terms of a unique polymorphic transition during crystallization [34]. Quite recently, on the basis of the mechanism of this polymorphic transition, we have successfully induced and inhibited the occurrence of preferential enrichment by controlling the mode of the polymorphic transition with appropriate seed crystals [35]. These results strongly support our proposed mechanism of preferential enrichment.

Consequently, we have recognized that preferential enrichment is ascribed to an event of a complexity system that can allow symmetry breaking. Namely, dissipative self-organization, emergence of new order and chaos, phase transition between chaos and/or order, and increasing returns in

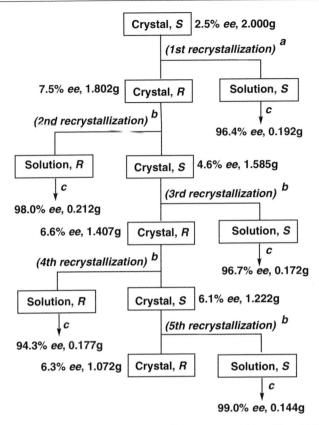

Fig. 2 Preferential enrichment of NNMe$_3$. Conditions: **a** EtOH (32 mL) at 25 °C for 4 days; **b** EtOH (32 mL) at 25 °C for 2 days; **c** removal of the solvent by evaporation. (Reprinted with permission from [34]. Copyright 2006 American Chemical Society)

a complexity system correspond to molecular aggregation, formation of metastable phases, polymorphic transition, and substantial enrichment of one enantiomer in preferential enrichment.

In this chapter we describe (1) the features of preferential enrichment as a spontaneous enantiomeric resolution phenomenon, (2) the mechanism of polymorphic transition causing preferential enrichment, (3) the relationship between molecular and crystal structures and preferential enrichment, (4) the origin of this unusual phenomenon, and (5) the induction and inhibition of preferential enrichment.

2
Features of Preferential Enrichment

Figure 2 shows a typical example of preferential enrichment for nearly racemic NNMe$_3$ (Fig. 3), which is a linear asymmetric secondary alcohol

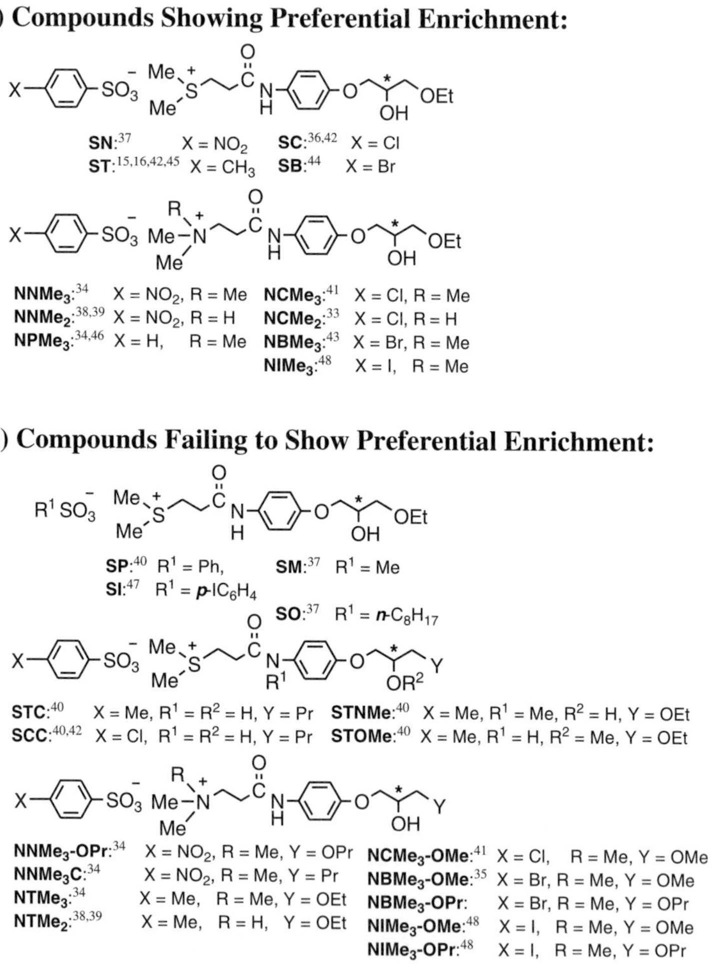

(a) Compounds Showing Preferential Enrichment:

SN:[37] X = NO$_2$ SC:[36,42] X = Cl
ST:[15,16,42,45] X = CH$_3$ SB:[44]   X = Br

NNMe$_3$:[34] X = NO$_2$, R = Me   NCMe$_3$:[41] X = Cl, R = Me
NNMe$_2$:[38,39] X = NO$_2$, R = H   NCMe$_2$:[33] X = Cl, R = H
NPMe$_3$:[34,46] X = H,   R = Me   NBMe$_3$:[43] X = Br, R = Me
                                     NIMe$_3$:[48] X = I, R = Me

(b) Compounds Failing to Show Preferential Enrichment:

SP:[40] R^1 = Ph,   SM:[37] R^1 = Me
SI:[47] R^1 = p-IC$_6$H$_4$
                          SO:[37] R^1 = n-C$_8$H$_{17}$

STC:[40] X = Me, R^1 = R^2 = H, Y = Pr   STNMe:[40] X = Me, R^1 = Me, R^2 = H, Y = OEt
SCC:[40,42] X = Cl, R^1 = R^2 = H, Y = Pr STOMe:[40] X = Me, R^1 = H, R^2 = Me, Y = OEt

NNMe$_3$-OPr:[34] X = NO$_2$, R = Me, Y = OPr   NCMe$_3$-OMe:[41] X = Cl, R = Me, Y = OMe
NNMe$_3$C:[34]   X = NO$_2$, R = Me, Y = Pr   NBMe$_3$-OMe:[35] X = Br, R = Me, Y = OMe
NTMe$_3$:[34]   X = Me, R = Me, Y = OEt   NBMe$_3$-OPr: X = Br, R = Me, Y = OPr
NTMe$_2$:[38,39]   X = Me, R = H, Y = OEt   NIMe$_3$-OMe:[48] X = I, R = Me, Y = OMe
                                                 NIMe$_3$-OPr:[48] X = I, R = Me, Y = OPr

Fig. 3 Relationship between molecular structures and preferential enrichment. Abbreviations of typical compounds are as follows. For ST, S stands for sulfonium ion and T represents the p-toluenesulfonate group. For NNMe$_3$, the first N stands for ammonium ion, the second N represents the p-nitrobenzenesulfonate group, and the Me$_3$ indicates the trimethyl-substituted ammonium structure. For NNMe$_3$ – OPr, OPr indicates the presence of a terminal propoxy group in place of a terminal ethoxy group in NNMe$_3$. For NNMe$_3$C, the last C shows the presence of a terminal propyl group in place of a terminal ethoxy group in NNMe$_3$

containing a glycerol moiety, an amide group, and a trimethylammonium *p*-nitrobenzenesulfonate structure [34]. This compound was synthesized from racemic epichlorohydrin. However, the crystalline sample obtained after the final synthetic procedure was no longer racemic but contained either enantiomer in small excess. Thus, preferential enrichment already occurred during the final synthetic procedure. By recrystallization of the obtained *S*-rich crystals of 2.5% ee from ethanol under highly supersaturated conditions, substantial enrichment of the same *S* enantiomer occurs in the mother liquor. At the same time, slight enrichment of the opposite *R* enantiomer occurs in the deposited crystals. Accordingly, by repeating recrystallization and filtration, alternating enrichment of the two enantiomers occurs largely in the mother liquors and slightly in the deposited crystals. These are the most important and fully reproducible features of preferential enrichment. Thus, by collecting the enantiomerically enriched mother liquors with the same handedness, very efficient separation of the two enantiomers (> 96% ee) has been easily achieved.

In addition, we could visually observe the occurrence of the polymorphic transition, which we discuss in Sect. 3. Namely, as can be seen in the photograph in Fig. 4, a lot of convective streams were evolved from the crys-

Fig. 4 Visual observation of polymorphic transition followed by redissolution of one enantiomer from the once-formed crystals during crystallization from the supersaturated solution of nearly racemic NNMe$_3$ (0.1 mol L^{-1}) in EtOH; **a** 15 min and **b** 60 min after crystallization began. (Reprinted with permission from [34]. Copyright 2006 American Chemical Society)

tal surface for 90 min at the beginning of crystallization of nearly racemic NNMe$_3$ at 25 °C [34]. These streams were found to be enriched by one enantiomer which dissolved from the just-made crystals into solution, resulting in enrichment of the same enantiomer in the mother liquor.

Accordingly, the technical and phenomenal features of preferential enrichment are summarized as follows (Fig. 2) [15, 16, 33–49]:

1. Usual standard recrystallization conditions are applied to the preferential enrichment experiment, except that approximately 4- to 25-fold supersaturated solutions are employed because the supersolubility (a solubility obtained by dissolving the sample in a solvent on heating followed by cooling) of the racemates showing preferential enrichment is considerably higher than that of the solubility at 25 °C.
2. Racemates or nonracemates of less than 10% ee are more suitable samples for the preferential enrichment experiment than those with higher ee values to achieve a very efficient resolution.
3. Recrystallization of the nonracemic sample of less than 10% ee from the supersaturated solution leads to a remarkable enrichment of the excess enantiomer up to 100% ee in the mother liquors (*a considerable enrichment of the excess enantiomer in the mother liquor*). At the same time, the resulting deposited crystals always display the opposite chirality at around 5% ee (*a slight enrichment of the opposite enantiomer in the deposited crystals*). These phenomena are fully reproducible.
4. A wide supersaturated concentration range exists for each compound to allow overdissolution of the excess enantiomer from the just-made crystals into solution until a slight enrichment of the opposite enantiomer in the deposited crystals occurs. At the lower supersaturated concentrations than the range, this essential phenomenon for preferential enrichment does not occur.
5. When the original supersaturated solution is strictly racemic, the probability for either the *R* or *S* enantiomer to be enriched in the mother liquor after crystallization was 50%. In the resulting deposited crystals, the opposite enantiomer is enriched at around 5% ee.
6. Only the racemic or nonracemic samples have to be crystalline to implement the preferential enrichment experiment efficiently. It does not matter whether the enantiomerically enriched materials with high ee values exist as solids or oils, in sharp contrast to preferential crystallization of *a racemic conglomerate*.
7. Seed crystals are not necessary at all. Addition of seed crystals may accelerate or inhibit the occurrence of preferential enrichment, depending on the kind of added seed crystals (Sect. 6).
8. To know whether preferential enrichment occurs or not for a given compound, one only has to repeat recrystallization of the racemic or nonracemic sample of less than 10% ee several times at 25, 0, or – 20 °C at

several different supersaturated concentrations, and measure the ee value in the supernatant solution by HPLC analysis etc. after each crystallization.

3
Mechanism of Polymorphic Transition Causing Preferential Enrichment

In general, a polymorphic transition frequently occurs during crystallization from the supersaturated solutions of organic compounds, particularly when the packing mode in the first-formed crystal is unstable as the crystal structure [50–54]. Although much less is known about the mechanism of the polymorphic transition during crystallization from solution, it has been believed that the phase transition should proceed through either a solvent-mediated dissolution–recrystallization mechanism according to "Ostwald's law of stages" [55] or a solid-to-solid transformation mechanism with free energy change [56], and that the rate of the polymorphic transition primarily depends on the free energy difference between the two crystalline phases [50–54].

Thus far, we have found three types of solvent-assisted solid-to-solid transformations of a kinetically formed metastable γ-form crystalline phase into a thermodynamically more stable one (α-, δ-, or ε-form) [15, 16, 33–49], which are relevant to the occurrence of preferential enrichment, in a process of crystal growth.

3.1
Assembly Mode of Enantiomers in Solution

To confirm the occurrence of the polymorphic transition and to elucidate the mechanism, it is primarily necessary to clarify the enantiomeric assembly mode in the first-formed metastable crystal prior to the polymorphic transition, and to compare it with the stable crystal structure after the polymorphic transition with respect to the racemic samples showing preferential enrichment. Since it is very possible that the stable molecular assembly structure in solution would be retained in the crystalline phase which is first formed by crystallization from the same solvent [57–59], we investigated the enantiomeric association mode in solutions of the racemates showing preferential enrichment. Consequently, in our study, the variable-temperature ^{1}H NMR technique proved to be inapplicable to deciding which molecular association mode is more stable in solution, homochiral or heterochiral [60–64]. Instead, the combined use of solubility and supersolubility measurements under various conditions and number-averaged molecular weight measurement by vapor pressure osmometry turned out to be a potent tool for this objective. Thus, it was concluded that a homochiral molecular assembly is essentially in preference to a heterochiral one in solution with respect to the racemic samples showing preferential enrichment, and that the homochiral supramolecular structure must be a 1D chain. These results are consistent with those obtained by molecular dynamics simulation of the oligomer models of ST and NNMe$_3$ (Fig. 3) [34].

3.2
Crystal Structure

3.2.1
General Crystalline Nature of a Racemic Mixed Crystal Showing Preferential Enrichment

The crystalline nature and physicochemical properties of the crystals belonging to *a racemic conglomerate* or *a racemic compound* have been thoroughly investigated [17, 28–32]. In contrast, other types of enantiomerically mixed crystals except these two are generally categorized as *a racemic mixed crystal*, namely, *a pseudoracemate* or *a solid solution* (Fig. 1) [17, 32]. Accordingly, *a racemic mixed crystal* refers to the crystal with an irregular alignment of the two enantiomers in defined positions to give both racemic and nonracemic crystals with a variety of ee values flexibly, depending on the molecular structure. Due to the diversity of its enantiomeric arrangement, the crystalline nature and physicochemical properties of this third class of crystals have not been sufficiently understood.

Whether a given racemic crystal is classified as a racemic mixed crystal or not is verified by (1) constructing the binary melting-point phase diagram and (2) comparing the X-ray diffraction pattern and the solid-state IR spectrum of its racemic sample with those of the nearly enantiopure one; almost identical patterns and spectra are obtained for a racemic mixed crystal. Ultimately, it is necessary to observe a molecular structure with the orientational disorder at the position of a substituent on the asymmetric center by the X-ray crystallographic analysis of the racemic and nonracemic single crystals. Depending on the degree of disorder of the enantiomeric arrangement, a racemic mixed crystal is further categorized into three types; (1) disordered, (2) fairly disordered (short-range ordered), and (3) ordered (long-range ordered), of which the ordered type can hardly be distinguished from a racemic compound solely by X-ray crystallographic analysis (Fig. 1) [17]. As described in the following section, the crystalline nature of the compounds showing preferential enrichment falls into *a fairly disordered racemic mixed crystal* that resembles a racemic compound (Fig. 1b).

3.2.2
Metastable γ-Form

We have searched for a compound that shows preferential enrichment and possesses a homochiral 1D chain structure in the metastable polymorphic form, because the homochiral molecular assembly structure in solution is supposed to be retained in the crystal which is first formed by crystallization from the same solvent [57–59]. As a result, the single crystal of racemic

Table 1 Relationship between racemic crystal structures and space groups, and preferential enrichment

Compound	Crystal structure[a]	Space group, Z	Preferential enrichment[b]
ST [42]	Stable α-form, ordered	$P\bar{1}, Z = 4$	Yes
ST [45]	Metastable δ-form, ordered	$P\bar{1}, Z = 2$	Yes
SB [44]	Stable δ-form, disordered (73 : 27)	$P\bar{1}, Z = 2$	Yes
SC [42]	Stable δ-form, ordered	$P\bar{1}, Z = 2$	Yes
NNMe$_2$ [38, 39]	Disordered δ-form, ordered	$P\bar{1}, Z = 2$	Yes
NNMe$_3$ [34]	Stable δ-form, ordered	$P\bar{1}, Z = 2$	Yes
NNMe$_3$-OMe [35]	Stable δ-form, disordered (72 : 28)	$P\bar{1}, Z = 2$	Yes
NCMe$_2$	Stable δ-form, disordered (74 : 26)	$P\bar{1}, Z = 2$	Yes
NCMe$_3$ [41]	Stable δ-form, disordered (77 : 23)	$P\bar{1}, Z = 2$	Yes
NBMe$_3$ [43]	Stable δ-form, disordered (67 : 33)	$P\bar{1}, Z = 2$	Yes
NIMe$_3$ [48]	stable δ-form, disordered (71 : 29)	$P\bar{1}, Z = 2$	Yes
NPMe$_3$ [46]	Stable ε-form, ordered	$P\bar{1}, Z = 2$	Yes
NPMe$_3$ [46]	Metastable γ-form, ordered	$P\bar{1}, Z = 2$	No
SI [47]	Nn, ordered	$P\bar{1}, Z = 2$	No
SP [40]	Nn, disordered (59 : 41)	$P2_1/c, Z = 4$	No
SCC [42]	Stable α-form, ordered	$P\bar{1}, Z = 2$	No
NNMe$_3$-OPr [34]	Stable γ-form, ordered	$P\bar{1}, Z = 2$	No
NNMe$_3$-C [34]	Stable α-form, ordered	$P\bar{1}, Z = 2$	No
NCMe$_3$-OMe [41]	Stable δ$_1$-form, ordered	$P\bar{1}, Z = 2$	No
NBMe$_3$-OMe [35]	Stable κ-form, ordered	$P2_1/c, Z = 4$	No
NBMe$_3$-OMe [35]	Metastable γ-form, ordered	$P\bar{1}, Z = 2$	No
NBMe$_3$-OPr	Stable γ-form, disordered (67 : 33)	$P\bar{1}, Z = 2$	No
NIMe$_3$-OMe [48]	Stable κ-form, ordered	$P2_1/c, Z = 4$	No
NIMe$_3$-OPr [48]	Stable γ-form, disordered (64 : 36)	$P\bar{1}, Z = 2$	No
NTMe$_2$ [38, 39]	Nn, disordered (65 : 35)	$P2_1/c, Z = 4$	No

[a] Each crystal structure form and the degree of disorder at the position of the hydroxy group on an asymmetric carbon atom; the constrained occupancy factors are indicated in parentheses. Nn means no name for the crystal structure form
[b] Indication of the occurrence of preferential enrichment upon the deposition of each crystal form from the supersaturated solution

NPMe$_3$ (Fig. 3) was found to possess the desired γ-form crystal structure that is classified as *an ordered racemic mixed crystal* and is composed of alternating alignment of homochiral 1D *R* and *S* chains in an antiparallel direction with space group $P\bar{1}$ ($Z = 2$) with the second lowest symmetry. Each homochiral chain comprises an alternating alignment of the long-chain ammonium ion and the sulfonate ion by two hydrogen bonds (1) between one oxygen atom of the sulfonate ion and the hydroxy group and (2) between the same sulfonate oxygen atom and the amide NH (Table 1 and Fig. 5) [34, 46]. There is no interchain interaction. This γ-form crystal structure is also seen in several other racemic samples such as NNMe$_3$ – OPr, NBMe$_3$ – OPr, NIMe$_3$ – OPr, and NBMe$_3$ – OMe (Table 1 and Fig. 3) which failed to show preferential en-

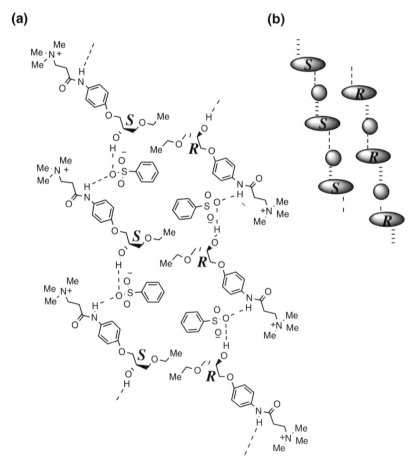

Fig. 5 a Intermolecular interactions in the γ-form crystals and **b** schematic representation of the homochiral 1D chain. The *ellipsoid* and *circle* indicate the long-chain cation and sulfonate ion, respectively. The *dashed lines* show the intermolecular hydrogen bonds. (Reprinted with permission from [46]. Copyright 2006 American Chemical Society)

richment [34, 35, 48], indicating that the subsequent polymorphic transition into the more stable α-, δ-, or ε-form is indispensable for preferential enrichment to occur. Actually, in the case of NPMe$_3$, the successive polymorphic transition of the γ-form into the ε-form occurred slowly to induce preferential enrichment [34, 46].

3.2.3
Stable δ-Form

Thus far, three different types of stable crystal structures, α-, δ-, and ε-forms, which are classified as *an ordered* or *fairly disordered racemic mixed crystal*, have been obtained for the racemic or nearly racemic samples of the compounds showing preferential enrichment. Among them, the δ-form crystal is most commonly found for the compounds showing preferential enrichment [33–44], while only one case is observed for each of the α- and ε-forms (Table 1) [15, 16, 34, 42, 46]. It has been indicated that the formation of the stable α-form crystal of ST is not essential to the occurrence of preferential enrichment and is caused by further polymorphic transition of the once-formed δ-form polymorph [45]. In this section, the crystal structure of the δ-form is described in detail, because this crystal structure provides crucial information on the mechanism of the polymorphic transition causing preferential enrichment.

For example, the ordered or fairly disordered crystal structure was observed in the δ-form single crystals of racemic NNMe$_3$ or NBMe$_3$ (Fig. 3), respectively, which were produced at much lower concentrations than the usual preferential enrichment conditions [34, 43, 49]. Furthermore, the crystal structure of the nonracemic (20% ee) NBMe$_3$ was found to be virtually isomorphous with that of the racemate and very similar to that of the pure enantiomer [34, 43, 49].

The crystal structure of nearly racemic NNMe$_3$ is characterized by two types of centrosymmetric cyclic dimers, types A and B. The head-to-head cyclic dimer of type A is formed by the hydrogen bonds between the hydroxy groups and the ethoxy oxygen atoms in a pair of *R* and *S* molecules, and another head-to-head cyclic dimer of type B is formed by (1) the hydrogen bond between one oxygen atom of a sulfonate ion and the amide NH and (2) the electrostatic interaction between another oxygen atom of the same sulfonate ion and the ammonium nitrogen atom in the neighboring long-chain cation (Fig. 6). By virtue of these intermolecular interactions, a heterochiral 1D chain is formed. Furthermore, each 1D chain interacts with two adjacent chains by another weak electrostatic interaction between the third oxygen atom of the same sulfonate ion and the ammonium nitrogen atom in the adjacent chains, eventually forming a weak 2D sheet structure [34]. This fragile 2D sheet structure is essential for crystal disintegration resulting in preferential enrichment, as discussed in Sect. 3.3.

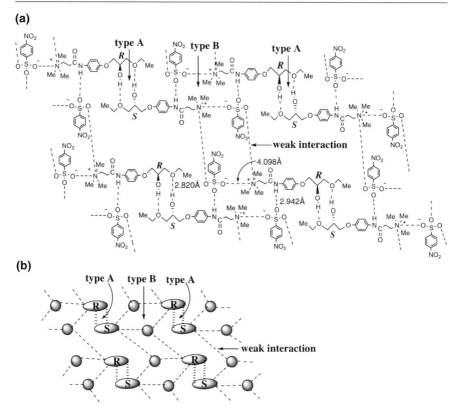

Fig. 6 a Intermolecular interactions in the δ-form crystal of racemic NNMe₃ and **b** schematic representation of the heterochiral 2D sheet structure. The *ellipsoid* and *circle* indicate the long-chain ammonium ion and sulfonate ion, respectively. The *dashed lines* show the intermolecular hydrogen bonds and electrostatic interactions. (Reprinted with permission from [34]. Copyright 2006 American Chemical Society)

In the crystals of racemic and nonracemic (20% ee) NBMe₃, the orientational disorder was observed at the position of the hydroxy group on an asymmetric carbon atom (Fig. 7). The degree of disorder per asymmetric unit (one salt) was estimated to be 65 : 35 and 75 : 25 (R vs S, or S vs R), respectively, by calculating the constrained occupancy factors of the hydroxy group, corresponding to a fairly disordered racemic mixed crystal [34, 43]. In other respects, the crystal structure is identical to that of the δ-form crystal of nearly racemic NNMe₃. Therefore, it is easily conceivable that such a nonracemic crystal with a low ee value, which is classified as a fairly disordered racemic mixed crystal and can accommodate excess enantiomers flexibly in the crystal lattice, is also produced by the preferential enrichment experiment in other cases. Namely, an ordered racemic mixed crystal such as NNMe₃ can also have fair disorder under the pref-

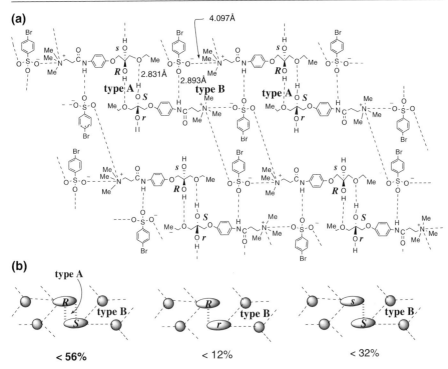

Fig. 7 a Intermolecular interactions in the δ-form crystal of S-rich NBMe$_3$ (20% ee) and **b** schematic representation of the heterochiral cyclic dimer and the homochiral non-cyclic ones in the crystal. The *ellipsoid* and *circle* indicate the long-chain ammonium ion and sulfonate ion, respectively. The *dashed lines* show the intermolecular hydrogen bonds and electrostatic interactions. The hydroxy group on the asymmetric carbon atom is disordered over two positions in **a**. The R and S enantiomers in the sites with higher occupancy factor (0.75) are designated R and S, and those in the sites with lower occupancy factors (0.25) are r and s. The contents of three dimer structures were estimated from the occupancy factors of the orientationally disordered hydroxy groups and the ee value (20%) of the crystal. (Reprinted with permission from [34]. Copyright 2006 American Chemical Society)

erential enrichment experimental conditions using a highly supersaturated solution.

3.3
Mode of Polymorphic Transition

Before discussing the overall mode of the polymorphic transition, let us consider the association mode of the homochiral 1D R and S chains in the metastable γ-form crystal formed kinetically from the highly supersaturated solution (Fig. 8). Supposing that the R enantiomer is slightly in excess in the supersaturated solution, then the resulting metastable γ-form crystal must

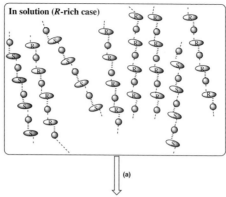
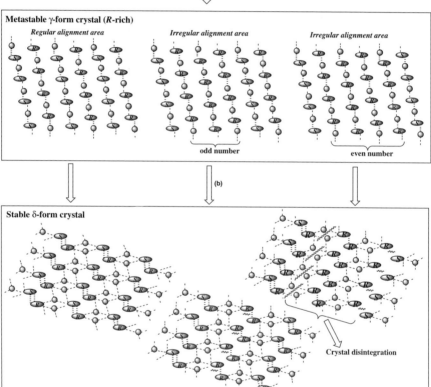

Fig. 8 Mechanism of preferential enrichment. **a** Association mode of homochiral 1D *R* and *S* chains in the metastable γ-form crystal formed kinetically from the highly supersaturated solution of slightly *R*-rich sample. **b** Modes of polymorphic transition of the γ-form into the δ-form in the regular and irregular alignment areas of the *R* and *S* chains in the crystal

become *R*-rich, too. This slightly *R*-rich crystal exists as a fairly disordered racemic mixed crystal which consists of three types of alignment modes of the homochiral *R* and *S* chains; the major regular alignment of the *R* and *S* chains, along with two minor irregular alignments in which an odd number of *R* chains and an even number of *R* chains are respectively surrounded by two *S* chains (Fig. 8a).

Now let us consider the mode of the polymorphic transition of the metastable γ-form into the δ-form. By comparing the crystal structures between the metastable γ-form composed of the homochiral 1D *R* and *S* chains and the stable δ-form comprised of the heterochiral 2D sheet, the mode of this single phase transition can be proposed (Fig. 8b).

Polymorphic transition in the regular alignment area of the *R* and *S* chains is initiated by rearrangement of the hydrogen bonds inside the crystal lattice. It is very possible for the nearest *R* and *S* molecules in the two adjacent chains in the γ-form to form new hydrogen bonds between the ethoxy oxygen atoms and the hydroxy groups by slight movement of the two molecules in the crystal. This rearrangement of hydrogen bonds, accompanied by slight movement of the sulfonate ions so as to form cyclic dimers of types A and B, occurs one after another in the crystal lattice and leads to the heterochiral 1D chains and then the weak 2D sheet structure.

The fashion of alignment of the homochiral *R* and *S* chains in the metastable γ-form crystal must define the type of the local δ-form crystal structure after the phase transition:

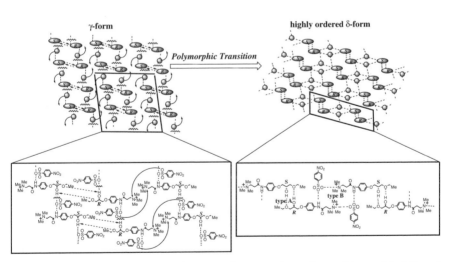

Fig. 9 Polymorphic transition in the case where equal numbers of homochiral *R* and *S* chains are alternately aligned in the γ-form crystal, leading to the ordered δ-form crystal

1. At the sites where equal numbers of homochiral R and S chains are alternately aligned in an antiparallel direction along one axis, an ordered δ-form crystal structure will be formed locally after the polymorphic transition (Figs. 8b and 9).

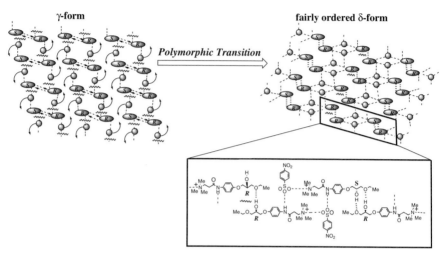

Fig. 10 Polymorphic transition in the case where an odd number of homochiral R chains are surrounded by two S chains in the γ-form crystal, leading to the fairly disordered δ-form crystal without crystal disintegration

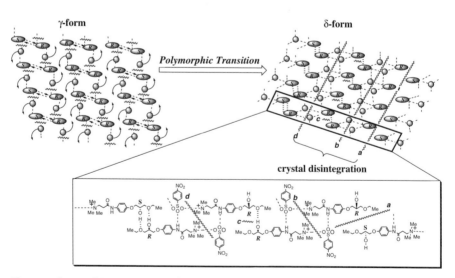

Fig. 11 Polymorphic transition in the case where an even number of homochiral R chains are surrounded by two S chains in the γ-form crystal, leading to local crystal disintegration in the δ-form crystal that occurs between the sites *a* and *b*

2. At the sites where an odd number of homochiral R chains are aligned between two S chains, after the polymorphic transition, the disordered δ-form crystal structure will be formed locally without disintegration of the crystal (Figs. 8b and 10).
3. At the sites where an even number of homochiral R chains are aligned between two S chains, after the polymorphic transition, partial disintegration occurs in the resulting δ-form crystal to release the R-rich area into solution, because the R-rich area was surrounded by the sites where the OH···OEt interaction cannot be formed as well as the sites with weak electrostatic interactions (Figs. 8b and 11).

This third case corresponds to the visually observed redissolution of the excess enantiomer from the just-made crystals into solution (Fig. 4) and must be responsible for the phenomenon characteristic of preferential enrichment, *a considerable enrichment of the excess enantiomer in the mother liquor.*

4
Relationship Between Molecular Structures and Preferential Enrichment

Typical compounds that have been found to show preferential enrichment are summarized in Fig. 3, together with the compounds failing to show this phenomenon. These are analogous linear asymmetric secondary alcohols containing a glycerol moiety, an amide group, and a sulfonium sulfonate or ammonium sulfonate structure. Their molecular weights are about 500. Figure 12 summarizes the structural requirements for the occurrence of preferential enrichment [34, 49]:

1. The amide, hydroxy, and terminal alkoxy groups are all indispensable for the formation of hydrogen bonds.
2. The terminal alkoxy group must be methoxy or ethoxy; longer or bulky alkoxy groups fail to show preferential enrichment due to the steric effect.
3. The onium sulfonate salt structure is advantageous for the occurrence of the polymorphic transition owing to the mobility of the anion in the crystal lattice.

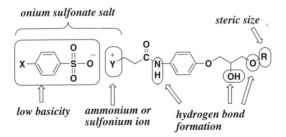

Fig. 12 Structural requirements for the occurrence of preferential enrichment

4. The selection of the sulfonate ion is crucial. Sulfonate ions with low basicity, which can form weak hydrogen bonds to allow the subsequent rearrangement of hydrogen bonds in the crystal lattice, i.e., polymorphic transition, are appropriate, whereas highly basic sulfonate ions, which can form strong hydrogen bonds, fail to induce preferential enrichment.

5
Origin of Preferential Enrichment

The origin of preferential enrichment can be interpreted in terms of an interplay of: (1) the polymorphic transition of the metastable γ-form into the stable α-, δ-, or ε-form; (2) the substantially higher solubility of the sample with high ee compared with that of low ee to allow *a considerable enrichment of the excess enantiomer in the mother liquor*; and (3) the presence of a wide supersaturated concentration range to allow the overdissolution of the excess enantiomer from the just-made crystals until *a slight enrichment of the opposite enantiomer in the deposited crystals* occurs.

The proposed overall mechanism of preferential enrichment illustrated in Fig. 8 is the case where a considerable enrichment of the R enantiomer occurs in the mother liquor by crystallization from the slightly R-rich (around 5% ee) supersaturated solution. In solution, homochiral 1D R and S chains are formed preferentially and aggregate to form a γ-form supramolecular cluster, which undergoes a phase transition to give the incipient metastable γ-form crystals with fair disorder. Since the R enantiomer is slightly in excess in solution, the resulting metastable γ-form crystals become slightly R-rich, too. These slightly R-rich crystals consist of three types of alignment modes of the homochiral R and S chains: the major regular alignment of the R and S chains, along with two minor irregular alignments in which an odd number of R chains and an even number of R chains are respectively surrounded by two S chains. The next step is the polymorphic transition initiated by the rearrangement of the hydrogen bonds in the crystal lattice, as discussed in Sect. 3.3. After the polymorphic transition, partial disintegration in the resulting δ-form crystals occurs at the sites where an even number of homochiral R chains are aligned between two S chains, releasing the R-rich area into solution. In contrast, at the sites where an even number of homochiral S chains are aligned between two R chains, the dissolution of the S-rich area into solution seemingly does not proceed. As soon as the dissolved 1D S chains meet with excess R chains in solution, crystallization of the R-rich γ-form crystals occurs, followed by redissolution of the excess R enantiomer into solution to result in the deposition of slightly S-rich δ-form crystals. Thus, the enantiomeric purity of the R enantiomer in the mother liquor is gradually raised until the crystallization ceases. More importantly, within the wide supersaturated concentration range, the substantially higher solubility of the materials with high ee values than those with low ee values allows the R enantiomer to be enriched in the mother liquor gradually until the opposite S enantiomer becomes in excess in the deposited crystals.

It should be stressed again that when the original supersaturated solution was strictly racemic (0.0% ee), the probability for either the R or the S enantiomer to be enriched in solution after crystallization was 50%. This is because initial capricious formation of the very first nonracemic metastable crystal nucleus should decide which enantiomer is enriched in the mother liquor, i.e., this is a typical feature in an event of a complexity system.

6
Induction and Inhibition of Preferential Enrichment

As shown in Sect. 4, it was found that a slight change in the molecular structures of a series of racemic substances showing preferential enrichment largely affects the mode of polymorphic transition during crystallization from solution, resulting in the formation of stable crystals with different crystal structures and thereby no occurrence of preferential enrichment. In this section, we describe two examples in which the racemic sample, which cannot show preferential enrichment by itself, comes to show this enantiomeric resolution phenomenon with the aid of appropriate seed crystals [35].

6.1
Significant Influence of Minor Molecular Modification on the Mode of Polymorphic Transition

Here we focus on the terminal methoxy derivatives [($\pm$)-NNMe$_3$ – OMe (**1a**), NCMe$_3$ – OMe (**1b**), and NBMe$_3$ – OMe (**1c**)] of the prototype ammonium sulfonates ($\pm$)-NNMe$_3$ (**2a**), NCMe$_3$ (**2b**), and NBMe$_3$ (**2c**) which showed preferential enrichment [34, 41, 43]. Whether ($\pm$)-**1a**, **1b**, and **1c** exhibit preferential enrichment or not depends only on the kind of electron-withdrawing group situated at the *para* position of the benzenesulfonate group. Namely, the *p*-nitro derivative ($\pm$)-**1a** successfully exhibited preferential enrichment [35], whereas the *p*-chloro and *p*-bromo derivatives [($\pm$)-**1b** and **1c**] failed to do so. It was concluded that all of ($\pm$)-**1a**–**1c** and ($\pm$)-**2a**–**2c** predominantly adopt a γ-form supramolecular structure in their supramolecular solutions in EtOH, by in situ ATR-FTIR spectroscopy and on the basis of our previous studies [34, 35, 49]. We compared the crystal structures of the stable forms of ($\pm$)-**1a**, **1b**, and **1c** with those of ($\pm$)-**2a**, **2b**, and **2c**, because it is possible to predict each mode of polymorphic transition during crystallization for these three racemates which cannot show preferential enrichment.

Similarly to the case of ($\pm$)-**2b** and **2c**, the crystal structure of ($\pm$)-**1a** is a fairly disordered δ-form which has the orientational disorder at the position of the hydroxy group on an asymmetric carbon atom with occupancy factors of 0.72 and 0.28 [35].

The stable crystal structure of (±)-**1b** was found to be partly similar to a δ-form which consists of a 1D chain composed of type A and type B heterochiral cyclic dimers by X-ray crystallographic analysis [41]. We call this new crystal structure a $δ_1$-form. The decisive difference in crystal structure between the δ- and $δ_1$-forms is the mode of interchain interactions between the analogous heterochiral 1D chains (Fig. 13). In the $δ_1$-form of (±)-**1b**, there are two interchain interactions: one is the relatively strong slipped-parallel π–π stacking between the benzene rings of the nearest two *p*-chlorobenzenesulfonate groups, and the other is weak $C(sp^2)H - Cl$ contacts between the chlorine atom of the *p*-chlorobenzenesulfonate group and two vicinal hydrogen atoms on the benzene ring of the neighboring long-chain cation. These fairly strong interchain interactions must be responsible for no occurrence of preferential enrichment of **1b**, because the subsequent partial crystal disintegration which is necessary to cause preferential enrichment does not occur.

With respect to (±)-**1c**, a metastable γ-form was crystallized from 2-PrOH, while a new, stable κ-form was obtained as a monophasic powder sample from EtOH [35]. The crystal structure of the κ-form, which was solved from its powder X-ray diffraction data by the direct-space approach using the Monte Carlo method with subsequent Rietveld refinement [65], consists of a fairly strong heterochiral 2D sheet structure particularly characterized by the Coulombic donor–acceptor attraction between the methoxy oxygen atom and the nearest bromine atom (Fig. 14). It is most likely that this O · · · Br in-

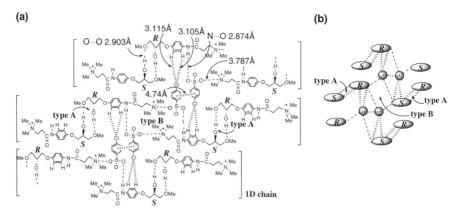

Fig. 13 a Intermolecular interactions in the $δ_1$-form crystal of (±)-**1b** and **b** schematic representation of the heterochiral 2D sheet structure. The *ellipsoid* and *circle* indicate the long-chain ammonium ion and sulfonate ion, respectively. The *dashed lines* show the intermolecular hydrogen bonds, electrostatic interactions, and π–π stackings. (Reprinted with permission from [35]. Copyright 2006 Wiley)

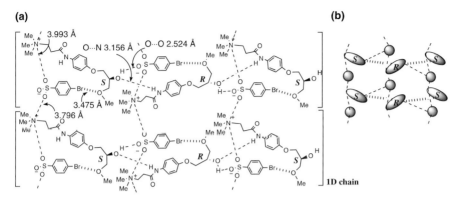

Fig. 14 a Intermolecular interactions in the κ-form crystal of (±)-**1c** and **b** schematic representation of the heterochiral 2D sheet structure. The *ellipsoid* and *circle* indicate the long-chain ammonium ion and sulfonate ion, respectively. The *dashed lines* show the intermolecular hydrogen bonds and electrostatic interactions. (Reprinted with permission from [35]. Copyright 2006 Wiley)

Table 2 Induction or inhibition of preferential enrichment by mutual seeding[a]

Compound	Solubility in EtOH (mg/mL)	Without seeding[b]	(±)-**1a** δ-form	Seed crystal[c] (±)-**1b** $δ_1$-form	(±)-**1c** κ-form
(±)-**1a**	12.9	Yes[d] (γ to δ)[g]	Acceleration[e] (γ to δ)[g]	Inhibition[f] (γ to $δ_1$)[g]	No effect
(±)-**1b**	59.1	No[h] (γ to $δ_1$)[g]	Induction[i,j] (γ to δ, then $δ_1$)[g]	–	No effect
(±)-**1c**	19.9	No[h] (γ to κ)[g]	No effect	Induction[i] (γ to δ, then $δ_1$)[g]	–

[a] Ten-, three-, and fivefold supersaturated EtOH solutions were used for recrystallization of (±)-**1a**, (±)-**1b**, and (±)-**1c**, respectively
[b] Recrystallization was carried out without seed crystals
[c] 5 wt % of seed crystals was added
[d] Preferential enrichment occurred
[e] Preferential enrichment was accelerated
[f] Preferential enrichment was completely inhibited by seeding
[g] Mode of polymorphic transition
[h] Preferential enrichment did not occur
[i] Preferential enrichment was induced by seeding
[j] See Fig. 16 (Reprinted with permission from [35]. Copyright 2006 Wiley)

teraction makes a significant contribution to building up the κ-form crystal structure, and therefore the polymorphic transition of the metastable γ-form into the δ-form causing preferential enrichment cannot occur.

Thus, it is quite plausible that the incipient formation of the individual metastable γ-form polymorphs and the subsequent polymorphic transition into the stable δ-form, $δ_1$-form, or κ-form should have occurred during crystallization of (±)-**1a**, (±)-**1b**, or (±)-**1c** from EtOH, respectively (Table 2).

6.2
Induction and Inhibition of Preferential Enrichment by Controlling the Mode of Polymorphic Transition with Seed Crystals

It is noteworthy that the δ-form of (±)-**1a**, the $δ_1$-form of (±)-**1b**, and the γ-form of (±)-**1c** have a similar columnar structure along one axis (Fig. 15a–c). This fact suggests the possibility of forced adsorption of the γ-form prenucleation aggregates of (±)-**1b** or (±)-**1c** on the columnar surface of the δ-form of (±)-**1a**, and the subsequent polymorphic transition of the incipient γ-form crystalline phase into the desired δ-form on the same crystalline surface of (±)-**1a**. If this assumption is correct, preferential enrichment must be induced or inhibited by the alteration of the mode of polymorphic transition by seeding with each other. Such was indeed the case [35].

By seeding the supersaturated solution of (±)-**1b** in EtOH with more than 3 wt % of the δ-form crystals of (±)-**1a**, crystallization of **1b** began quickly and preferential enrichment was distinctly induced (Fig. 16 and Table 2). The crystal structure of the deposited crystals was found to be a fairly disordered (at the position of the hydroxy group on an asymmetric carbon atom with occupancy factors of 0.7 : 0.3) $δ_1$-form by using the direct-space approach with the Monte Carlo method (Fig. 15e). It is quite plausible that the successive polymorphic transitions of the metastable fairly disordered γ-form of (±)-**1b** into the fairly disordered δ-form and then the fairly disordered $δ_1$-form should occur on the columnar surface of the δ-form seed crystal of (±)-**1a**.

In contrast, by seeding the supersaturated solution of (±)-**1c** in EtOH with the δ-form crystals of (±)-**1a**, neither the acceleration of crystallization nor the formation of another polymorphic form was observed, nor was the induction of preferential enrichment noted. Instead, more interestingly, seeding

Fig. 15 Similar columnar crystal structures of **a** the δ-form of (±)-**1a** viewed down the *b* axis, **b** the $δ_1$-form of (±)-**1b**, crystallized without seed crystals, viewed down the *a* axis, **c** the γ-form of (±)-**1c** viewed down the *a* axis, **d** the $δ_1$-form of (±)-**1a** viewed down the *a* axis, **e** the $δ_1$-form of (±)-**1b**, crystallized with seed crystals (5 wt %) of the δ-form of (±)-**1a**, viewed down the *a* axis, and **f** the $δ_1$-form of (±)-**1c** viewed down the *a* axis. The C, O, N, S, Cl, and Br atoms are represented by *gray, red, blue, yellow, green,* and *purple sticks*, respectively. All H atoms in **c** are omitted. (Reprinted with permission from [35]. Copyright 2006 Wiley)

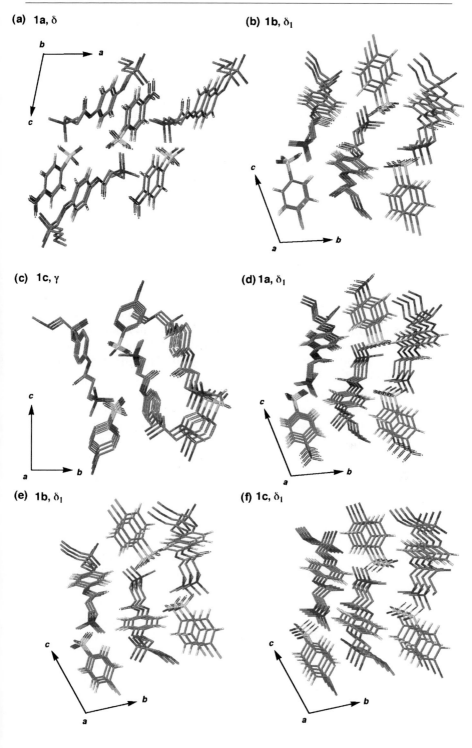

(a) **1a, δ**

(b) **1b, δ₁**

(c) **1c, γ**

(d) **1a, δ₁**

(e) **1b, δ₁**

(f) **1c, δ₁**

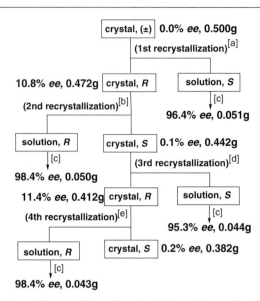

Fig. 16 Preferential enrichment of (±)-**1b** induced by seeding with the δ-form crystal of (±)-**1a** (5 wt %) in EtOH. Conditions: **a** in EtOH (2.5 mL) with 0.025 g of (±)-**1a** at – 16 °C for 7 days; **b** in EtOH (2.4 mL) with 0.024 g of (±)-**1a** at – 16 °C for 7 days; **c** removal of the solvent by evaporation; **d** in EtOH (2.20 mL) with 0.021 g of (±)-**1a** at – 16 °C for 7 days; **e** in EtOH (2.0 mL) with 0.020 g of (±)-**1a** at – 16 °C for 7 days. (Reprinted with permission from [35]. Copyright 2006 Wiley)

with the $δ_1$-form crystal (5 wt %) of (±)-**1b** in EtOH induced prompt crystallization to provide a fairly disordered (at the position of the hydroxy group on an asymmetric carbon atom with occupancy factors of 0.7 : 0.3) $δ_1$-form of **1c** (Fig. 15f) and distinctly induced preferential enrichment for (±)-**1c** (Table 2). Therefore, it is again likely that successive polymorphic transitions of the fairly disordered γ-form of (±)-**1c** into the fairly disordered δ-form and then the fairly disordered $δ_1$-form should occur on the columnar surface of the $δ_1$-form seed crystal of (±)-**1b**. This is a quite interesting example of the successful induction of preferential enrichment by using the seed crystals of another compound that cannot show preferential enrichment by itself.

Conversely, the occurrence of preferential enrichment of (±)-**1a** in EtOH was completely inhibited by seeding with the $δ_1$-form crystals (more than 3 wt %) of (±)-**1b** (Table 2). In this case, it was shown that a single phase transition of the γ-form into the $δ_1$-form of (±)-**1a** should occur on the columnar surfaces of the $δ_1$-form seed crystal of (±)-**1b**.

With respect to the compounds showing preferential enrichment by themselves, such as (±)-**1a** and (±)-**2a**–(±)-**2c**, addition of the individual seed crystals to their own supersaturated solutions in EtOH was found to in-

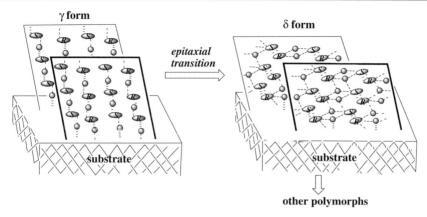

Fig. 17 Schematic representation of "epitaxial transition" of the metastable γ-form polymorph into the δ-form, causing preferential enrichment on the surface of a δ- or $δ_1$-form seed crystal. (Reprinted with permission from [35]. Copyright 2006 Wiley)

duce prompt crystallization and thereby accelerate the rate of preferential enrichment (Table 2). This acceleration effect can be accounted for by an "epitaxial transition" phenomenon that involves (1) the adsorption of the γ-form prenucleation aggregates, (2) the heterogeneous nucleation and crystal growth of the incipient metastable γ-form polymorph, and (3) the subsequent polymorphic transition into the more stable δ-form; these multistage processes occur on the same surface of the δ-form seed crystal (Fig. 17). Likewise, in the mechanism of preferential enrichment induced by seeding the supersaturated solution of (±)-**1b** or (±)-**1c** in EtOH with the δ-form of (±)-**1a** or the $δ_1$-form of (±)-**1b**, respectively, a similar epitaxial transition phenomenon must operate (Fig. 17). These results strongly support our proposed mechanism of preferential enrichment including polymorphic transition.

7
Conclusions and Scope

The mechanism of preferential enrichment has been interpreted in terms of a symmetry-breaking complexity system including three major elements: (1) the polymorphic transition of an incipient metastable crystalline form into the more stable form, (2) the substantial higher solubility of the sample with high ee than that of low ee to allow *a considerable enrichment of the excess enantiomer in the mother liquor*, and (3) the presence of a wide supersaturated concentration range to allow the overdissolution of the excess enantiomer from the just-made crystals into solution until *a slight enrichment of the opposite enantiomer in the deposited crystals* occurs. Therefore, preferential enrichment is not only an excellent chiral separation method but also a good research model for a complexity system.

Successful induction and inhibition of preferential enrichment, based on a new concept of epitaxial transition, strongly support the proposed mechanism of preferential enrichment. If the crystal lattices and crystal structures of the two compounds are partly similar, it is possible to control the mode of polymorphic transition mutually by seeding with each other. Accordingly, once one finds a leading racemic sample that can show preferential enrichment, it is possible to induce preferential enrichment for analogous compounds that cannot show preferential enrichment by themselves, on the basis of this epitaxial transition protocol.

It is quite conceivable that preferential enrichment should be applicable to other chiral organic substances that exist as a fairly disordered racemic mixed crystal, if the above requirements are satisfied. Furthermore, by carefully choosing the kinetic crystallization conditions or seeding with appropriate crystals, preferential enrichment might be applicable to even certain racemic compounds that satisfy the structural requirements for supramolecular association in solution and in the crystalline state.

References

1. Pasteur L (1848) Ann Chim Phys 24:442
2. Lough WL, Wainer IW (eds) (2002) Chirality in natural and applied science. Blackwell Science, Oxford
3. Cintas P (2002) Angew Chem Int Ed 41:1139
4. Quack M (2002) Angew Chem Int Ed 41:4618
5. Avalos M, Babiano R, Cintas P, Jiménez JL, Palacios JC (2000) Chem Commun 887
6. Buschmann H, Thede R, Heller D (2000) Angew Chem Int Ed 39:4033
7. Poclech J (1999) Angew Chem Int Ed 38:477
8. Lee T, Yang C (1956) Phys Rev 104:254
9. Wu C, Ambler E, Hayward R, Hoppes D, Hudson R (1957) Phys Rev 105:1413
10. Avalos M, Babiano R, Cintas P, Jiménez JL, Palacios JC, Barron LD (1998) Chem Rev 98:2391
11. Mainzer K (2005) Symmetry and complexity: the spirit and beauty of nonlinear science. World Scientific, Singapore
12. Kauffman S (2000) Investigations. Oxford University Press, Oxford
13. Waldrop MM (1992) Complexity: the emerging science at the edge of chaos. Simon & Schuster, New York
14. Pérez-García L, Amabilino DB (2002) Chem Soc Rev 31:342
15. Ushio T, Tamura R, Takahashi H, Yamamoto K (1996) Angew Chem Int Ed 35:2372
16. Ushio T, Tamura R, Azuma N, Nakamura K, Toda F, Kobayashi K (1996) Mol Cryst Liq Cryst 276:245
17. Jacques J, Collet A, Wilen SH (1994) Enantiomers, racemates and resolutions. Krieger, Malabar, FL
18. Kozma D (ed) (2001) CRC handbook of optical resolutions via diastereomeric salt formation. CRC, Boca Raton, FL

19. Kinbara K, Saigo K (2003) Top Stereochem 23:207
20. Vries T, Wynberg H, van Echten E, Koek J, ten Hoeve W, Kellog RM, Broxterman QB, Minnaard A, Kaptein B, van der Sluis S, Hulshof L, Kooistra J (1998) Angew Chem Int Ed 37:2349
21. Nieuwenhuijzen JW, Grimbergen RFP, Koopman C, Kellogg RM, Vries TR, Pouwer K, van Echten E, Kaptein B, Hulshof LA, Broxterman QB (2002) Angew Chem Int Ed 41:4281
22. Gervais C, Grimbergen RFP, Markovits I, Ariaans GJA, Kaptein B, Bruggink A, Broxterman QB (2004) J Am Chem Soc 126:655
23. Sakai K, Sakurai R, Yuzawa A, Hirayama N (2003) Tetrahedron Asymmetry 14:3713
24. Sakai K, Sakurai R, Hirayama N (2004) Tetrahedron Asymmetry 15:1073
25. Sakai K, Sakurai R, Nohira H, Tanaka R, Hirayama N (2004) Tetrahedron Asymmetry 15:3495
26. Toda F (1987) Top Curr Chem 140:43
27. Toda F (2004) In: Toda F (ed) Enantiomer separation. Kluwer, Dordrecht, p 1
28. Gernez M (1866) C R Hebd Seances Acad Sci 63:843
29. Collet A, Brienne M-J, Jacques J (1980) Chem Rev 80:215
30. Collet A (1996) Optical resolution. In: Reinhoudt DN (ed) Comprehensive supramolecular chemistry, Vol. 10. Pergamon, Oxford, p 113
31. Kinbara K, Hashimoto Y, Sukegawa M, Nohira H, Saigo K (1996) J Am Chem Soc 118:3441
32. Eliel E, Wilen SH, Mander LN (1994) Stereochemistry of organic compounds. Wiley, New York, p 297
33. Tamura R, Takahashi H, Hirotsu K, Nakajima Y, Ushio T, Toda F (1998) Angew Chem Int Ed 37:2876
34. Tamura R, Fujimoto D, Lepp Z, Misaki K, Miura H, Takahashi H, Ushio T, Nakai T, Hirotsu K (2002) J Am Chem Soc 124:13139
35. Tamura R, Mizuta M, Yabunaka S, Fujimoto D, Ariga T, Okuhara S, Ikuma N, Takahashi H, Tsue H (2006) Chem Eur J 12:3515
36. Tamura R, Ushio T, Nakamura K, Takahashi H, Azuma N, Toda F (1997) Enantiomer 2:277
37. Tamura R, Ushio T, Takahashi H, Nakamura K, Azuma N, Toda F, Endo K (1997) Chirality 9:220
38. Takahashi H, Tamura R, Ushio T, Nakajima Y, Hirotsu K (1998) Chirality 10:705
39. Tamura R, Takahashi H, Hirotsu K, Nakajima Y, Ushio T (2001) Mol Cryst Liq Cryst 356:185
40. Tamura R, Takahashi H, Ushio T, Nakajima Y, Hirotsu K, Toda F (1998) Enantiomer 3:149
41. Tamura R, Takahashi H, Miura H, Lepp Z, Nakajima Y, Hirotsu K, Ushio T (2001) Supramol Chem 13:71
42. Takahashi H, Tamura R, Lepp Z, Kobayashi K, Ushio T (2001) Enantiomer 6:57
43. Takahashi H, Tamura R, Fujimoto D, Lepp Z, Kobayashi K, Ushio T (2002) Chirality 14:541
44. Takahashi H, Tamura R, Yabunaka S, Ushio T (2003) Mendeleev Commun p 119
45. Miura H, Ushio T, Nagai K, Fujimoto D, Lepp Z, Takahashi H, Tamura R (2003) Cryst Growth Des 3:959
46. Fujimoto D, Tamura R, Lepp Z, Takahashi H, Ushio T (2003) Cryst Growth Des 3:973
47. Takahashi H, Tamura R, Yabunaka S, Mizuta M, Ikuma N, Tsue H, Ushio T (2004) Mendeleev Commun p 239
48. Fujimoto D, Takahashi H, Ariga T, Tamura R (2006) Chirality 18:188

49. Tamura R, Ushio T (2004) In: Toda F (ed) Enantiomer separation. Kluwer, Dordrecht, p 135
50. Bernstein J (2002) Polymorphism in molecular crystals. Oxford University Press, Oxford
51. Brittain HG (ed) (1999) Polymorphism in pharmaceutical solids: drugs and the pharmaceutical sciences, vol. 95. Marcel Dekker, New York
52. Bernstein J, Davey RJ, Henck J-O (1999) Angew Chem Int Ed 38:3440
53. Dunitz JD, Bernstein J (1995) Acc Chem Res 28:193
54. McCrone WC (1965) Polymorphism. In: Fox D, Labes MM, Weissberger A (eds) Physics and chemistry of the organic solid state, vol. 2. Interscience, New York, p 726
55. Ostwald W (1899) Grundriss der Allgemeinen Chemie. Leipzig
56. Parkinson GM, Thomas JM, Williams JO, Goringe MJ, Hobbs LW (1976) J Chem Soc Perkin Trans 2 836
57. Maruyama S, Ooshima H (2000) J Cryst Growth 212:239
58. Kitamura M, Ueno S, Sato K (1998) Molecular aspects of the polymorphic crystallization of amino acids and lipids. In: Ohtaki H (ed) Crystallization processes. Wiley, Chichester, p 99
59. Kitamura M (1989) J Cryst Growth 96:541
60. Dobashi A, Saito N, Motoyama Y, Hara S (1986) J Am Chem Soc 108:307
61. Jursic BS, Goldberg SI (1992) J Org Chem 57:7172
62. Cung MT, Marraud M, Neel J (1978) Biopolymers 17:149
63. Harger MJP (1977) J Chem Soc Perkin Trans 2 1882
64. Harger MJP (1978) J Chem Soc Perkin Trans 2 326
65. Harris KDM, Tremayne M, Kariuki BM (2001) Angew Chem Int Ed 40:1626

Racemization, Optical Resolution and Crystallization-Induced Asymmetric Transformation of Amino Acids and Pharmaceutical Intermediates

Ryuzo Yoshioka

Process Chemistry Department CMC Research Laboratories, Tanabe Seiyaku Co., LTD., 532-8505 Osaka, Japan
yoshioka@tanabe.co.jp

1	Introduction	84
2	Racemization and Asymmetric Transformation	85
2.1	Racemization of Free Amino Acids	86
2.1.1	Racemization of Various Free Amino Acids	87
2.1.2	Effect of Various Aldehydes and Low Aliphatic Acids	88
2.1.3	Racemization Mechanism of Free Amino Acids	90
2.1.4	Racemization of Racemic Mixtures and Racemic Compounds	92
2.2	Racemization of Amino Acid Salts	93
2.2.1	Racemization of Various Amino Acid Salts	94
2.2.2	Effect of Various Aldehydes and Free Amino Acids	95
2.3	Conclusions	97
3	Asymmetric Transformation of Enantiomers	98
3.1	Enantiomeric Resolution and Its Asymmetric Transformation	98
3.2	Asymmetric Transformation of Acyl-DL-Amino Acids	100
3.3	Asymmetric Transformation of DL-Amino Acid Salts	102
3.3.1	Asymmetric Transformation of DL-HPG·oTS	102
3.4	Reports and Aspect	104
4	Asymmetric Transformation of Diastereomers	106
4.1	Diastereomeric Resolution and Its Asymmetric Transformation	107
4.2	Asymmetric Transformation of Amino Acid Salts Using Chiral Reagents . .	107
4.2.1	Asymmetric Transformation of DL-HPG with (+)-PES	108
4.2.2	Asymmetric Transformation of L-Asp(OMe) with (–)-PES	112
4.3	Reports and Aspect of Pharmaceutical Intermediates	114
4.3.1	Substrates	116
4.3.2	Resolving Agents	126
4.3.3	Reaction Conditions	126
5	Conclusions	127
References		128

Abstract Tanabe Seiyaku has been investigating an efficient optical resolution method for the production of optically active amino acids since the 1950s. As one of the practical applications of the resolution methods, we focused on crystallization-induced asymmetric transformation, with which it is possible to obtain more than 50% of one enantiomer of a racemate. In order to achieve the asymmetric transformation, an elegant method for racemization of optically active amino acids and their salts was developed. This successful racemization procedure led to efficient and economical preparation paths for various optically active amino acids by the two crystallization-induced asymmetric transformations, one of which is a combination of enantiomeric resolution and simultaneous racemization and the other is a combination of diastereomeric resolution and simultaneous epimerization. Here, many examples of our studies and recent reports of pharmaceutical intermediates are presented.

Keywords Optical resolution · Asymmetric transformation · Racemization · Chiral compound · Amino acid

Abbreviations
HPG   p-hydroxyphenylglycine
PES   1-phenylethanesulfonic acid
Asp(OMe) aspartic acid β-methyl ester

1
Introduction

Optical resolution via crystallization, although seen as a classical method, has been used for over 150 years since the first discovery by Pasteur [1, 2] and is recognized now as one of the convenient procedures available for separating enantiomers at both laboratory and industrial scales [3, 4]. Recent important developments in chiral pharmaceuticals include optical resolution of synthesized racemic substrates (*synthesis-resolution*), which is of great importance for developing single enantiomers because it is often used in the first stage of chiral drugs development [5–7, 10]. However, the present synthesis-resolution process, especially when industrial applications are contemplated, has an unavoidable disadvantage. That is, the wanted isomer (denoted as e.g. S) is ordinarily only one enantiomer of the corresponding racemate (RS), and the maximum theoretical yield by this resolution is 50% (half of the RS-isomer). Accordingly, if the unwanted isomer (e.g. R) can be racemized into the RS-isomer it should be reused in the next resolution process, but the recycled RS-isomer is also theoretically only of half use. Thus, it is tedious and difficult to perfectly use a racemate. Such an alternative recycling process is not only complicated but also ineffective, expensive and impractical compared with other simple chiral-technologies [8–14], such as asymmetric synthesis and fermentation procedures.

In order to overcome the disadvantages of the synthesis-resolution process, we have been studying asymmetric transformation via crystallization, especially for industrial applications of α-amino acids, since the early 1980s [15–25]. The present work is part of the results of our investigations. It describes examples of our experiments with asymmetric transformation and views future problems and prospects.

2
Racemization and Asymmetric Transformation

Until quite recently, asymmetric transformation via crystallization has been used with a very limited range of substrates, i.e. optically labile compounds. Since about 1980, asymmetric transformation via crystallization began to be used aggressively, especially with the rise in demand for optically active compounds. The technique had traditionally been known as "second-order asymmetric transformation" (or "asymmetric transformation of the second kind") [26–28], "optical activation" [29] and "isomerization-crystallization" [30], but now is often referred to as "*crystallization-induced asymmetric transformation*" [3, 28]. The principle of this technique, as shown in Fig. 1a (the left side), is that when optical resolution of a racemate (*RS*) takes place continuously under conditions of simultaneous racemization of the undesired isomer (e.g. *S*) in solution, the system greatly shifts to crystallization of the desired isomer (e.g. *R*) as a whole. This process, classified as a type of one-pot asymmetric reaction with crystallization, is expected to be highly practical because all (100%) of the racemate will theoretically be transformed into the desired isomer through a simple process. If elegant reaction conditions can be designed, this technique should be as highly practical as asymmetric synthesis or fermentation procedures. Therefore, crystallization-induced asymmetric transformation is on its way to becoming distinguished from the traditional second-order asymmetric transformation. Alternatively, although dynamic kinetic resolution [31–36], which is characterized by a difference in the rate of chemical reaction between the two isomers (Fig. 1b), is said to be similar to crystallization-induced asymmetric

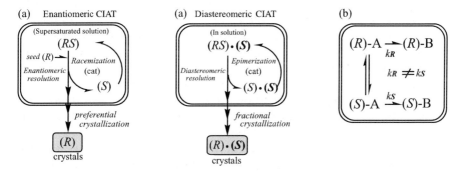

Fig. 1 Crystallization-induced asymmetric transformation (CIAT) (**a**) and dynamic kinetic resolution (**b**)

transformation, a marked difference between these two methods exists in the crystallization step. However, both methods are considered economically very attractive.

There are two types of crystallization-induced asymmetric transformation (Fig. 1a). One is a combination of enantiomeric resolution and simultaneous racemization (Sect. 3), and the other is a combination of diastereomeric resolution and simultaneous epimerization (Sect. 4). Details of the principals of each type are given in the respective sections. For these asymmetric transformations, since racemization (or epimerization) and crystallization are continuously repeated in a system, the reaction rate is controlled by both the rate of racemization (or epimerization) and the rate of crystallization. Therefore, racemization (or epimerization) is an extremely important process in the development of practical crystallization-induced asymmetric transformation. However, in many cases the conditions of racemization (or epimerization) are poorly set, leading to decomposition. Accordingly, it is generally recognized that the racemization reaction is difficult, especially under simple and mild conditions.

We have developed a rational and economic procedure for the racemization of optically active amino acids by combining various existing methods.

2.1
Racemization of Free Amino Acids

In general, it is recognized that racemizations of free amino acids are not easier than those of N-acyl amino acids and amino acid ester derivatives [37]. For example, optically active free amino acids can be racemized:
(i) By heating with water in the presence of a strong base or strong acid [38].
(ii) By heating with water in the absence of a strong base or strong acid in a sealed vessel at 150–250 °C [39].
(iii) By heating with water in the presence of an aldehyde and a metal ion under neutral or weakly alkaline conditions [40].
(iv) By heating with water in the presence of pyridoxal or its analogues, including the so-called resin catalyst, and a metal ion [41, 42].
(v) By heating with a lower aliphatic acid such as acetic acid [43, 44].

On economic grounds, however, these methods are still unsatisfactory since some decomposition of the amino acids often occurs at high temperature and/or the rate of racemization frequently is not practical.

On the other hand, we paid attention to both effects of an aldehyde and a lower aliphatic acid in the above examples. As a result, these combinations allowed the successful development of a convenient method for the racemization of free amino acids and their salts by heating in a lower aliphatic acid with catalytic amounts of an aldehyde. Later, because of this finding we started the study of asymmetric transformation with crystallization.

2.1.1
Racemization of Various Free Amino Acids

Racemization of a wide variety of optically active α-amino acids, including neutral amino acids, acidic amino acids, basic amino acids, and imino acids, was found to be greatly accelerated in acetic acid solution in the presence of catalytic amounts of aldehyde [18]. For example, when various amino acids were heated in a medium of acetic acid at 80–100 °C for 1 h with or without catalytic amounts of salicylaldehyde, their racemizations accelerated slightly in the acetic acid but dramatically after the addition of 0.2 molar equivalent of salicylaldehyde (Table 1). Except for some discoloration in the case of serine and tryptophan, no significant decomposition was observed. As L-aspartic acid, L-glutamic acid, and L-tyrosine were hardly soluble even in acetic acid solution containing 20% or 5% water, the reaction was carried out under

Table 1 Racemization of various amino acids in acetic acid [a]

	Racemization degree % without salicylaldehyde	with salicylaldehyde
L-alanine	13	100
L-arginine	9	100
L-arginine hydrochloride	11	100
L-aspartic acid [b]	4	6
L-glutamic acid [b]	14 (51) [c]	36 (81) [c]
L-histidine hydrochloride	11	100
L-isoleucine [d]	4	93
L-leucine	23	100
L-lysine	9	100
L-lysine hydrochloride [b]	9	67
L-methionine	24	100
L-phenylalanine	35	100
L-proline [e]	3	91
L-serine [f]	1	81
L-tryptophan [f]	0	6
L-tyrosine [b]	0	39
L-valine	4	87

[a] A mixture of L-amino acid (1.5 mmol), salicylaldehyde (0.3 mmol), and acetic acid (6 ml) was heated in a sealed tube in both at 100 °C for 1 h
[b] Aqueous 95%(v/v) acetic acid
[c] Aqueous 80%(v/v) acetic acid, for 2 h
[d] A mixture of L-isoleucine and D-alloisoleucine was obtained
[e] Acetic acid (0.6 ml)
[f] Aqueous 95%(v/v) acetic acid, at 80 °C. Some decomposition took place

heterogeneous conditions. However, the degree of racemization was lower than that of the other amino acids. This phenomenon seems to suggest that racemization occurred only in the liquid phase because the insoluble part of the L-isomer in the reaction mixture could not dissolve continuously in the reaction mixture saturated with the DL-form. This result is supported experimentally by the following facts: (i) the precipitates separated from the final reaction mixture were almost optically pure L-isomer; (ii) under the conditions of racemization, DL-aspartic acid and DL-glutamic acid crystallized as a *racemic mixture* (or *conglomerate*) but not as a *racemic compound*, and their optically active isomers were insoluble in a saturated solution of their racemic mixture [3].

2.1.2
Effect of Various Aldehydes and Low Aliphatic Acids

The effect of varying the aldehyde on the racemization is shown in Table 2. This shows that various aliphatic or aromatic aldehydes can serve as catalysts

Table 2 Effect of kinds of aldehyde in racemization reactions[a]

Aldehyde	Reaction Temp, °C	Racemization degree %			
		L-Ala	L-Met	L-Phe	L-Pro
None	80	7	0	35	0
	100	13	24	35	3
Formaldehyde	100	83	95	100	63
Acetaldehyde	100	97	100	100	98
Propionaldehyde	100	78	100	100[b]	87
n-butyraldehyde	80	97	95[b]	100[b]	99
n-heptylaldehyde	80	100	100[c]	100[c]	100
Acrolein	100	76	100	100[c]	100
Benzaldehyde	100	72	100	100	72
Salicylaldehyde	80	100	100	100	91
p-hydroxybenzaldehyde	100	75	100	100	48
p-anisaldehyde	80	80	100	100	56
o-nitrobenzaldehyde	80	100[b]	100[b]	100	34
5-nitrosalicylaldehyde	80	100	100	100	91
Furfural	100	100[c]	100[c]	100[c]	100[c]

[a] A mixture of L-amino acid (1.5 mmol), aldehyde (0.3 mmol), and acetic acid (6 ml) was heated in a sealed tube in a oil bath at 100 °C for 1 h
[b] A ninhydrin-positive spot of degradation products was slightly observed on TLC of the reaction mixture
[c] Considerable decomposition was detected by thin-layer chromatography of the reaction mixture

Table 3 Effect of molar ratio of salicylaldehyde to L-amino acids on their racemization [a]

Molar ratio of salicylaldehyde	Racemization degree %	
	L-Ala	L-Met
0	13	24
0.001	46	44
0.005	87	78
0.01	100	95
0.05	100	100
0.1	100	100

[a] The reaction was carried out at 100 °C for 1 h

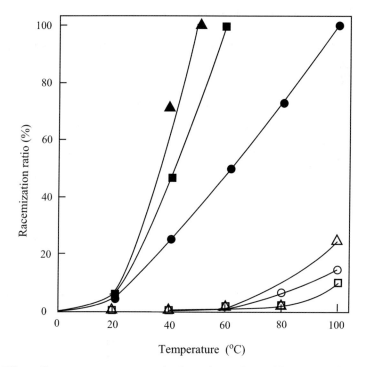

Fig. 2 Effect of temperature on racemization of L-amino acids. ●,○: L-alanine; ▲,△: L-methionine; ■,□: L-lysine. *Black mark*: with salicylaldehyde. *White mark*: without salicylaldehyde. A mixture of L-amino acid (0.1 g), salicylaldehyde (0.2 molar equiv), and acetic acid (3 ml) was heated in a sealed tube in both at 100 °C for 1 h

in the place of salicylaldehyde. Table 3 shows the effect of varying the amount of aldehyde. To complete the racemization of L-alanine in acetic acid within 1 h at 100 °C, it is sufficient to use only 0.01 mol of salicylaldehyde/mol of the amino acid. The racemization reaction was also markedly accelerated by

Table 4 Comparison of kinds of solvent in the racemization reaction

Aliphatic acid	Racemization degree[a] %			
	L-Ala	L-Lys	L-Met	L-Phe
Formic acid	81	43[b]	49	100
	(53)	(19)[b]	(18)	(95)
Propionic acid	9[c]	99[b]	96[b]	100[b]
	(2)[c]	(15)[b]	(19)[b]	(100)[b]
Acetic acid	100	100	100	100
	(13)	(9)	(24)	(35)

[a] Numbers in parentheses indicate the results when the racemization was carried out without salicylaldehyde
[b] A ninhydrin-positive spot of degradation products was slightly observed on TLC of the reaction mixture
[c] Because of low solubility, the reaction was carried out under heterogeneous conditions

increasing the reaction temperature (Fig. 2). L-Lysine and L-methionine were completely racemized even at 60 °C for 1 h without substantial decomposition. Racemization also proceeded when formic or propionic acid was used instead of acetic acid as shown in Table 4, although racemization in acetic acid was most rapid. In addition, other organic acids, organic solvents and their mixtures will be substantially able to be used as a medium.

2.1.3
Racemization Mechanism of Free Amino Acids

As can be seen from Scheme 1, the mechanism of the present racemization is suggested to progress by way of initial protonation of the imine (Schiff base), followed by proton abstraction from the α-carbon atom by acetate anion [18, 45]. The rate of racemization seems to be dependent on the overall reaction rate of the two steps. From the dissociation constants of the aliphatic acids, the order of protonation should be formic acid > acetic acid > propionic acid while the order of proton abstraction is propionic acid > acetic acid > formic acid. Thus, the overall reaction rate may be largest in acetic acid. When L-phenylalanine was racemized in the presence of salicylaldehyde, the extent of racemization decreased sharply with an increase in the water content of acetic acid (Fig. 3). The imine formation from amino acids and an aldehyde, the first step in Scheme 1, might be inhibited by the presence of water.

Traditional racemization procedures using an aldehyde also employed a metal ion which forms a chelate compound with the initially formed Schiff base, and the reaction was carried out under neutral or weakly alkaline conditions. In the method developed by us, racemization of α-amino

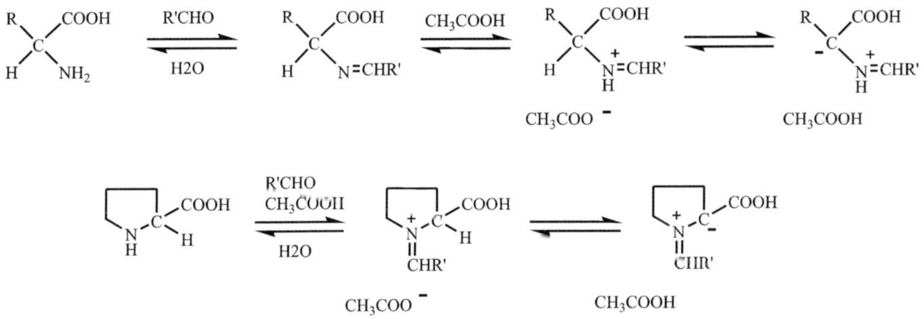

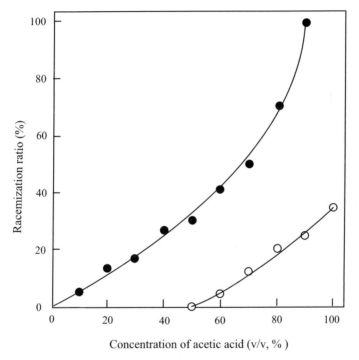

Scheme 1 Racemization mechanism of amino acids

Fig. 3 Effect of water content of acetic acid on racemization of L-phenylalanine. A mixture of L-Phe (0.1 g), salicylaldehyde (0.2 molar equiv), and aqueous acetic acid (3 ml) was heated in a sealed tube in both at 100 °C for 1 h. ●: with salicylaldehyde; ○: without salicylaldehyde

acids, including that of proline, is carried out under acidic conditions without metal ions. There is also no restriction on the type of aldehyde which can be used. Roughly speaking, there is no great difference in their catalytic ability to induce racemization. Thus, aldehydes are not restricted to pyridoxal analogues or ortho-substituted benzaldehyde derivatives. Conse-

Table 5 Racemization of amino acids under various conditions

Media	Catalyst[a]	Temp °C	Racemization degree % L-Ala	L-Lys[b]	L-Met	L-Phe	L-Pro
AcOH	none	60	0	0	0	25	0
	SA	60	55	100	97	86	3
	SA, Cu^{2+}	60	38	68	39	42	0
AcOH	mone	80	0	0	0	35	0(3)[c]
	SA	80	72	100	100	100	21(91)[c]
	SA, Cu^{2+}	80	98	97	100	100	0
H_2O buffered at pH 4[d]	mone	100	0	0	0	0	0
	SA	100	0	0	0	0	0
	SA, Cu^{2+}	100	0	0	0	0	0
H_2O buffered at pH 10[e]	mone	100	0	0	0	0	0
	SA	100	4	0	0	0	0
	SA, Cu^{2+}	100	31	58	37	27	0

[a] Salicylaldehyde (SA; 0.2 molar equiv) and 0.1 molar equiv of $CuCl_2$ (Cu^{2+}) were used
[b] L-Lysine base was used in an acetic acid solution, and L-lysine hydrochloride was used in a buffered aqueous solution
[c] Numbers in parentheses indicate the results when the racemization was carried out at 100 °C by using 0.6 mL of acetic acid
[d] 1 M CH_3COONa/HCl
[e] 0.2 M Na_2CO_3/$NaHCO_3$

quently, the mode of action of the aldehyde must be different from that of the previous methods. Moreover, as shown in Table 5, the racemization efficiency of the present method was superior to the previous methods or other known methods that used acetic acid without a catalyst. As a practical example, L-alanine, L-lysine, L-methionine, L-phenylalanine, L-proline, L-phenylglycine, L-*p*-hydroxyphenylglycine, and L-serine were racemized by the present method and the respective racemic amino acids were separated and analyzed. The results are shown in Table 6. The racemized DL-amino acids were obtained in high yield (85–96%) and in such high purity that no byproduct was detected by thin-layer chromatography. Isolation of these racemic modifications could be easily achieved by simple filtration.

2.1.4
Racemization of Racemic Mixtures and Racemic Compounds

As can also be seen from Table 6, phenylglycine, *p*-hydroxyphenylglycine, and serine were completely racemized even in systems in which large amounts of the amino acids were insoluble. These results were contrasted with the case of insoluble aspartic acid or glutamic acid shown in Table 1. This phenomenon may be explained by the fact that aspartic acid or glutamic acid crystallized

Table 6 Preparation of DL-amino acids by racemization

Amino acid	Composition[i] AcOH mL	SA[a] mL	Reaction Temp °C	Reaction Time h	Separated crystals[b] Yield g	Separated crystals[b] Yield %	$[\alpha]_D^{25}$	Optical purity[c] %
Alanine	180	0.35	100	1	5.74	96	0[d]	0
Lysine	180	0.22	100	1	6.64[e]	89[e]	0[d]	0
Methionine	180	0.21	100	1	5.62	94	0[d]	0
Phenylalanine	180	0.19	100	1	5.62	94	0[f]	0
Proline	36	0.28	100	1	5.10	85	−8.5[f]	9.9
Phenylglycine	20[g]	0.20	100	2	5.36	89	+1.6[h]	1.0
p-hydroxyphenylglycine	20[g]	0.20	100	2	5.58	93	+3.3[h]	2.1
Serine	20[g]	0.10	80	5	5.14	86	0[i]	0

[a] Salicylaldehyde
[b] The byproduct was not detected by thin-layer chromatography
[c] Based on the values in the literature
[d] c 2 in 5 N HCl
[e] Lysine monohydrochloride
[f] c 1 in H$_2$O
[g] AcOH/H$_2$O, 95/5(v/v)
[h] c 1 in 1 N HCl
[i] A 6 g amount of the L form of the compound was used throughout

as a *racemic mixture*, whereas DL-phenylglycine, DL-p-hydroxyphenylglycine, and DL-serine crystallized as a *racemic compound* under the racemization conditions. That is to say, since optically active isomers of these amino acids could be dissolved into a saturated solution of the respective racemic compounds, their racemizations were accomplished even in a heterogeneous system by repeating the following (a) → (b) → (c) processes: (a) dissolution of the optically active isomer into a saturated solution of the racemized amino acids, (b) racemization of the dissolved optically active isomer, and (c) deposition of the racemized amino acids from the saturated solution.

2.2
Racemization of Amino Acid Salts

In general, optical resolution is carried out in the form of salts of racemic modification, because the resolved isomer salt can be readily separated to the free isomer by treatment with neutralization or anionic exchange resin. For instance, physico-chemical resolutions of many amino acids have been achieved through the formation of enantiomeric salts [46–55] and diastereomeric salts [56–62]. In these cases, the undesired enantiomers are also obtained in the form of the same salts. If the racemization of the undesired enantiomers can be easily accomplished in the form of the salts, the racemized salts themselves can be advantageously reused as the starting material for the next resolution. Consequently, the racemization of amino acid salts is economically of great importance, especially from the viewpoint of industrial reuse. It is, however, known that the racemization of salts of amino acids is generally more difficult than that of free amino acids, although only a few salts of optically active amino acids could be racemized by heating in water at a much higher temperature under high pressure [49, 55].

We have developed a new practical method for racemization of various amino acid salts by application of the above racemization of free amino acids [19].

2.2.1
Racemization of Various Amino Acid Salts

In a manner similar to the racemization of free amino acid, a wide variety of salts of optically active α-amino acids, including neutral amino acids, basic amino acids and imino acids with sulfonic acids or mineral acids were

Table 7 Racemization of various amino acid salts [a]

Amino acid salt [b]	Racemization degree [c] % in the presence of		
	None	SA [d]	SA [d] and DL-amino acid
L-alanine·HCl	3	55	100
L-alanine·BS	0	13	88
L-leucine·BS	0	7	64
L-lysine·ABS [e]	0	59	59
L-methionine·HCl	0	41	100
L-phenylalanine·HCl [e]	14	59	100
L-phenylalanine·pXS	0	19	100
L-proline·BZ	15	28	32
L-serine·mXS [f]	0	5	65
L-valine·HCl	0	0	20

[a] The reactions were carried out in acetic acid at 100 °C for 1 h. The amounts of salicylaldehyde and DL-amino acid used were 0.1 molar equivalents
[b] BS, benzenesulfonate; ABS, p-aminobenzenesulfonate; pXS, p-xylene-2-sulfonate; BZ, benzoate; mXS, m-xylene-4-sulfonate
[c] The degrees were calculated from the difference between initial and final optical rotation
[d] SA, salicylaldehyde
[e] The reactions were carried out in a solid–liquid heterogeneous state
[f] The reaction was carried out at 80 °C

heated in a medium of acetic acid at 80–100 °C for 1 h in the presence of catalytic amounts of an aldehyde. These racemizations were also accelerated by the addition of 0.1 molar equivalent of salicylaldehyde, but the rate of racemization was smaller than that in the case of free amino acids. To improve this lower rate, 0.1 molar equivalent of free DL-amino acid corresponding to the L-amino acid moiety of salt was added to the reaction system together with salicylaldehyde. As a result (shown in Table 7) the racemization of the amino acid salts was greatly accelerated, and in the reaction mixtures the content of amino acid was almost 100% without signification decomposition. The racemization mechanism of amino acid salts seems to be essentially the same as that of free amino acids. Namely, a part of the amino acid moiety of optically active amino acid salt leads to the free state by adding free DL-amino acid and the free optically active amino acid is racemized by an imine formation from the amino acid and aldehyde.

Interestingly, in the case of L-lysine p-aminobenzenesulfonate, the racemization was accelerated by only the addition of salicylaldehyde, and the addition of free DL-amino acid was not effective. This seems to be because the α-amino group in lysine p-aminobenzenesulfonate exists already in the free state (uncharged form) predominantly without the addition of free amino acid.

2.2.2
Effect of Various Aldehydes and Free Amino Acids

The effect of the kind of aldehyde on the racemization of L-alanine benzenesulfonate and L-methionine hydrochloride was examined at 100 °C for 1 h in glacial acetic acid. As a result, various aldehydes, such as aliphatic or aromatic aldehyde, accelerated efficiently the racemization of amino acid salts under the coexistence of free DL-amino acid (Table 8). Also, the effect of the amount of salicylaldehyde on the racemization of L-alanine benzenesulfonate and L-methionine hydrochloride was examined at 100 °C for 1 h under the coexistence of 0.2 molar equivalent of DL-alanine or DL-methionine in glacial acetic acid. In addition, the effect of the amount of free DL-leucine on the racemization of L-leucine benzenesulfonate was examined at 100 °C for 30 min or 3 h in the presence of 0.1 molar equivalent of salicylaldehyde in glacial acetic acid. The results are shown in Tables 9 and 10. Salicylaldehyde and free DL-leucine exerted their effects in the use of 0.001 molar equivalent and gave a sufficient racemization degree in the use of 0.05 molar equivalent.

As a practical example, L-alanine benzenesulfonate, L-alanine p-toluenesulfonate, L-lysine benzenesulfonate, L-methionine p-chlorobenzenesulfonate and L-p-hydroxyphenylglycine o-toluenesulfonate were racemized by the present method, and the respective racemic amino acid salts were directly separated from the reaction mixtures and analyzed. The results are shown

Table 8 Effect of kinds of aldehyde in racemization reactions[a]

Aldehyde	Racemization degree %			
	L-alanine·BS		L-methionine·HCl	
	with aldehyde	with aldehyde and DL-alanine	with aldehyde	with aldehyde and DL-methionine
None	0	0	0	5
Formaldehyde	14	46[b]	20[c]	46[b,c]
Propionaldehyde	0	40[b]	12	44[b]
n-butyraldehyde	0	45	12	49
n-heptylaldehyde	0	79[b]	27	73[b]
Acrylaldehyde	0	48	16	38[b]
Benzaldehyde	0	48	10	49
Salicylaldehyde	13	88	41	100
p-hydroxybenzaldehyde	4	35	9	44
o-nitrobenzaldehyde	0	39	17	54
5-nitrosalicylaldehyde	14	76[c]	42	86[e]
Furfural	23	91[b]	74	100[b,c]

[a] The reaction were carried out at 100 °C for 1 h in acetic acid. The amounts of aldehyde and DL-amino acid used were 0.1 molar equivalents
[b] A 0.2 molar equivalent of DL-amino acid was used
[c] A ninhydrin positive spot of degradation product was detected by TLC of the reaction mixture

Table 9 Effect of molar ratio of salicylaldehyde[a]

Molar ratio of salicylaldehyde	Racemization %	
	L-Ala·BS	L-Met·HCl
0	0	9
0.001	13	15
0.005	48	53
0.01	70	71
0.05	100	100
0.1	100	100

[a] The reactions were carried out in the presence of 0.2 molar ratio of DL-amino acid at 100 °C for 1 h

in Fig. 4. The racemized DL-amino acid salts were obtained in high yield (89–92%) and in such high purity that no byproduct was detected by thin-layer chromatography.

Table 10 Effect of amount of DL-leucine on the racemization of L-leucine·BS [a]

Molar ratio of DL–leucine	Racemization degree %	
	30 min	3 h
0	2	27
0.001	4	34
0.005	6	43
0.01	8	47
0.05	24	85
0.1	40	100
0.2	61	100

[a] The reactions were carried out in the presence of 0.1 molar ratio of salicylaldehyde at 100 °C

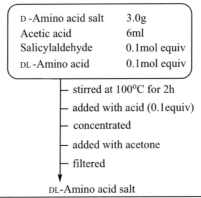

Amino acid salt	Separated amino acid salts	
	Yield g (%)	$[\alpha]_D^{25}$ (c 1, water)
Alanine·BS	2.95 (89.4)	0 (c 1, EtOH)
Alanine·pTS	3.05 (92.4)	0
Leucine·BS	2.85 (86.4)	0
Methionine·pCBS	3.00 (90.9)	0

Fig. 4 Racemization of amino acid salts. BS: Benzenesulfonate, pTS: *p*-Toluenesulfonate, pCBS: *p*-Chlorobenzenesulfonate

2.3
Conclusions

The racemization demonstrated in this section is a simple and convenient method since various optically active α-amino acids including imino acid could be easily racemized by heating in a medium of a low fatty acid at

80–100 °C for 1–2 h in the presence of catalytic amounts of aldehyde. Furthermore, although racemization of salts of these amino acids is ordinary very difficult, they could be achieved by the use of both catalytic amounts of aldehyde and free DL-amino acid. As aldehyde catalysts, both various aliphatic and aromatic aldehydes can be used, and their catalytic amounts are sufficient with 1–10% of substrate. As mediums, various low fatty acids (especially acetic acid) can be efficiently used.

As an interesting finding, when amino acid (or its salt) is a *racemic compound* (not as a *racemic mixture*), the optically active isomer can be racemized without dissolving completely. This fact has an advantage since many free amino acids and their salts are *racemic compounds*, although their enantiomeric resolutions by preferential crystallization are not possible.

Since the racemization presented here is a simple and practical method, it is promising for both laboratory and industrial applications, and today it is often used in crystallization-induced asymmetric transformation (see next Section).

3
Asymmetric Transformation of Enantiomers

The preferential crystallization procedure [63–65] is a useful resolution path for industrial purposes since it enables the desired optically active isomer to crystallize preferentially from a supersaturated solution of the racemate by the simple inoculation of the same isomer. On the contrary, the asymmetric transformation of enantiomer by a combination of preferential crystallization and simultaneous racemization is unique and rare, and practical examples have not been reported except for the case of the α-amino-ε-caprolactam complex [66]. The present method, therefore, has been generally of more limited use than asymmetric transformation of diastereomers (Sect. 4).

We have developed the unique manufacturing methods for enantiomer asymmetric transformation of acyl-amino acids [15–17] and amino acid salts [19–25].

3.1
Enantiomeric Resolution and Its Asymmetric Transformation

A comparison of principles of enantiomeric resolution and its asymmetric transformation is illustrated in Fig. 5.

Figure 5a shows the principle of the ordinary enantiomeric resolution by a preferential crystallization procedure: When seed crystals of the desired isomer (e.g. D) are inoculated in a supersaturated solution of racemate (DL), crystallization of D-isomer arise slowly and then the precipitated D-isomer is separated before spontaneous nucleation of the opposite L-isomer. In addition, the same operation (seed → crystallization → separation cycles) is

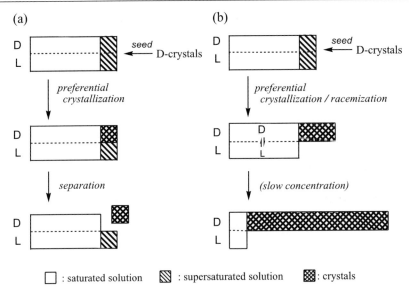

Fig. 5 Enantiomeric resolution (a) and its asymmetric transformation (b)

repeated to give reciprocally L- and D-isomer, adding the racemate to each of the separated mother solutions. This procedure is known to be a useful method for industrial purposes, because it has features suitable for easy automation and continuous production, and an optically active source is not needed (e.g. resolving agent or chiral catalyst) although little seed crystals are necessary. Conversely, the drawback of the ordinary preferential crystallization is that the resolution yield is limited by the density and stability of the supersaturated solution and is usually low (5–15% yield based in DL) because a supersaturated solution is relatively unstable.

On the other hand, Fig. 5b shows the principle of the preferential crystallization with asymmetric transformation: When seed crystals of the desired D-isomer are inoculated in a supersaturated solution of the DL-isomer under the racemization condition, preferential crystallization of D-isomer and simultaneous racemization of L-isomer take place continuously in a system. As a consequence, almost all of the DL-isomer in a supersaturated part in solution can be transformed into D-isomer.

The differences between the above procedures are clear. In the ordinary preferential crystallization, unseeded L-isomer remains in the liquid phase as an unstable supersaturated state (L-rich state). Consequently, the resolution process is not finished until the crystallized D-isomer is separated from the mother liquor containing the enriched L-isomer. To the contrary, the reaction system in the asymmetric transformation is greatly shifted to the crystallized D-isomer as a whole before separating it and then the supersaturated state is cancelled. Here, if during the reaction the solvent is gradually removed,

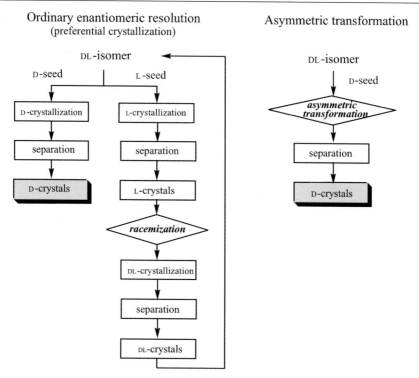

Fig. 6 Comparison of producing processes

almost all of the racemate will be theoretically converted to the crystalline D-isomer. In this case, the yield obtained by the operation can be rapidly improved over that of the ordinary preferential crystallization. The present method is a simple and reasonable one-pot resolution process since both operation of alternative separation and treatment of the undesired isomer are not needed (Fig. 6).

3.2
Asymmetric Transformation of Acyl-DL-Amino Acids

It is well known that some N-acylamino-acids crystallize as a true *racemic mixture* suitable for the preferential crystallization procedure and are actually resolvable by this procedure. Firstly, we have chosen such N-acylamino acids as a test compound for asymmetric transformation. As a result of extensive studies, n-butyrylproline (N-nBu-Pro) [52], N-acetylleucine (N-Ac-Leu), and N-benzoylphenyglycine were found to be greatly racemized in acetic acid solution by heating in the presence of catalytic amounts of acetic anhydride, and preferential crystallization of desired isomer could take place under the racemization conditions. In all cases, the yield of the asymmetric transformation was rapidly increased compared to that of the ordinary preferential crystallization [16].

In the case of N-Ac-Leu [17], when a supersaturated solution of N-Ac-DL-Leu (150 g) in acetic acid containing catalytic amounts of acetic anhydride was seeded with the crystals of N-Ac-L-Leu and cooled at a rate of 10 °C/h from 100 °C to 40 °C, almost optically pure N-Ac-L-Leu (112 g) was obtained in a yield of 70% (Fig. 7).

Interestingly, the asymmetric transformation of N-nBu-DL-Pro could be achieved by melting it with acetic anhydride without solvent, although in 63% yield and 61.4%ee. This result without solvent suggests the first type of asymmetric transformation, referred to as *zonemelted asymmetric transformation*, occurs by means of a combination of preferential crystallization from the super-cooled melt and in situ racemization by the melt.

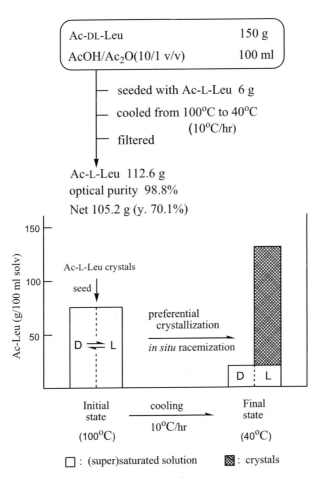

Fig. 7 Enantiomer asymmetric transformation of Ac–DL–Leu

3.3
Asymmetric Transformation of DL-Amino Acid Salts

In general, the optical resolution of DL-amino acids is carried out in the form of their salts and the undesired enantiomers are also obtained in the form of salts. For instance, enantiomeric resolution by the preferential crystallization procedure can be applied to many amino acids by converting them to suitable salts such as DL-alanine p-chlorobenzenesulfonate [46], DL-serine m-xylene-4-sulfonate [53, 54], DL-tryptophan p-phenolsulfonate [50], DL-chlorotryptophan methanesulfonate [50], and DL-p-hydroxyphenylglycine o-toluenesulfonate (DL-HPG·oTS) [55], etc. Among these amino acids, we focused on HPG, since its D-isomer is very important and useful worldwide as a side chain of semisynthetic penicillins or cephalosporins.

3.3.1
Asymmetric Transformation of DL-HPG·oTS

Using the racemization method for amino acid salts as described in the previous Sect. 2, we determined the most essential requirements and conditions for the asymmetric transformation by a combination of the optical resolution of DL-HPG·oTS and simultaneous racemization of the undesired isomer salt, as follows:

(i) DL-HPG·oTS, existing as a *racemic mixture*, was easily resolved to the corresponding enantiomer pure salts by preferential crystallization in acetic acid or water (Table 11).
(ii) The undesired L-HPG·oTS was easily racemized into the corresponding DL-isomer by heating at 100 °C in aqueous 95%(v/v) acetic acid in the presence of small amounts of salicylaldehyde and free DL-HPG (Table 12).
(iii) DL-HPG·oTS crystallized as a *racemic mixture* suitable for the preferential crystallization also under such racemizing conditions.
(iv) The crystals of D-HPG·oTS were not racemized under such conditions for racemization in a liquid phase.

The optimum condition controlled by a combination of both (i) and (ii) in a reaction system led to a successful asymmetric transformation as shown in Fig. 8: A heterogeneous mixture consisting of DL-HPG·oTS, salicylaldehyde and free DL-HPG in aqueous 95%(v/v) acetic acid was seeded with the crystals of D-HPG·oTS at 100 °C, and then poured for 20 h at a rate of 5.0 ml/h with the solution consisting of o-toluenesulfonic acid dihydrate and acetic anhydride, in order to provide continuously the supersaturated state of DL-HPG·oTS as a driving force. This reaction was continuously controlled for a total of 30 h, recycling the crystallization of D-HPG·oTS and simultaneous racemization of L-HPG·oTS together with salt formation of DL-HPG with o-toluenesulfonic acid in a system. The change in the composition of both enantiomers by the reaction is shown in Fig. 8 (the right side), and the re-

Table 11 Properties of HPG·oTS

$$HO-\langle\ \rangle-\overset{*}{C}H\cdot NH_3^+ \quad ^-O_3S-\langle\ \rangle$$
$$\qquad\qquad\ \ \ \overset{|}{COOH}\qquad\qquad H_3C$$

p-Hydroxyphenylglycine *o*-Toluenesulfonate

		DL-HPG·oTS	D-HPG·oTS
Mp (°C)		213–215	222–224
$[\alpha]_D^{25}$ (c 1, water)		0.0	– 66.6
Solubility (g/100ml solv.)	20 °C (water)	12.3	7.7
	40 °C (water)	21.0	12.2
	80 °C (AcOH)	1.80	–
	100 °C (AcOH)	3.06	–

Table 12 Effects of salicylaldehyde and DL-HPG on racemization of L-HPG·oTS

$$\text{L-HPG·oTS} \xrightarrow[100\,°C,\ 3\,h]{cat/AcOH} \text{DL-HPG·oTS}$$

Additive (0.2 molar equiv)	Racemization degree (%)		
	100% AcOH	95% AcOH	90% AcOH
–	20	18	18
Salicylaldehyde	55	30	24
DL-HPG	76	66	49
Salicylaldehyde and DL-HPG	99	94	62

95% AcOH: Water/AcOH = 5/95(v/v)

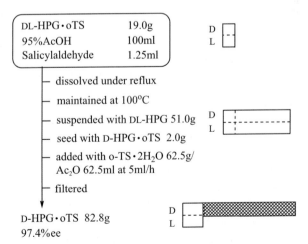

Fig. 8 Continuous asymmetric transformation of DL-HPG·oTS

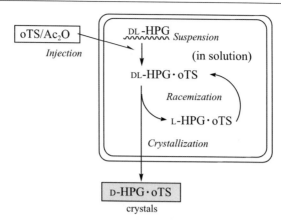

Fig. 9 Enantiomer asymmetric transformation of DL-HPG·oTS

sult is that 77.2% of suspended DL-HPG was transformed to D-HPG via the pathway as shown in Fig. 9.

3.4
Reports and Aspect

Recently reported examples of preferential crystallization with asymmetric transformation are summarized in Table 13. Although aggressive investigation has been undertaken by our laboratory [15–17] and Arai's group [30, 61] since the 1980s, the number of reported examples using this method is comparatively small in the literature. In some cases, catalysts to accelerate the racemization reaction were devised, with findings of mild conditions that allow simultaneous resolution. For instance, the basic catalysts for compounds 1 and 2 are alkoxides, such as NaOMe, and those for compounds 3, 4, 7 and 8 are 1,8-diazabicyclo-[5.4.0]undec-7-ene (DBU). Especially DBU and related compounds, such as DBN, DABCO, and TBD, are useful as basic catalysts for this type of asymmetric transformation. On the contrary, both compounds 5 and 6, which were configurationally labile, were achieved in solution without a basic catalyst, although it is suggested that racemization of 5 proceeds in MeOH containing acetic acid through intramolecular tautomer formation of a labile iminium salt. Surprisingly, enantiomeric transformation of 2 was achieved by half-melting only with neither a catalytic agent nor a solvent. In addition, the α-amino-ε-caprolactam 1, which is useful as a precursor for preparing lysine, was asymmetrically transformed via formation of a metal complex with nickel. This is known as a practical study of the type of enantiomer-asymmetric transformation, and was examined at a micro-pilot scale as a tool for preparing D-lysine [8].

Since the driving force for the preferential crystallization of a seeded isomer is provided only by the supersaturation degree of the racemate, the yield

Table 13 Asymmetric transformation of enantiomers

	Substrate	Conditions	Results	Refs.
1	(caprolactam-NH$_2$)$_3$·NiCl$_2$	EtOK/EtOH 78 °C, 28 h	76% 97%ee	[66]
2	MeO-naphthyl-CH(Me)-COOMe	MeONa/MeOH 50 → 25 °C, 20 h	89%	[67–71]
3	Ph-CH(Me)-COO-naphthyl	DBU/i-Pr$_2$O 28 → 10 °C, 2 h	64% 98%ee	[67–71]
4	Ph-CH$_2$-CH(N=CH-C$_6$H$_4$-Cl)-COOMe	DBU/i-PrOH 38 → 5 °C, 7 h	76% 99%ee	[67–71]
5	Br-benzodiazepine-F	MeOH, 40 °C	85% 99%ee	[72]
6	Cl-C$_6$H$_4$-CH$_2$-CH(triazole)-C(O)-C(Me)$_3$	MeOH/H$_2$O/NaBH$_4$ 5 → 25 °C, 2 h	85% 95%ee	[73]
7	Ph-dioxolanone(Me,Me)	DBU/hexane-IPE 32 → 10 °C	82.3% 90.4%ee	[74, 75]
8	benzothiophene-dioxolanone(Me,Me)	DBU/t-Amylalcohol 50 → 25 °C	87.5% 96.2%ee	[74, 75]
9	benzodioxole-CH(COO^{t}Bu)(O-N-phthalimide)	DBN/Et$_2$CO 40 °C, 3 h → rt, 1 h	79% 85%ee	[76, 77]
10	N-Me-OMe bicyclic	Et$_3$N/EtOH 68 → 40 °C, 16 h, → 25 °C	84% >99%ee	[78]

DBU: 1,8-Diazabicyclo-[5.4.0]undec-7-ene
DBN: 1,5-Diazabicyclo-[4.3.0]non-5-ene

of the desired enantiomer produced by the present asymmetric transformation is dependent on the extent of the supersaturated state of the racemate dissolved in the reaction mixture. To gain a higher yield of the desired enantiomer in one operating cycle, first an attempt to increase the supersaturation degree is generally performed. This tool, however, is occasionally impossible to resolve safely because it becomes difficult to control spontaneous nucleation of the opposite antipode of the seeded isomer. As a counterplan to this, addition of co-existing salts, such as organic salts or surface-active agents may be effective for increasing the stability of the supersaturated state [79, 80], but it is still difficult to make up to 20% or more of saturated solution. Consequently, since improvement of the yield is limited only by the extent of the supersaturated state, it may be necessary to continuously create a supersaturated state during crystallization. If the amount of solvent corresponding to the amount of precipitated crystals can gradually evaporate, it should become possible to crystallize while maintaining the supersaturated state. Alternatively, it is also a useful tool to continuously but slowly supply the racemate in limited amounts as in the above-mentioned case of DL-HPG·oTS where it was difficult to maintain a high degree of supersaturation due to the low solubility.

Asymmetric transformation of enantiomers has attractive economical applications as it does not require the use of chiral auxiliary, but it has serious disadvantages due to the fact that the target compound is limited to a *racemic mixture (conglomerate)* and because it is difficult to maintain the supersaturated state under harsh conditions for racemization. Therefore, the present asymmetric transformation is a reasonable tool to overcome the principal disadvantage of enantiomeric resolution (< 50% yield), but it is not widely used as expected in industrial application of products with added value (chiral pharmaceuticals).

4
Asymmetric Transformation of Diastereomers

Second-order asymmetric transformation (via crystallization) was first observed in a diastereomeric resolution system [28, 83], and it is now occasionally accepted as an effective tool corresponding to asymmetric synthesis. In this section, our studies and recent reports, particularly focusing on the practical applications of the diastereomer asymmetric transformation, are mainly described with principles.

4.1
Diastereomeric Resolution and Its Asymmetric Transformation

The principle behind optical resolution by ordinary diastereomeric formation [81], as shown in Fig. 10a, is that when a racemate (DL-A) is reacted with a resolving agent (e.g. (+)-B) in a solvent, the diastereomeric pair of D- and L-A·(+)-B occur and then the least soluble diastereomer (e.g. D-A·(+)-B) is separated by fractional crystallization from the resolution solution. In this case, the maximum yield of the desired (D-A) is theoretically only half of the corresponding racemate (DL-A).

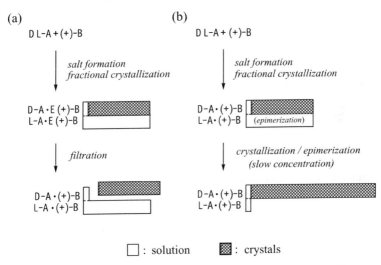

Fig. 10 Diastereomeric resolution (a) and its asymmetric transformation (b)

On the other hand, asymmetric transformation of diastereomers, as shown in Fig. 10b, is when DL-A is reacted with a resolving agent (+)-B in solution under epimerizing conditions, fractional crystallization of the less soluble D-A·(+)-B and the epimerization of the more soluble L-A·(+)-B in the solution proceed simultaneously and consequently is greatly shifted to the crystalline D-A·(+)-B as a whole. In addition, if during the reaction the solvent can be gradually removed, it will result in a more favorable yield. In the present case, almost all of DL-A·(+)-B should be theoretically transformed into the desired crystalline D-A·(+)-B.

4.2
Asymmetric Transformation of Amino Acid Salts Using Chiral Reagents

An elegant example of diastereomer asymmetric transformation of amino acid esters was reported by Clark's group, who have successfully achieved asymmetric transformation of diastereomeric salts of substituted DL-phenylglycine esters with (+)-tartaric acid [82].

First our approach to asymmetric transformation of diastereomers, using the above-mentioned racemization method of amino acid salts, was undertaken for the well-known diastereomeric resolution of DL-phenylglycine (PG) with (+)-camphorsulfonic acid (CS) [62, 84, 85]. Namely, a mixture of DL-PG (15.1 g), (+)-CS (22.1 g), and salicylaldehyde (0.24 g) was stirred in glacial acetic acid (70 ml) at 80 °C for 4 h, and the precipitated crystals were collected to give D-PG·(+)-CS (26.1 g), optical purity 95.9%, yield 68.1% (based on the DL-salt) [19]. This result shows that the selective precipitation of the less soluble D-PG·(+)-CS and the epimerization of the more soluble L-PG·(+)-CS took place at the same time. This encouraging result motivated us to take up asymmetric transformations of the other amino acid salts.

4.2.1
Asymmetric Transformation of DL-HPG with (+)-PES

The preparation process for D-HPG, which is useful as a side chain of semisynthetic penicillins or cephalosporins, has already been developed by the foregoing preferential crystallization [55] and its asymmetric transformation (Sect. 3.3.1). Alternatively, in the course of our studies of diastereomeric resolutions [56–62], an optically active 1-phenylethanesulfonic acid (PES) [57] was found to be an efficient resolving agent for the optical resolution of DL-HPG. Using this (+)-PES, the first approach to the diastereomer asymmetric transformation of DL-HPG [23] took place as follows:

(i) When DL-HPG was resolved by fractional crystallization of its diastereomeric salt with (+)-PES in water or acetic acid, the crystals of less soluble D-HPG·(+)-PES precipitated, and then more soluble L-HPG·(+)-PES remained in its mother liquor (Table 14).
(ii) More soluble L-HPG·(+)-PES could be easily epimerized into DL-HPG·(+)-PES by heating it at 100 °C in glacial acetic acid in the presence of a small amount of salicylaldehyde (Fig. 11).
(iii) (+)-PES as a resolving agent was optically and chemically stable in hot water or acetic acid.

When the optimum condition of a combination of both (i) and (ii) in a reaction system was controlled, its asymmetric transformation accelerated successfully. For instance, a mixture of DL-HPG (10g), (+)-PES (11.2 g), and salicylaldehyde (0.63 ml) was stirred in glacial acetic acid (200 ml) at 100 °C for 5 h. The reaction proceeded continuously in a slurry state of the solid-liquid heterogeneous system, as the reaction pathway shown in Fig. 12. That is, when the crystals of D-HPG·(+)-PES were fractionally precipitated from a salt solution consisting of DL-HPG and (+)-PES, the epimerization of the more-soluble L-HPG·(+)-PES proceeded simultaneously in the liquid phase, and D-HPG·(+)-PES was also thereupon immediately precipitated from the

Table 14 Properties of HPG·(+)-PES

p-Hydroxyphenylglycine(+)-1-Phenylethanesulfonate

		D-HPG·(+)-PES	L-HPG·(+)-PES
Mp (°C)		251–252	204–205
$[\alpha]_D^{25}$ (c 1, MeOH)		−78.9	+98.6
Solubility			
(g/100ml water)	20 °C	1.1	77
	40 °C	1.7	–
(g/100ml AcOH)	20 °C	0.2	1.0
	70 °C	0.2	4.1

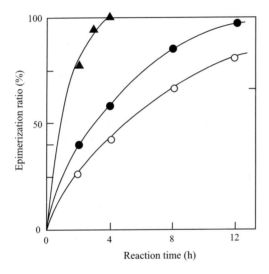

▲ L-HPG·(+)-PES/AcOH, Salicylaldehyde (0.2 molar ratio, at 100°C)
○ L-HPG·(+)-PES/ Water, at 140°C
● L-HPG·(+)-PES/ Water, DL-HPG (0.1 molar ratio), at 140°C

Fig. 11 Epimerization of L-HPG·(+)-PES

epimerizing solution. Since these steps were repeated continuously in the reaction system, the equilibrium finally shifted in favor of D-HPG·(+)-PES as a whole. As a result, 80% of DL-HPG·(+)-PES was asymmetrically transformed into D-HPG·(+)-PES. Thus, the yield of D-HPG·(+)-PES in the asymmetric transformation resulted in nearly twice the yield of the ordinary diastereomeric resolution (Fig. 13).

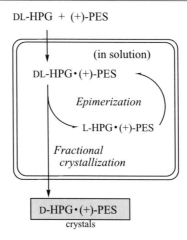

Fig. 12 Reaction pathway of asymmetric transformation of DL-HPG·(+)-PES

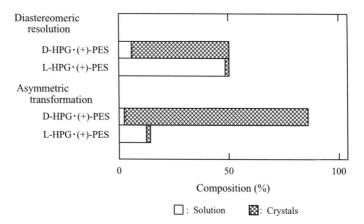

Fig. 13 A comparison of the results of diastereomeric resolution and asymmetric transformation

In the course of our further studies [24], fortunately the following was also noted:
(a) The solubility of L-HPG·(+)-PES in water at 20 °C is about 80 times greater than that of D-HPG·(+)-PES.
(b) More soluble L-HPG·(+)-PES can be easily epimerized into DL-HPG·(+)-PES by heating it in water (Fig. 11).

A rational combination of both (a) and (b) conditions in a reaction system led to an improved asymmetric transformation, the reaction of which could be achieved by heating merely with water in an autoclave. For example, when a mixture of DL-HPG, (+)-PES·NH4, sulfuric acid, water, and xylene in an autoclave was agitated at 125 °C for 20 h, the reaction proceeded in the slurry state of the solid-liquid heterogeneous system, and crystals of D-HPG·(+)-PES were obtained in yield 91.2% and optical purity 98%. As illustrated in Fig. 14, this result indicated that first the composition of D-: L-HPG·(+)-PES = 50 : 50 shifted greatly to that of almost D-HPG·(+)-PES only. The present method, confirmed by reproducibility at the pilot scale with a 20 L autoclave, is a very simple and economically promising process for practical production of D-HPG since it was possible to give D-HPG·(+)-PES 1.5 kg to water 1 kg (Fig. 15).

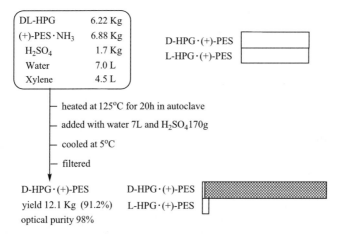

Fig. 14 Preparation of D-HPG by diastereomer asymmetric transformation at the pilot scale

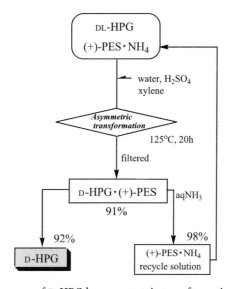

Fig. 15 Preparation process of D-HPG by asymmetric transformation

4.2.2
Asymmetric Transformation of L-Asp(OMe) with (−)-PES

Aspoxicillin, which was developed as a semisynthetic penicillin by our company, has a unique chemical structure with two D-amino acids, D-HPG and N^4-methyl-D-asparagine (N^4-Me-D-Asn), in its side chain [86].

We have investigated the manufacturing method of a precursor of N^4-Me-D-Asn of another raw material, D-aspartic acid β-methyl ester [D-Asp(OMe)], and fortunately have also found that optically active PES was an efficient resolving agent for the optical resolution of DL-Asp(OMe). Using this (−)-PES, the approach to the asymmetric transformation of L-Asp(OMe) [25] was practiced as follows:

(a) When the salt formation of DL-Asp(OMe) took place with (−)-PES in acetonitrile, D-Asp(OMe)·(−)-PES was fractionally crystallized as the less soluble diastereomeric salt in good yield (40%, based on DL salt), and the more-soluble L-Asp(OMe)·(−)-PES remained in the mother liquor.

(b) The more soluble L-Asp(OMe)·(−)-PES could be easily epimerized into DL-Asp(OMe)·(−)-PES by heating it in acetonitrile in the presence of catalytic amounts of aldehyde.

(c) (−)-PES itself was chemically and optically stable under such epimerizing conditions.

A rational combination of both (a) and (b) conditions was optimized in a reaction system, and consequently its asymmetric transformation was accomplished. As a typical example, a mixture of L-Asp(OMe)·(−)-PES, salicylaldehyde (0.05–0.2 mol equiv), and L-Asp(OMe) (0.05–0.1 mol equiv) in acetonitrile was stirred at 80 °C for 2 h, and after seeding with D-Asp(OMe)·(−)-PES the reaction mixture was stirred at 70 °C for 4 h (Fig. 16). The former reac-

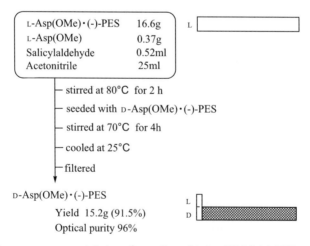

Fig. 16 Diastereomer asymmetric transformation of L-Asp(OMe)·(−)-PES

tion condition is required for rapid epimerization of L-Asp(OMe)·(−)-PES to DL-Asp(OMe)·(−)-PES, and the latter condition is for epimerization-fractional crystallization of DL- to D-Asp(OMe)·(−)-PES (Fig. 17). This reaction proceeded in the slurry state of a solid-liquid heterogeneous system, and the resulting crystalline D-Asp(OMe)·(−)-PES was in yield 91.5% and optical purity 96.0%. As an outcome, L-Asp(OMe)·(−)-PES could be inverted into D-Asp(OMe)·(−)-PES in up to ca. 90% yield, as illustrated in Fig. 16.

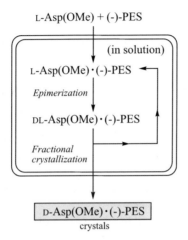

Fig. 17 Reaction pathway of asymmetric transformation of L-Asp(OMe) with (−)-PES

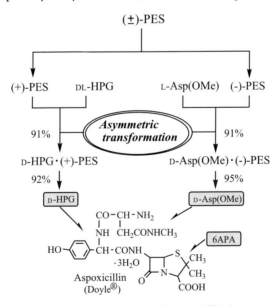

Fig. 18 Preparation process of starting materials of Aspoxicillin by asymmetric transformation

In the final result, we fortunately achieved the rational preparation of both D-HPG and D-Asp(OMe) by asymmetric transformation using antagonistic chiral PES as the resolving agent. The combination of two asymmetric transformations is a rational and advantageous procedure, since the two D-amino acids required to synthesize Aspoxicillin can be prepared by the use of one resolving agent pair (Fig. 18). The developed processes were, therefore, proposed to be advantageous procedures for economical production of Aspoxicillin.

4.3
Reports and Aspect of Pharmaceutical Intermediates

Asymmetric transformation of diastereomers has been used more widely than the preceding asymmetric transformation of enantiomers, because the reaction system of the former is comparatively stable owing to a simple system consisting of only the solubility difference of the diastereomeric pair. Therefore, asymmetric transformation of diastereomers has been used in a number of recently reported examples, as shown in Tables 15–18, many of which are chiral substrates of raw materials, pharmaceutical intermediates, and atropisomers. In these cases, some of the racemic substrates are asymmetrically transformed by diastereomeric salt formation using a chiral resolving agent, whereas more than half of the substrates are accomplished by application of diastereomeric intramolecular relation (covalent diastereomers) through two asymmetric centers. Of these methods, the above-mentioned racemization of amino acid salts and its related conditions, i.e. heating in a medium of low fatty acids or organic solvent in the presence of catalytic amounts of an aldehyde, has been widely used as in examples of compounds **1, 4, 6, 15, 16, 17**, and **20** without being limited to ordinary amino acids.

Furthermore, as shown in Tables 16–17, the present asymmetric transformation has begun to be widely used in pharmaceutical intermediates due to the rise in recent years of demand for chiral drugs of a single enantiomer [5–7]. For example, **15**, which is an aminobenzodiazepinon derivative of a CCK antagonist, was first prepared by Reider's group [102]. In this case, a combination of a general resolving agent (CSA) and a salicylaldehyde derivative used as an epimerizing catalyst was devised and resulted in a yield of more than 91% and 98%ee. Similar compounds, i.e. **16**, and **17**, were elegantly achieved in the same manner, and **16** was successfully manufactured at the ton scale. In other cases, the Naproxen derivative anti-inflammatory agents **24–26**, the antibiotics **33–36**, which have an inversed phenylglycine moiety as the side chain, the more recently prepared NK1 receptor antagonists **19** and **28**, the anticancer agent **29**, which interestingly was obtained by stereoselective glycosidation, and the pyrethroid-insecticides **37** and **38** [8] have all been prepared by this method.

Table 15 Asymmetric transformation of diastereomers (1): chiral intermediates

Entry	Substrate	Resolving Agent	Reaction Conditions	Results	Refs.
1	HO-C6H4-CH(NH2)-COOH	(+)-BCS	SA, AcOH, H2O 70 °C, 12 h	(R)-(+) 99%, 98%de Kg scale	[81]
2	Ph-C(CH2(CH3)3)(OH)-CN	brucine	MeOH r.t., 24 h	(+)-(−) 100%, 94%ee	[88]
3	biphenyl-F, Me-CH-COOH	(−)-PEA	Petroleum 125 °C, 72 h	(+)-(−) 76%, 74%ee Kg scale	[89, 90]
4	thiazolidine-COOH	(S)-TTA	Formaldehyde/AcOH 80 → 5 °C, 2 h	(R)-(S) 82%, 99%ee	[91]
5	Ph-CH(CN)(NHCH2Ph)	(R)-MA	EtOH 23 °C, 20 h	(S)-(R) 90%, >97%de	[92]
6	Cl-C6H4-CH(NH2)-COOMe	L-TTA	SA, MeOH reflux, overnight	(R)-L 68%, 98%ee	[93]
7	Ph-CH(CONH2)-HN-C(CN)(tBu)		NaCN/AcOH/H2O 70 °C, 24 h	(R)-(S) 93%, >98%de	[94]
8	NC-CH(iBu)-NH-CH(Me)(Ph)		H2O/MeOH r.t., 3 days	(S)-(S) 98%, 94%de	[95]
9	Ph-CO-CH2-CH(NH-Bz)-CO-N(H)-CH(Ph)-COOH		H2O/BzNH2 20–25 °C, 7 days	(S)-(S) 81%, >90%de	[96]
10	Ph-CH(Me)-NH-CO-CH(Cl)(Ph)		aqNH3/MeOH r.t., 7 days	(S)-(S) 95%, >94%de	[97]
11	Me-Ar-furanone-HCl·HN-CH(Ph)		4 M HCl 50 °C, 30 h	cis(3R5R)-(R) 70%, 98%de	[98]
12	MeO-C6H4-CO-CH(COOH)-NH-CH(Me)(Ph)		EtOH 60 °C, 16 h	(1S)-(αS) 90%, 97%de	[99]

Table 15 (continued)

Entry	Substrate	Resolving Agent	Reaction Conditions	Results	Refs.
13	(imidazolidinone with Me, Ph, Br, 2'-Me substituents)	n-Bu$_4$NBr/THF evaporate, > 24 h	(4R,5S)-(2'R) 91%, > 98%de	[100]
14	(biphenylcarbonyl amide with Br-alkyl ester)		MeSO$_3$H/Et$_2$O 40 °C, 15 days	90%, 92%de	[101]

BCS: 3-bromocamphor-9-sulfonic acid, PEA: 1-phenylethylamine, TTA: tartaric acid, MA: mandelic acid, CSA: camphor-10-sulfonic acid, BNHP: 1,1'-binaphthalene-2,2'-dihydrogen phosphate, DTT: di-p-toluoyltartaric acid, SA: salicylaldehyde, DIBA: diisobutylamine, DCSA: 3,5-dichlorosalicylaldehyde, DBU: 1,8-diazabicyclo(5.4.0)undec-7-ene, DBN: 1,5-diazabicyclo(4.3.0)non-5-ene, DMOK: 3,7-dimethyl-3-octanol potassium salt, DMAP: 4-(dimethylamino)pyridine, DHN: 2,7-dihydroxynaphthalene, CHA: 1-cyclohexylamine

The majority of these examples have been accomplished by megapharmaceutical companies (e.g, Merck, Novartis, J&J, AstraZeneca, Takeda), and major chiral chemical (e.g. DSM, Kaneka, Sumitomo) and venture (e.g. Chirotech, Chiroscience, Sepracor) companies. Therefore, a number of chiral chemical companies worldwide are presently interested in asymmetric transformation of diastereomers as an attractive technique for the preparation of chiral drug intermediates at the pilot or industrial scale.

From the many examples shown in Tables 15–18, common technical aspects for the use of asymmetric transformation can be summarized as described below.

4.3.1
Substrates

The chemical structure of a racemic substrate is very important for the asymmetric transformation of diastereomers. Of course, the best desired substrate, as shown in Table 18, should be configurationally labile, however, other various substrates have been achieved by devising reaction conditions for epimerization. Generally, since many of the substrates are formed via Schiff bases and tautomerism for epimerization, it is important that the substrate structure has amino groups and neighboring carbonyl groups in the chiral center. In addition, other neighboring functional groups, carboxy, nitrile, halogen, phosphate, and their cyclic structures are a more favorable choice, because they can enhance acidity of the chiral center by their electrophilic action, leading to an enhancement of epimerization.

Table 16 Asymmetric transformation of diastereomers (2): pharmaceutical intermediates

Entry	Pharmacology	Substrate	Resolving Agent	Reaction Conditions	Results	Refs.
15	CCK antagonist		(S)-CSA	DCSA i-C$_3$H$_7$OAc-MeCN 20–25 °C, 16 h	(3S)-(S) 91%, 98%ee 10 kg scale	[102]
16	CCK antagonist		L-TTA	SA/EtOH 40 °C, 3.5 h	(S)-L 84.2%, 95.5%ee 1.5 t scale	[103]
17	K channel blocker		(R)-MA	DCSA H$_2$O/i-PrOAc r.t., 24 h	(+)-(R) 92%, 99.4%ee	[104]
18	Growth hormone secretagogue			Mineral spirite 140 °C, 16 h	(S)-(S) 88%, >98%de 100 g scale	[105]
19	NK1 receptor antagonist			EtOH 37 °C, 60 h	(R)-(9R) 69%, 96%ee	[106]

Table 16 (continued)

Entry	Pharmacology	Substrate	Resolving Agent	Reaction Conditions	Results	Refs.
20	Antiplatelet drug		(R)-MA	SA/i-PrOH reflux, 24 h → 22 °C	(S)-(R) 78%, 98.7%ee	[107]
21	Neurokinin-1 receptor		(S)-BCS	i-PrOAc 89 °C, 3 days	(S)-(S) 90%, 99.4%ee multi-kg scale	[108]
22	Antidiabetic agent		(−)-PEA	AcOEt r.t., 24 h	(S)-(−) 79.4%, > 99%ee	[109]
23	Analgesic anti-inflammatory			DBN/AcOEt-hexane r.t. → cool, 2 h	(−)-(R) 88%, 83%ee	[110]
24	Anti-inflammatory (naproxen)			DBU/DMF 50–55 °C, 28 h → 0 °C	(S)-(S) 81.4%, 97.2%de	[111]

Table 16 (continued)

Entry	Pharmacology	Substrate	Resolving Agent	Reaction Conditions	Results	Refs.
25	Anti-inflammatory (naproxen)	MeO—[naphthyl]—CH(Me)-CONH-CH(Et)-CH₂OH		MeONa/toluene-MeOH $105 \to 40 \to 5\,°C$	(−)-(+) 94%, > 99%ee	[112, 113]
26	Anti-inflammatory (naproxen)	MeO—[naphthyl(Br)]—CH(Me)-CONH-CH(COOH)-CH₂COOH		MeONa/n-BuOH reflux, 1 h	(−)-(S) 85.6%	[114, 115]

BCS: 3-bromocamphor-9-sulfonic acid, PEA: 1-phenylethylamine, TTA: tartaric acid, MA: mandelic acid, CSA: camphor-10-sulfonic acid, BNHP: 1,1′-binaphthalene-2,2′-dihydrogen phosphate, DTT: di-p-toluoyltartaric acid, SA: salicylaldehyde, DIBA: diisobutylamine, DCSA: 3,5-dichlorosalicylaldehyde, DBU: 1,8-diazabicyclo(5.4.0)undec-7-ene, DBN: 1,5-diazabicyclo(4.3.0)non-5-ene, DMOK: 3,7-dimethyl-3-octanol potassium salt, DMAP: 4-(dimethylamino)pyridine, DHN: 2,7-dihydroxynaphthalene, CHA: 1-cyclohexylamine

Table 17 Asymmetric transformation of diastereomers (3): pharmaceutical intermediates

Entry	Pharmacology	Substrate	Resolving Agent	Reaction Conditions	Results	Refs.
27	Endothelin antagonist		Cinchonine	i-PrOH/MeOH 70 °C, 20 h, 40 °C, 5 h	(R)-(+) 83%, 99%de	[116]
28	NK1-receptor antagonist			DMOK/heptane −5–10 °C, 5 h	(R)-(R) 85%, > 99%de	[117]
29	Anticancer			BF$_3$OEt$_2$/MeCN −10 °C, 5 h	(β) 79% kg scale	[118]
30	Anti-inflammatory			TsOH/cyclohexane Dean–Stark, 18 h, reflux, 8 h	(S)-(S) 91%, > 99%de	[119]

Table 17 (continued)

Entry	Pharmacology	Substrate	Resolving Agent	Reaction Conditions	Results	Refs.
31	Antiallergy		(R)-CSA	MeCN 80 °C, 8 h	(+)-(R) 97%, 96%ee kg scale	[120, 121]
32	FXa inhibitor			NaOEt/EtOH	(S)-(S) 88%, > 99.5%ee	[122]
33	Antibiotic			Pyridine/dioxane r.t., 14 days	(6R,7R,4'R) 75%, > 99%ee	[123]
34	Cephalosporin antibiotic			Pyridoxal, DHN/H$_2$O r.t., 24 h	(R)-(R,R) 86%	[124]

Table 17 (continued)

Entry	Pharmacology	Substrate	Resolving Agent	Reaction Conditions	Results	Refs.
35	Antibiotic	(structure)		NH$_3$/acetone	(S) 90%, 94%ee	[125, 126]
36	Antibiotic	(structure)		Pyridine/AcOEt r.t., overnight	(S) 75%	[127, 128]
37	Insecticide	(structure)		DIBA/iPrOH 20–23 °C, 9 days	88%	[129, 130]
38	Insecticide	(structure)		NH$_3$/MeOH 32 °C, 14 h → 10 °C → −10 °C, 1 h	(S)-(S) 84%, 99%ee	[131, 132]

BCS: 3-bromocamphor-9-sulfonic acid, PEA: 1-phenylethylamine, TTA: tartaric acid, MA: mandelic acid, CSA: camphor-10-sulfonic acid, BNHP: 1,1′-binaphthalene-2,2′-dihydrogen phosphate, DTT: di-p-toluoyltartaric acid, SA: salicylaldehyde, DIBA: diisobutylamine, DCSA: 3,5-dichlorosalicylaldehyde, DBU: 1,8-diazabicyclo(5.4.0)undec-7-ene, DBN: 1,5-diazabicyclo(4.3.0)non-5-ene, DMOK: 3,7-dimethyl-3-octanol potassium salt, DMAP: 4-(dimethylamino)pyridine, DHN: 2,7-dihydroxynaphthalene, CHA: 1-cyclohexylamine

Table 18 Asymmetric transformation of diastereomers (4) atropisomers and others

Entry	Pharmacology	Substrate	Resolving Agent	Reaction Conditions	Results	Refs.
39			(−)-Sparteine	CuCl$_2$/MeOH r.t., 20 h	(−)-(−) 94%, 80%ee	[133]
40			(−)-BNHS	EtOH 0 °C, 3 h	(+)-(−) 93%	[134]
41	Acetylcholinesterase inhibitor		(+)-DTT	EtOH reflux, 1 h → 40 °C, 16 h	(−)-(+) 92%, 97%ee	[135]
42	Anti-Alzheimer drug		Cat. (−)-narwedine	EtOH/Et$_3$N	(−)-(−) 80%, kg scale	[136]
43			(−)-CHA	Solid/diazomethane −40 °C, $h\nu$	(−)-(R) 92%, 90%ee	[137]

Table 18 (continued)

Entry	Pharmacology	Substrate	Resolving Agent	Reaction Conditions	Results	Refs.
44				Without solvent 110 °C, 4 days	(S)-(3′S) 99%de	[138]
45				Hexane hot → r.t.	(M) 95%, 90%rt	[139]
46				DCC/DMAP r.t., >1 day	(−)-(S) 82%, 99%de	[140]
47				MeCN 55 °C, 10 min → r.t., 30 h	(S)-(R) 83%, 91%ee	[141]

Table 18 (continued)

Entry	Pharmacology	Substrate	Resolving Agent	Reaction Conditions	Results	Refs.
48		ROOC-CH=CH-COOR	R=(−)-L-menthyl	Et$_3$N/pentane −20 °C, 2 days	(−)-(R) 90%, 98%de	[142]
49		(MeO-phenyl phosphine with menthyl ester)		EtOH r.t., crystallization	(Sp) 85%, 97%de	[143]
50		(Cl-OCO cyclopentane phosphate)		^{n}Bu$_3$N/MeCN −5 °C, 23 h	(α)-(D) 92%, 97%de	[144]
51		(Ph-B-F oxazaborine)		Hexane evaporate, r.t.	80%, 95%de	[145]

BCS: 3-bromocamphor-9-sulfonic acid, PEA: 1-phenylethylamine, TTA: tartaric acid, MA: mandelic acid, CSA: camphor-10-sulfonic acid, BNHP: 1,1′-binaphthalene-2,2′-dihydrogen phosphate, DTT: di-p-toluoyltartaric acid, SA: salicylaldehyde, DIBA: diisobutylamine, DCSA: 3,5-dichlorosalicylaldehyde, DBU: 1,8-diazabicyclo(5.4.0)undec-7-ene, DBN: 1,5-diazabicyclo(4.3.0)non-5-ene, DMOK: 3,7-dimethyl-3-octanol potassium salt, DMAP: 4-(dimethylamino)pyridine, DHN: 2,7-dihydroxynaphthalene, CHA: 1-cyclohexylamine

4.3.2
Resolving Agents

In general, since the epimerization reaction is almost always executed at high temperature in the presence of a catalyst, it is of importance to select an enduring resolving agent (or covalent agent) under such severe conditions. Some commonly used acidic or basic resolving agents [146–148] are listed in Tables 15–18. Of these agents, tartaric acid, mandelic acid, camphor-10-sulfonic acid, 1-phenylethylamine, and their derivatives are commonly used due to their relative stability, commercial availability, and cost. The alkaloid analogs brucine, cinchonine, and sparteine are also useful as lower basic resolving agents, although they are not available when the desired isomer is the opposite isomer.

In order to obtain high yield and ee%, the resolving agent used in the present asymmetric transformation should be chemically and optically stable. Previously, we have prepared chemically and optically stable synthetic chiral sulfonic acid resolving agents, which led to successful resolution and asymmetric transformation of various amino acids (Sect. 4.2.1) [23, 25, 57]. In the future we hope to develop more environmentally friendly resolving agents and expand substrate application range.

4.3.3
Reaction Conditions

Most of the present reactions were conducted under simple and mild conditions taking into account various paths for effective epimerization. Beside substrates, the major factors that affect the reaction system are solvents, catalysts, and temperature of the system. Useful mediums are alcohols, low fatty acids, acetates, ethers, ketones, hydrocarbons (carbon 4–7), acetonitrile, and their mixtures, all of which are inert solvents aimed at lowering solubility in the reaction system. Of these solvents, low fatty acids and ketones can be conveniently used as both solvents and catalysts. Also, as useful catalysts that promote epimerization, aldehydes, ketones, and general acids as well as bases are used. Aromatic aldehydes are more readily activated than aliphatic aldehydes, particularly salicylaldehyde and its derivatives (e.g. **15, 17**) can effectively be used. Interestingly, pyridoxal (**34**), which is a salicylaldehyde analog, has been reported to be useful for epimerization in aqueous solution under mild conditions. Examples with acidic catalysts are few, whereas basic catalysts, such as alkoxides, DBU, quaternized amines and their derivatives are widely used. The use of these basic catalysts expands the range of substrates applied to asymmetric transformation. As a noteworthy case, the weak base pyridine is effectively used in the preparation of **33, 36**, although the reaction is very slow.

The reaction temperature is a very important factor in controlling epimerization rate. Whereas high temperatures invite the risk of a decrease in yield and decomposition, low temperatures result in slow reactions. Therefore, most reactions are carried out at a temperature below the boiling point of the solvent used.

As in the case of enantiomer transformation, control of solubility in the reaction system is particularly important for asymmetric transformation of diastereomers. As a useful tool, removal of solvent or addition of inert solvents can lead to high yield, although this may influence epimerization rate and optical purity. Consequently, the secret for success in this process is to moderately adjust the conditions of fractional crystallization, while considering epimerization rate.

5
Conclusions

In this piece of work, we reviewed the principles, features, and recently reported examples of asymmetric transformation via crystallization while describing our research in this field. The use of this transformation method, which is a chiral technology closely related to classical optical resolution methods, seems to be increasing steadily as the demand and interest in chiral substrates rise. With the development of new racemization reactions other than those described in this report for amino acid salts, asymmetric transformation via crystallization can be applied to not only new compounds but also to the already existing resolvable compounds. Therefore, since the driving force for this transformation method is the efficiency of racemization, it is critical to find new efficient methods for racemization. Recently, the results of an extensive research on racemization carried out between 1967–1996 and reported by Ebbers's [149] group have shown that the majority of object substrates were amino acids and their derivatives and that almost all of the reactions were ordinary thermal, acid-, base-, or Schiff base-catalyzed racemization or epimerization.

In general, the racemization reaction needs a large amount of energy and is often limited by the characteristics of the substance itself. In some cases, however, racemization can be achieved by devising a catalyst. For instance, it is suggested that mild reaction conditions achieved by the use of biocatalysts [150–155] and/or photochemicals [156–159] are an attractive and promising tool for future racemizations. In addition to these methods, further advancement of asymmetric transformation via crystallization is dependent on the development of new methods for racemization.

Acknowledgements The author wishes to express gratitude to all the work and effort of Tanabe Seiyaku's researchers, and in particular to Drs. Hongo C, Yamada S, and Chibata I for their interest and encouragement.

References

1. Pasteur LM (1848) Campt Rend 26:535
2. Pasteur LM (1848) Ann Chem Phys [3]24:442
3. Jacques J, Collet A, Wilen SH (1981) Enantiomers, Racemates, and Resolutions. Wiley, New York
4. Eliel EL, Wilen SH, Mander LN (1994) Stereochemistry of Organic Compounds. Wiley, New York
5. Stinson SC (2001) Chemical & Engineering News, Vol. 79, 36. ACS, p 79
6. Rouhi AM (2003) Chemical & Engineering News, Vol. 81, 18. ACS, p 45
7. Rohhi AM (2004) Chemical & Engineering News, Vol. 82, 24. ACS, p 47
8. Collins AN, Sheldrake GN, Crosby J (1992) Chirality in Industry. Wiley, New York
9. Sheldon RA (1993) Chirotechnology. Marcel Dekker, New York
10. Collins AN, Sheldrake GN, Crosby J (1997) Chirality in Industry. Wiley, New York
11. Kurihara N, Miyamoto J (eds) (1998) Chirality in Agrochemicals. Wiley, New York
12. Ager DJ (ed) (1999) Handbook of Chiral Chemicals. Marcel Dekker, New York
13. Subramanian G (ed) (2001) Chiral Separation Techniques, a Practical Approach. Wiley, New York
14. RocSearch Ltd (2004) Chiral Technolgies: Market Potentialities and Technological Development
15. Yamada S, Hongo C, Chibata I (1980) Chem Ind (London) 5:539
16. Hongo C, Yamada S, Chibata I (1981) Bull Chem Soc Jpn 54:3286
17. Hongo C, Yamada S, Chibata I (1981) Bull Chem Soc Jpn 54:3291
18. Yamada S, Hongo C, Yoshioka R, Chibata I (1983) J Org Chem 48:843
19. Hongo C, Yoshioka R, Tohyama M, Yamada S, Chibata I (1983) Bull Chem Soc Jpn 56:3744
20. Hongo C, Yoshioka R, Tohyama M, Yamada S, Chibata I (1984) Bull Chem Soc Jpn 57:1328
21. Hongo C, Tohyama M, Yoshioka R, Yamada S, Chibata I (1985) Bull Chem Soc Jpn 58:433
22. Yoshioka R, Ohtsuki O, Tosa T (1986) Bunri Gijutsu 16:350
23. Yoshioka R, Tohyama M, Ohtsuki O, Yamada S, Chibata I (1987) Bull Chem Soc Jpn 60:649
24. Yoshioka R, Tohyama M, Yamada S, Ohtsuki O, Chibata I (1987) Bull Chem Soc Jpn 60:4321
25. Yoshioka R, Ohtsuki O, Senuma M, Tosa T (1989) Chem Pharm Bull 37:883
26. Buchanan C, Graham SH (1950) J Chem Soc, p 500
27. Harris MM (1958) Klyne W, Mare PBD (eds) Progress in Stereochemistry, vol 2. Butterworths Scientific Publications, London, p 157
28. McNaught AD, Wilkinson A (1997) IUPAC Compendium of Chemical Terminology, 2nd edn. Blackwell Science, Oxford
29. Suzuki K, Kiyooka S, Miyagawa T, Kawai A (1980) Nippon Kagaku Kaishi 2:287
30. Arai K (1986) J Synth Org Chem Jpn (Yuki Gosei Kagaku Kyokai Shi) 44:486

31. Dynamic kinetic resolution is occasionally confused with asymmetric transformation via crystallization. Alternatively, dynamic kinetic resolution via crystallization is rarely referred to as "Crystallization-induced dynamic resolution" [97, 100]
32. Noyori R, Tokunaga M, Kitamura M (1995) Bull Chem Soc Jpn 68:36
33. Ward RS (1995) Tetrahedron: Asymmetr 6:1475
34. Caddick S, Jenkins K (1996) Chem Soc Rev, p 447
35. Faber K (2001) Chem Eur J 7:5004
36. Pellissier H (2003) Tetrahedron 59:8291
37. Kaneko T, Izumi Y, Chibata I, Itoh T (1974) Synthetic Production and Utilization of Amino Acids. Kodansha, Tokyo and Wiley, New York
38. Neuberger A (1948) In: Anson ML, Edsall JT (eds) Advance in Protein Chemistry, vol 4. Academic Press, New York, p 339
39. Sasaji I, Hara M, Tatsumi S, Seki K, Akashi T, Ohno K (1965) US Patent 3 213 106, Ajinomoto Co., Inc., Japan
40. Ogasawara H, Tatemichi H, Suzuki S (1967) JP Patent 42-13445
41. Metzler DE, Ikawa M, Snell EE (1954) J Am Chem Soc 76:648
42. Toi K, Izumi Y, Akabori S (1962) Bull Chem Soc Jpn 35:1422
43. Sakieki I, Mitsuno M (1959) J Chem Soc Chem Jpn 80:1035
44. Matsuo H, Kawazoe Y, Sato M, Ohnishi M, Tatsuno T (1970) Chem Pharm Bull 18:1788
45. Grigg R, Gunaratne HQN (1983) Tetrahedron Lett 24:4457
46. Chibata I, Yamada S, Yamamoto M, Wada M (1968) Experientia 24:638
47. Yamada S, Yamamoto M, Chibata I (1973) Chem Ind (London), (issue II), p 528
48. Yamada S, Yamamoto M, Chibata I (1973) J Agr Food Chem 21:889
49. Yamada S, Yamamoto M, Chibata I (1973) J Org Chem 38:4408
50. Yamada S, Yamamoto M, Hongo C, Chibata I (1975) J Agr Food Chem 23:653
51. Yamada S, Hongo C, Yamamoto M, Chibata I (1976) Agr Biol Chem 40:1425
52. Hongo C, Shibazaki M, Yamada S, Chibata I (1976) J Agr Food Chem 24:903
53. Hongo C, Yamada S, Chibata I (1981) Bull Chem Soc Jpn 54:1905
54. Hongo C, Yamada S, Chibata I (1981) Bull Chem Soc Jpn 54:1911
55. Yamada S, Hongo C, Chibata I (1978) Agr Biol Chem 42:1521
56. Yamada S, Hongo C, Chibata I (1977) Agr Biol Chem 41:2413
57. Yoshioka R, Ohtsuki O, Da-te T, Okamura K, Semuma M (1994) Bull Chem Soc Jpn 67:3012
58. Yamada S, Hongo C, Yoshioka R, Chibata I (1979) Agr Biol Chem 43:395
59. Yamada S, Yoshioka R, Shibatani T (1997) Chem Pharm Bull 45:1922
60. Yoshioka R, Okamura K, Yamada S, Aoe K, Da-te T (1998) Bull Chem Soc Jpn 71:1109
61. Yamada S, Tsujioka I, Shibatani T, Yoshioka R (1999) Chem Pharm Bull 47:146
62. Yoshioka R, Hiramatsu H, Okamura K, Tsujioka I, Yamada S (2000) J Chem Soc Perkin Trans 2, p 2121
63. Secor RM (1963) Chem Rev 63:297
64. Wilen SH, Collet A, Jacques J (1997) Tetrahedron 33:2725
65. Collet A, Brienne Mj, Jacques J (1980) Chem Rev 80:215
66. Boyle WJ Jr, Sifiniades S, Van Peppen JF (1979) J Org Chem 44:4841
67. Arai K, Obara Y, Iizumi T, Takakuwa Y (1985) 47th Yuki Gosei Symposium (Japanese), p 64
68. Takahashi Y, Arai K, Obara Y, Matsumoto H (1986) JP 61 103852 (Nissan Chemical Industries, Ltd., Japan)

69. Takahashi Y, Arai K, Obara Y, Matsumoto H (1986) Chem Abstr 105:114749t
70. Obara Y, Matsumoto H, Arai K, Tsuchiya S (1986) JP 61 1652 (Nissan Chemical Industries, Ltd., Japan)
71. Obara Y, Matsumoto H, Arai K, Tsuchiya S (1986) Chem Abstr 105:43327r
72. Okada Y, Takebayashi T, Sato S (1989) Chem Pharm Bull 37:5
73. Black SN, Williams LJ, Davey RJ, Moffatt F, Jones RVH, McEwan DM, Sadler DE (1989) Tetrahedron 45:2677
74. Nakano J, Taya N, Chaki H, Yamafuji T, Momonoi K (1994) JP 6-9615 (Toyama Chemical Co., Ltd., Japan)
75. Nakano J, Taya N, Chaki H, Yamafuji T, Momonoi K (1994) Chem Abstr 120:106751g
76. Murakami K, Aizawa N, Mochida S (1991) JP 91-56482 (Mochida Seiyaku, Co., Ltd., Japan)
77. Murakami K, Aizawa N, Mochida S (1991) Chem Abstr 115:56482q
78. Shieh W-C, Carlson JA (1994) J Org Chem 59:5463
79. Akashi T (1962) Nippon Kagaku Kaishi 83:421
80. Nohira H, Miura H (1975) Nippon Kagaku Kaishi 6:1122
81. Kozma D (2001) CRC Handbook of Optical Resolution via Diastereomeric Salt Formation. CRC, Boca Raton, Florida
82. Clark JC, Phillipps GH, Steer MR (1976) J Chem Soc Perkin Trans 1, p 475
83. Leuchs H, Wutke J (1913) Ber 46:2420
84. Betti M, Mayer M (1908) Ber 41:2071
85. Ingersoll AW (1925) J Am Chem Soc 47:1168
86. Wagatsuma M, Seto M, Miyagishima T, Kawazu M, Yamaguchi T, Ohshima S (1983) J Antibiot 36:147
87. Bhattacharya A, Araullo-Mcadams C, Meier MB (1994) Syn Commun 24:2449
88. Toda F, Tanaka K (1983) Chem Lett 661
89. Nicholson JS, Tantum JG (1987) JP 62-6536, DE 2 809 794 (Boots Co., Ltd.)
90. Nicholson JS, Tantum JG (1987) Chem Abstr 90:22610j
91. Shiraiwa T, Kataoka K, Sakata S, Kurokawa H (1989) Bull Chem Soc Jpn 62:109
92. Hassan NA, Bayer E, Jochims JC (1998) J Chem Soc Perkin Trans 1, p 3747
93. Maryanoff CA, Scott L, Shah RD, Villani FJ Jr (1998) Tetrahedron:Asymmetr 9:3247
94. Boesten WHJ, Seerden J-PG, de Lange B, Dielemants HJA, Elsenberg HLM, Kaptein B, Moody HM, Kellogg RM, Broxterman QB (2001) Org Lett 3:1121
95. Resnick L, Galante RJ (2006) Tetrahedron:Asymmetr 17:846
96. Jakubec P, Berkeš D, Považanec F (2004) Tetrahedron Lett 45:4755
97. Lee S-K, Lee SY, Park YS (2001) Synlett 1941
98. Berkeš D, Kolarovič A, Manduch R, Baran P, Považanec F (2005) Tetrahedron:Asymmetr 16:1927
99. Yamada M, Nagashima N, Hasegawa J, Takahashi S (1998) Tetrahedron Lett 39:9019
100. Caddick S, Jenkins K (1996) Tetrahedron Lett 37:1301
101. Brunetto G, Gori S, Fiaschi R, Napolitano E (2002) Helv Chem Acta 85:3785
102. Reider PJ, Davis P, Hughes DL, Grabowski EJJ (1987) J Org Chem 52:955
103. Kotera T, Watanabe M, Katori G, Ichihara M, Kkara K (1998) 63th Kagaku Kogaku Kai Nenkai (Japanese), p 123
104. Shi Y-J, Wells KM, Pye PJ, Choi W-B, Churchill HRO, Lynch JE, Maliakal A, Sager JW, Rossen K, Volante RP, Reider PJ (1999) Tetrahedron 55:909
105. Shieh W-C, Carlson JA, Zaunius GM (1997) J Org Chem 62:8271
106. Ikeura Y, Ishimaru T, Doi T, Kawada M, Fujishima A, Natsugari H (1998) Chem Commun 2141

107. Colson P-J, Przbyla CA, Wise BE, Babiak KA, Seaney LM, Korte DE (1998) Tetrahedron:Asmmetr 9:2587
108. Alabaster RJ, Gibson AW, Johnson SA, Edwards JS, Cottrell IF (1997) Tetrahedron:Asymmetr 8:447
109. Sohda T, Mizuno K, Kawamatsu Y (1984) Chem Pharm Bull 32:4460
110. Hagmann WK (1986) Syn Commun 16:437
111. Lopez FJ, Ferriño SA, Reyes MS, Román R (1997) Tetraheron:Asymmetr 8:2497
112. Cannata V, Tamerlani G (1986) JP 61-129148, EP 182 279 (Alfa Chemicals Italiana S.p.A.)
113. Cannata V, Tamerlani G (1986) Chem Abstr 105:78693j
114. Cannata V, Tamerlani G, Morotti M (1986) JP 61-289057, EP 204 911 (Alfa Chemicals Italiana S.p.A.)
115. Cannata V, Tamerlani G, Morotti M (1987) Chem Abstr 107:6951k
116. Konoike T, Matsumura K, Yorifuji T, Shinomoto S, Ide Y, Ohya T (2003) J Org Chem 67:7741
117. Brands KMJ, Payack JF, Rosen JD, Nelson TD, Candelario A, Huffman MA, Zhao MM, Li J, Craig B, Song ZJ, Tschaen DM, Hansen K, Devine PN, Pye PJ, Rossen K, Dormer PG, Reamer RA, Welch CJ, Mathre Dj, Tsou NN, McNamara JM, Reider PJ (2003) J Am Chem Soc 125:2129
118. Silverberg LJ, Kelly S, Vemishetti P, Vipond DH, Gibson FS, Harrison B, Spector R, Dillon JL (2000) Org Lett 2:3281
119. Napolitano E, Farina V (2001) Tetrahedron Lett 42:3231
120. Ataka K, Okamura S (1997) JP 97 221 485 (Ube Industries, Ltd., Japan)
121. Ataka K, Okamura S (1997) Chem Abstr 127:248097s
122. Yokoyama Y (1998) Medicinal Chemistry Symposium No18 (Japanese), p 41
123. Cooper J, Humber DC, Long AG (1986) Syn Commun 16:1469
124. Kemperman GJ, Zhu J, Klunder AJH, Zwanenburg B (2000) Org Lett 2:2829
125. Yang KS (1981) JP 1-18077, EP 18 811 (Lilly, Eli, and Co.)
126. Yang KS (1981) Chem Abstr 94:180670r
127. Onoue H, Ohtani M, Watanabe F (1984) JP 59-007193, EP 98 545 (Shionogi and Co., Ltd.)
128. Onoue H, Ohtani M, Watanabe F (1984) Chem Abstr 101:23214r
129. Fuchs R, Wittig A (1987) JP 62-456, EP 206 149 (Bayer AG)
130. Fuchs R, Wittig A (1987) Chem Abstr 107:40134f
131. Takuma K, Suzuki Y, Morino H, Kakimizu A (1984) JP 58-183 662, EP 206 149 (Sumitomo Chemical Co., Ltd.)
132. Takuma K, Suzuki Y, Morino H, Kakimizu A (1984) Chem Abstr 100:116506g
133. Smrčina M, Lorenc M, Hanuš V, Sedmera P, Kočovsvský P (1992) J Org Chem 57:1917
134. Wilen SH, Qi JZ, Williard PG (1991) J Org Chem 56:485
135. Chaplin DA, Johnson NB, Paul JM, Potter GA (1998) Tetrahedron Lett 39:6777
136. Czollner L, Frantsits W, Küenburg B, Hedenig U, Fröhlich J, Jordis U (1998) Tetrahedron Lett 39:2087
137. Xia W, Scheffer JR, Patrick BO (2005) Cryst Eng Comm 7:728
138. Kanomata N, Ochiai Y (2001) Tetrahedron Lett 42:1045
139. Ates A, Curran DP (2001) J Am Chem Soc 123:5130
140. Ihara M, Taniguchi N, Yasui T, Hosoda S, Fukumoto K (1991) 33th Symposium on Natural Organic Compounds (Japanese), p 124
141. Vedejs E, Donde Y (2000) J Org Chem 65:2337
142. Node M, Nishide K, Fujiwara T, Ichihashi S (1998) Chem Commun 2363

143. Vedejs E, Donde Y (1997) J Am Chem Soc 119:9293
144. Komatsu H, Awano H (2002) J Org Chem 67:5419
145. Vedejs E, Chapman RW, Müller SLM, Powell DR (2000) J Am Chem Soc 122:3047
146. Allinger NL, Eliel EL (1971) Top Stereochem 6:107
147. Newman P (1978) Optical Resolution Procedures for Chemical Compounds, vol 1. Optical Resolution Information Center, New York
148. Newman P (1981) Optical Resolution Procedures for Chemical Compounds, vol 2. Optical Resolution Information Center, New York
149. Ebbers EJ, Ariaans GJA, Houbiers JPM, Bruggink A, Zwanenburg B (1997) Tetrahedron 53:9417
150. Faber K (1997) Biotransformations in Organic Chemistry, 3rd edn. Springer, Berlin Heidelberg New York
151. Strauss UT, Felfer U, Faber K (1999) Tetrahedron:Asymmetr 10:107
152. Fulling G, Sih CJ (1987) J Am Chem Soc 109:2845
153. van der Deen H, Cuiper AD, Hof RP, van Oeveren A, Feringa BL, Kellogg RM (1996) J Am Chem Soc 118:3801
154. Larsson ALE, Persson BA, Backvall J-E (1997) Angew Chem Int Ed Engl 36:1211
155. Strauss UT, Felfer U, Faber K (2005) Tetrahedron:Asymmetr 16:1927
156. Inoue Y (1992) Chem Rev 92:741
157. Leibovitch M, Olovsson G, Scheffer JR, Trotter J (1998) J Am Chem Soc 120:12755
158. Ramamurthy V, Schanze KS, Dekker M (1998) Molecular and Supramolecular Photochemistry, vol 2. Marcel Dekker, New York
159. Horspool WM, Lenci F (2004) CRC Handbook of Organic Photochemistry and Photobiology, 2nd edn. CRC, Boca Raton, FL

Advantages of Structural Similarities of the Reactants in Optical Resolution Processes

Ferenc Faigl (✉) · József Schindler · Elemér Fogassy

Research Group and Department of Organic Chemical Technology,
Hungarian Academy of Sciences, Budapest University of Technology and Economics,
Budafoki út 8., 1111 Budapest, Hungary
ffaigl@mail.bme.hu

1	Introduction	134
2	Optical Resolution of Structurally Similar Racemates with the Same Resolving Agent	137
2.1	Investigation of a Series of 1-Substituted-Alkylamines	137
2.2	Optimization of the Resolving Agent for a Series of Atropisomeric Quinazolinone Derivatives	139
3	Separation of Enantiomers by Distillation in the Presence of a Mixture of Structurally Similar Resolving Agents . .	140
4	Resolution with an Optically Active Derivative of the Racemate	141
4.1	Use of a Half Equivalent of Derivative Resolving Agents Under pH Control	142
4.2	Application of One or More Equivalents of a Derivative Resolving Agent . .	143
5	Resolution of Structurally Similar Racemates with Homologous Series of their Optically Active Derivatives	144
5.1	Resolution of 1-Arylethylamines with Homologous Series of Derivative Resolving Agents	144
5.2	Effects of Achiral Additives on the Results of Resolutions	147
5.3	Solvent-Free Resolutions with Homologous Series of Resolving Agents . . .	148
6	Resolution of the Dicarboxylic Acid Monoester Derivatives of 1-(Substituted Phenyl)ethyl Alcohols	150
6.1	Resolution of Dicarboxylic Acid Monoesters in the Presence of Another Similar Racemic or Achiral Monoester	152
6.2	Resolution of Structurally Similar Dicarboxylic Acid Monoesters with Mixtures of Optically Active and Achiral Amines	153
7	Parallel Kinetic Resolution with a Quasi-racemate-Type Resolving Agent Mixture	155
8	Conclusions and Outlook	155
	References	156

Abstract Advantages and limitations of the use of structurally similar compounds in racemate resolution via diastereoisomeric salt formation are discussed in this review. An effective conception on "derivative resolving agents" (use of the optically active derivatives of a racemate as resolving agents) is presented by examples and the method is extended to the homologous series of the derivative resolving agents and/or achiral additives. A recently developed distillation version of the family approach to optical resolution and the novel, solvent-free resolution methods using a half equivalent amount of resolving agent are also discussed.

Keywords Optical resolution · Diastereoisomeric salt · Derivative resolving agent · Structural similarity · Homo- and heterochiral assemblies

1
Introduction

Optical resolution via diastereoisomeric salt formation is one of the most frequently used methods in the fine chemical and pharmaceutical industries for manufacture of pure enantiomers. However, selection of the optimum resolving agent and the other chiral or achiral additives have remained a game of trial and error. In order to reduce this time-consuming experimental work numerous methods are offered in the recent reviews [1, 2]. A simple and usually efficient approach is to take a survey of the known resolutions of compounds which are structurally similar to our racemate. Thus, for example, the efficiency and the absolute configuration of the crystallizing diastereoisomeric salts could be predicted by computation in a series of 2-phenylglycine derivatives [3]. In a recently developed method, the authors used a small library of structurally similar resolving agents for the diastereoisomeric salt formation with a given racemate [4]. The same authors have developed the reverse system too: mixtures of structurally similar racemic compounds have been resolved with one resolving agent. In most cases synergistic effects of such mixtures on the efficiencies of enantiomer separations have been observed.

Another approach for finding efficient resolving agents is the application of optically active derivatives of the racemic mixture to be resolved. The basis of this method is the observed strong tendency of enantiomers to form heterochiral assemblies in solution (Scheme 1) and racemate-type molecular compounds in crystalline form.

The heterochiral packing usually results in more stable supramolecular structures and it is assumed that at least 85% of the known enantiomers crystallize out in true racemate form [5, 6]. Our recently developed enantiomeric enrichment methods: the selective precipitation, selective extraction, crys-

$$\begin{aligned} D+D &\rightleftharpoons DD \rightleftharpoons (DD)_m \\ L+L &\rightleftharpoons LL \rightleftharpoons (LL)_n \end{aligned} \Bigg\} \text{homochiral}$$

$$D+L \rightleftharpoons DL \rightleftharpoons (DL)_o \quad \text{heterochiral}$$

Scheme 1 Formation of homo- and heterochiral supramolecules in the solutions of D and L enantiomers ($m, n, o \geq 0$)

tallization or distillation of chiral acids and bases [7–9] have provided experimental evidence of this phenomenon. The same tendency to form quasi-symmetrical complexes can also be observed in such resolutions where the resolving agent and the racemate are structurally similar compounds. The first, 50-year-old example is the optical resolution of racemic tartaric acid by mixed-crystal formation with malic acid, a structurally closely related diacid. (S) Malic acid preferentially crystallizes with (R,R)-tartaric acid forming a quasi-racemate [10].

Crystallization of the more stable diastereoisomeric salt of a racemic compound with a quasi-enantiomeric resolving agent can also be used for separation of enantiomers. It has to be mentioned, however, that the salt formation with inorganic or structurally diverse chiral reagents may increase the homochiral salt-forming tendency [5]. Regarding similar structures of the racemates and resolving agents, only few examples were known two decades ago for the resolution of protected α-amino acids with optically active amino acid derivatives (Table 1, [11]) and half of the examples described quasi-racemate type crystalline diastereoisomeric salts.

Table 1 Resolution of α-amino acids with optically active α-amino acid derivatives

Resolving agent	Racemate		Solvent	Crystallized salt	Yield (%)
(S)-1		2	Ethanol	(S)-2.(S)-1	70
		3	Ethanol	(S)-3.(S)-1	60
		4	Methanol	(S)-4.(S)-1	94
		5	Ethanol	(S)-5.(S)-1	51
		6	Ethanol	(S)-6.(S)-1	78
		7	Ethanol	(S)-7.(S)-1	46

Table 1 (continued)

Resolving agent	Racemate		Solvent	Crystallized salt	Yield (%)
(S)-8: HO-C6H4-CH2-CH(NH2)-C(O)-NHNH2	CH3-CH(N-Z)-COOH	9	Ethanol	(R)-9.(S)-8	75
	N-Z pyrrolidine-2-COOH	10	Methanol	(R)-10.(S)-8	93
	HOOC-CH2-CH(NH2)-COOH	4	Water	(R)-4.(S)-8	67
	(CH3)2CH-CH(NHZ)-COOH	11	Ethanol	(R)-11.(S)-8	96
	azetidine-2-COOH (NH)	12	Methanol	(R)-12.(S)-8	55
(S)-13: 3,5-Br2-4-HO-C6H2-CH2-CH(NHAc)-COOH	H2N-(CH2)3-CH(NH2)-COOH	14	Water	(S)-14.(S)-13	78
(S)-15: Ph-CH2-CH(NH-COOCH2Ph)-COOH	Ph-CH2-CH(NH2)-COOtBu	16	Diethylether	(S)-16.(S)-15	100
(S)-4: HOOC-CH2-CH(NH2)-COOH	H2N-(CH2)3-CH(NH2)-COOH	14	Water	(S)-14.(S)-4	86

In these examples, one could consider the structural similarities around the stereogenic carbons of the reactants but, of course, the further parts of these compounds are quite different from each other. In spite of this fact, the reported efficiencies of the resolutions were usually good or excellent. (In order to characterize the result of a resolution with a dependent variable the S-value (S = yield × optical purity or ee) was postulated by us [12].)

All the above mentioned and the following literature data illustrate the unambiguous advantages of the use of structurally similar reagents in chiral discriminating processes. Application of these approaches would help chemists to find resolving agents much faster than usually happens by the traditional trial and error game, and highly efficient enantiomer separations should be accomplished. There are no well-defined rules for exact determination of sufficient structural similarities among the racemic compounds, resolving agents, and achiral additives. However, this comprehensive discussion of the recently developed enantiomer separation methods using single or homologous series of "derivative resolving agents", mixtures of structurally similar racemates and/or "families of resolving agents", as well as homologous series of chiral and achiral reagents can show a convenient way to design efficient optical resolutions.

2
Optical Resolution of Structurally Similar Racemates with the Same Resolving Agent

Application of the hypothesis that structurally similar racemates can be resolved with the same resolving agent (and in the same solvent) can often reduce radically the number of preliminary experiments. It is therefore useful to review known procedures for the resolution of analogs and the conditions used for the structurally most similar racemate should be tried first. There are several data on the systematic investigation of the applicability of a given resolving agent in a family of structurally similar racemates.

2.1
Investigation of a Series of 1-Substituted-Alkylamines

The optical isomers of isopropylideneglycerol-monophtalate (**17**) has been described recently as efficient resolving agents of racemic 1-substituted-alkylamines (**18–35**, Scheme 2) [13, 14]. The diastereoisomeric salt formation reactions were carried out in methanol/2-propanol mixtures with an equivalent amount of the resolving agent, and the (*S*) enantiomers of the amines gave crystalline salt with (*S*)-**17** in every cases.

Comparison of the results (*S* parameters) as a function of the structure of racemates showed the limitations of similarities within that series of amines. Good separation of the enantiomers could be achieved when an aromatic ring (or a vinyl group, **29**) was attached to the stereogenic center. Saturation of the ring, its substitution with an alkyl chain (compounds **33, 34**) or insertion of a methylene group between the phenyl group and the asymmetric carbon atom (**31**) hindered the crystallization of any diastereoisomeric salt. That is because of the sensitivity of the supramolecular assemblies (formed in the supersaturated solutions before appearing the solid product) against such types of structural deviations. However, modification of the alkyl chain bound to the stereogenic carbon was not as critical as the previous modifications. Com-

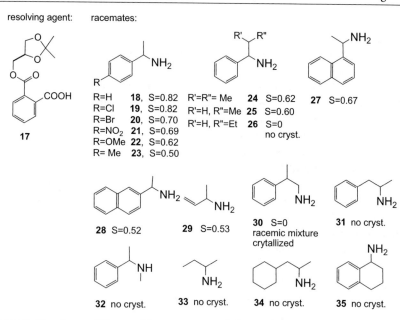

Scheme 2 Structures of the resolving agent 17 and the resolved chiral ethylamine derivatives 18–35 (S efficiency of the resolution)

pounds 24 and 25 could be resolved, in contrast to compounds 26 and 35, in which the alkyl groups were further extended.

Optical resolutions of structurally similar piperazine (36) and piperidine derivatives (37) were also accomplished with the same resolving agent (38) in our laboratory [15].

Since compound 36 was a very new lead molecule of antipsychotic agents, only a small amount of the racemic sample was available that time. Therefore, test reactions for selecting resolving agents were carried out using the analogous compound 37. The optimum resolving agent (38) was then applied for separation of the enantiomers of 36. Efficiencies of the two resolutions were practically the same (Scheme 3).

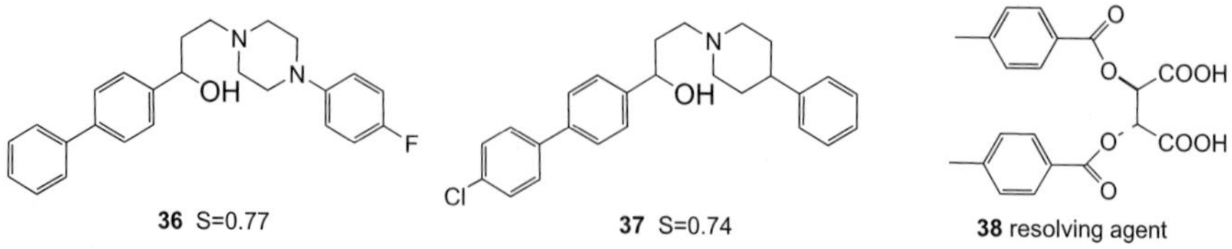

Scheme 3 Structures of the amino alcohols 36, 37 and the resolving agent 38 (S efficiency of the resolution)

2.2
Optimization of the Resolving Agent for a Series of Atropisomeric Quinazolinone Derivatives

Dai and coworkers synthesized several new quinazolinone groups containing phosphine ligands (**39–41**) and performed the resolution of the atropisomers with (1S)-10-camphorsulfonic acid [16]. The diastereoisomeric salts were separated as well-developed crystals, but the decomposition produced only the starting racemates. The structure of the crystalline salt of **39** was determined by X-ray crystallography. It seemed that the carbonyl group, which distinguishes the two sides of the resolving agent, is not bulky enough to differentiate between the enantiomers of the bases. Therefore, the ketone function was transformed into its benzenesulfonylhydrazone derivative (compound **42**), which proved to be efficient with all the three racemates (Scheme 4). It has to be mentioned that these resolutions were the first examples of using **42** as resolving agent. Regarding the structural similarities of the atropisomeric racemates, we can conclude that the multipoint interaction system (the second order bonds between the base and the chiral acid within the salt and among the ion pairs [12]) is practically resistent to change of the substituents R^1 and R^2 in compounds **39–41**.

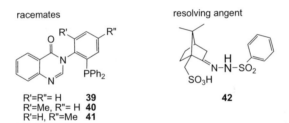

Scheme 4 Atropisomeric quinazolinone group containing phosphine ligands (**39–41**) and the optimum resolving agent **42**

A well-selected resolving agent can also be used to resolve structurally similar racemates via complex formation. An excellent example is Noyori's method for the synthesis of optically active BINAP via resolution of BINAP oxide (**43**, Scheme 5) with O,O'-dibenzoyl-(R,R)-tartaric acid [17]. The process is simple and very efficient. Solutions of the racemate and the resolving agent (made in chloroform or ethylacetate/chloroform mixture) were reacted and the crystalline diastereoisomeric complex was separated and recrystallized in the same solvent. Then it was treated with an aqueous base solution to liberate the optically active phosphine oxide in enantiomerically pure form. In the last two decades, Noyori's method has been successfully used to resolve numerous chiral diphosphine oxides such as **44** [18], **45** [19], **46** [20], and **47** [21, 22].

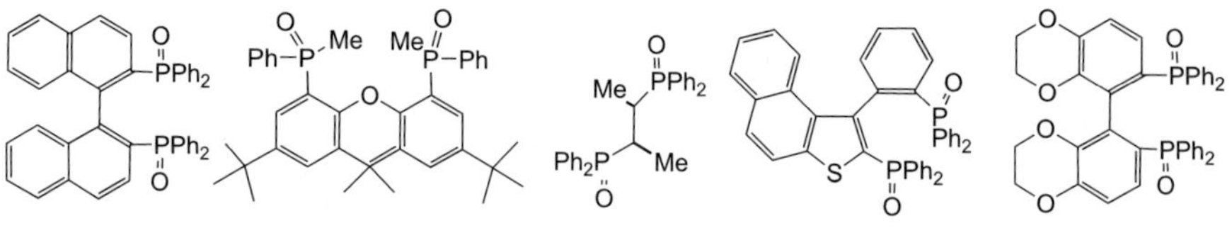

Scheme 5 Chiral diphosphine oxides resolved by Noyori's diastereoisomeric complex formation method

Table 2 Data on the resolution of acids 48–50

Racemate		Resolving agent	Solvent	Enantiomer in the crystallized salt
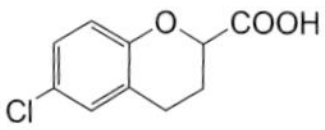	48	Chinconidine	Ethyl acetate	(+)-48
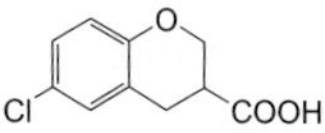	49	(R)-2-Amino-1-phenylpropane	Ethanol/water	(+)-49
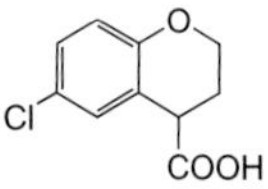	50	(R)-1-Phenylethylamine	Ethanol	(+)-50

Further examples of the selection of a resolving agent using the principle of maximum similarity can be found in the literature [1, 23]. However, there are limitations to this simple rule when, for example, the position of the asymmetric carbon atom is changed. Despite their structural similarities, acids **48–50** required different resolving agents and solvents for efficient enantiomer separation (Table 2, [24]). An explanation might be the fact that the carboxylic group was bound to different positions in the racemates and, consequently, the relative position of the stereogenic center changed in each case.

3
Separation of Enantiomers by Distillation in the Presence of a Mixture of Structurally Similar Resolving Agents

Several years ago, a new method was developed by Dutch researchers to find the optimum resolving agent using libraries of structurally similar compounds [4, 25]. Details of the "Dutch resolution" is discussed in another chapter of this volume written by R. Kellogg. Starting from their observations,

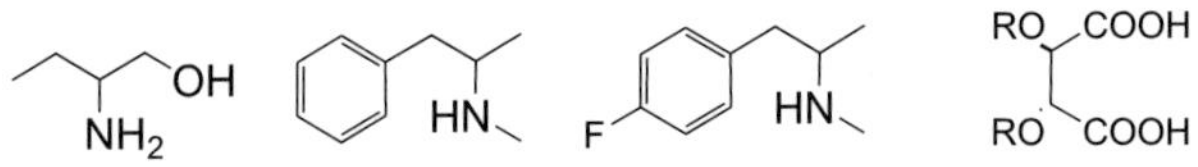

Scheme 6 The racemates (53–55) and the family of resolving agents (38, 51, 52) used for optical resolution by distillation

our laboratory has developed a novel, distillation version of the Dutch resolution [26]. According to our method, the model compounds (53–55) were reacted with a half equivalent of resolving agents (38, 51, and 52) without any solvent, and the excess of the free enantiomers were separated by distillation (Scheme 6). Resolutions with mixtures of tartaric acid derivatives gave the same, or (in several cases) better, results than were obtained with individual resolving agents. A calculation method was also developed for counting the synergetic effect. It is important to mention that an individually inactive resolving agent could also contribute to successful resolutions in a mixture of resolving agents. We supposed that certain supramolecular structures (formed among the acidic resolving agents and the base enantiomers in solution) were responsible for the enantiomer discrimination.

4
Resolution with an Optically Active Derivative of the Racemate

The strong tendency of chiral compounds to form heterochiral assemblies over homochiral ones demonstrates the advantages of the economic spatial arrangement of symmetrical structures in nature. It is because, by complementing each other, a pair of mirror image molecules can usually form more compact (i.e. lower energy) crystals than the homochiral dimers. Starting from these facts one can treat a non-racemic enantiomeric mixture (in solution) as a mixture of two dimers: the heterochiral molecular compound (racemate) and the homochiral aggregate (Scheme 1). Of course, higher member associates can also form before crystallization as combinations of homo- and heterochiral dimers. In other words, the excess of the major enantiomer works in these cases as a resolving agent of the racemate. In practice, optical resolution of a racemic mixture can be carried out with a structurally modified optically active derivative of the racemic compound instead of the free enantiomer. An important condition for the formation of quasi-racemic crystals is that the structures of the reaction partners should be as similar as possible.

4.1
Use of a Half Equivalent of Derivative Resolving Agents Under pH Control

A few early resolutions using derivative resolving agents of amino acids have been mentioned in the introduction of this chapter. Recent examples of such resolutions were mainly developed in our laboratory (Table 3, [1, 6, 27]).

The molar ratio of the racemate and resolving agents has big practical importance in these resolutions. The best results could be achieved using a half

Table 3 Optical resolutions performed with the optically active derivatives of the racemic substrates

Enantiomer in the precipitated salt		Resolving agent		ee (%)	Yield (%)	S-factor
	(R)-56		(S)-57	95	90	0.86
	(S)-58		(R)-59	100	80	0.80
	(R)-60		(S,S)-61	81	190[a]	1.54[a]
	(S)-62		(R)-63	100	86	0.86
	(S,S)-64		(S,S)-65	100	85	0.86
	66		67	86	89	0.76
	(R)-68		(S)-69	75	85	0.64

[a] The yields were calculated as half of the racemate. Therefore, in the case of second order asymmetric transformation, 200% yield would be the theoretical maximum. © (2006) from CRC Handbook of Optical Resolution via Diastereoisomeric Salt Formation from David Kozma (ed). Reproduced by permission of Routledge/Taylor and Francis Group, LLC.

equivalent of the resolving agents in the presence of achiral additives. The amounts of these additives (having the same basic or acidic character as for the resolving agent) strongly influence the pH value of the reaction mixtures and, in this way, can modify the S-values (for theoretical background see the equilibrium model of resolution in [28]). Thus, for example, resolution of β-phenylalanine (**58**) could be accomplished with the use of a half equivalent of its optically active N-benzoyl derivative (**59**) in aqueous methanol solution. The yield and optical purity of the crystalline diastereoisomeric salt strongly depended on the pH value of the reaction mixture. The best result (yield 80%, ee 100%, S-factor 0.80) could be achieved by addition of an equivalent amount of hydrochloric acid (as side reagent, calculated to the amount of the racemate) to the reaction mixture. Similar results were obtained during resolution of 1-phenylglycine (**56**) with its optically active derivative (**57**).

4.2
Application of One or More Equivalents of a Derivative Resolving Agent

Aminoketone **60** is configurationally stable in acidic medium but suffers racemization by enolization at about pH 7.5, in protic solvents. This fact let us to convert the whole amount of racemic **60** into the desired (R)-**60** isomer by second order asymmetric transformation using (S,S)-**61** as derivative resolving agent [2] (Table 3).

Separation of the enantiomers of **68** represents a special example of resolution with derivative resolving agents (Table 3). Interestingly enough, good separation could be achieved using hexane as solvent and the crystalline diastereoisomeric salt contained (R)-**68** and (S)-**69** resolving agent in 1 : 2 molar ratio. Though the resolving agent did not dissolved in hexane, the salt formation was complete within 24 h producing the diastereoisomeric salt in 75% ee and 85% yield [27]. This result was the best among all the resolution trials carried out with other resolving agents [29].

Drauz and coworkers [30] attempted the optical resolution of *tert*-butylleucinol (**70**) with 14 different optically active acids. Four among them were the N-acylated derivatives (**71–73**). They used a half equivalent of resolving agent in 2-propanol. Comparison of the experimental data demonstrated that the most effective resolutions could be accomplished with the derivative

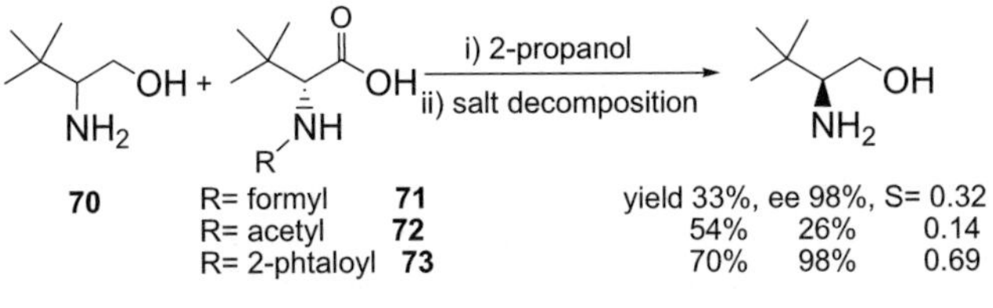

R= formyl **71**	yield 33%, ee 98%,	S= 0.32
R= acetyl **72**	54% 26%	0.14
R= 2-phtaloyl **73**	70% 98%	0.69

Scheme 7 Resolution of *tert*-butylleucinol (**70**) with its optically active derivatives (**71–73**)

resolving agents (Scheme 7). One among them, N-(2-phtaloyl)-L-*tert*-leucine (**73**), provided the best separation (70% yield and 98% ee).

5
Resolution of Structurally Similar Racemates with Homologous Series of their Optically Active Derivatives

The above mentioned results confirmed our theory that the best resolving agent of a given racemic compound should be found among the optically active derivatives of the same compound. In addition, these derivatives should be as similar to the parent compound (racemate) as possible. However, scarcely anything is known about the proper structure of the derivative-forming moiety of the molecule. Prediction of the effects of chiral or achiral additives on the outcome of such resolutions has also been missing from the chemist's toolkit. Therefore, a well-designed series of experiments were carried out in our laboratory to test the structural limitations of similarity between racemates and a group of derivative resolving agents. These investigations were extended to the resolutions of mixtures of racemates with structurally similar resolving agents. In addition, the influence of achiral additives on the efficiency of enantiomer separation was also investigated.

5.1
Resolution of 1-arylethylamines with Homologous Series of Derivative Resolving Agents

In order to construct multicomponent systems of structurally similar racemates and resolving agents, 1-phenylethylamine (**18**) and its acidic derivatives (prepared by acylation of (R)-**18** with different dicarboxylic acid anhydrides or esters [31]) were chosen as model compounds. Resolutions of racemic **18** with an equivalent amount of the amide derivatives ((R)-**74–77**, Scheme 8) were accomplished in acetone or ethylacetate [31]. The yields, enantiomer excesses of the crystallized diastereoisomeric salts, and the S-factors are given in Table 4.

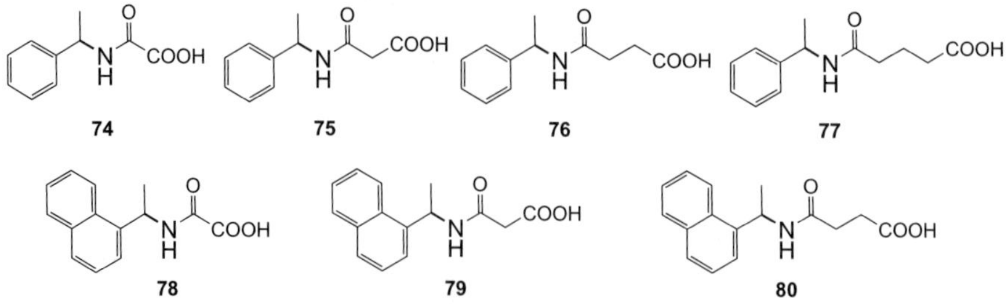

Scheme 8 Optically active dicarboxylic acid amide derivatives of amines **18** and **27**

Table 4 Results of the resolutions of **18** with a series of its dicarboxylic acid amide derivatives

Racemate	Resolving agent	ee (configuration) (%)	Yield [a] (%)	S-factor
18	(R)-74	98 (S)	31	0.37
18	(R)-75	99 (S)	56	0.55
18	(R)-76	100 (S)	61	0.61
18	(R)-77	77 (S)	58	0.35

[a] Diastereoisomeric salts were recrystallized twice from acetone or ethyl acetate before workup

In that series of experiments, the quasi-racemate-type salts crystallized in all cases and compound **76** was the optimal resolving agent among the homologous derivatives.

In another series of experiments the dicarboxylic acid amide derivatives of (R)-1-naphtylethylamine (**78–80**, Scheme 8) were used for resolution of racemic 1-naphtylethylamine (**27**)(Schindler and Fogassy, unpublished results) and an equivalent amount of (R)-**78** preformed the best separation. After two recrystallizations from ethyl acetate, optically pure heterochiral diastereoisomeric salt ((S)-**27**(R)-**78**) could be obtained in 60% yield (Table 5). Compound (R)-**80** behaved similarly under similar conditions but the efficiency of the resolution was approximately half of the previous one. While compounds **78** and **80** crystallized with their quasi-enantiomer pair ((S)-**27**), the malonate derivative ((R)-**79**) provided a solvent-dependent discrimination between the two enantiomers of compound **27**. Resolutions were accomplished in these cases with a half equivalent of (R)-**79** in water or water/cyclohexane mixture instead of ethyl acetate. In the latter case, half of the racemate remained in the apolar cyclohexane. In aqueous solution, a half equivalent amount of methanesulfonic acid was used in order to form a water-

Table 5 Results of the resolutions of **27** with a series of its optically active dicarboxylic acid amide derivatives (**78–80**)

Racemate	Resolving agent (solvent)	ee (configuration) (%)	Yield (%)	S-factor
27	(R)-78 (ethyl acetate)	100 (S)	60 [a]	0.59
27	(R)-79 (water/cyclohexane)	12 (R)	86	0.10
27	(R)-79 (water)	15 (S)	81	0.13
27	(R)-80 (ethyl acetate)	100 (S)	100 [a]	0.31

[a] In these experiments the salts were decomposed after recrystallization from ethyl acetate

soluble salt with the non-crystallizing part of the amine. In the last two cases the optically active amines were liberated from the crude diastereoisomeric salts (Table 5).

The presented experimental facts confirmed that the resolution trials with derivative resolving agents represent a useful method for finding a good resolving agent of a racemic compound. It is also clear from these experiments that the difference between the aromatic moieties of the two racemates (**18** and **27**) decreases the similarity of the amines with respect to their choice among the homologous derivatives.

The effects of the substituents connected to the phenyl group of **18** have also been investigated using the above-mentioned resolving agents ((R)-**74–77**). In these series of experiments seven ring-substituted 1-phenylethylamines (**19, 20, 22, 23, 81–83**) were subjected to diastereoisomeric salt formation reactions, separately (Scheme 9) [32]. All experiments were carried out at a 50 mmol scale using an equivalent amount of resolving agents in acetone or ethylacetate. The results of the successful experiments are collected in Table 6.

The crystalline diastereoisomeric salts contained the (S) enantiomer of the amines, in all cases showing the preference of the heterochiral assemblies in these systems. On the basis of the experimental data one can conclude that the ring substitution of **18** has a crucial role in the selection of the appropriate derivative resolving agent. Thus, the oxalic acid amide (**74**) proved to be the best resolving agent of the *ortho*-substituted amine **81**, similar to the findings for the resolution of compound **30**. On the other hand, only the succinic acid derivative (**76**) crystallized with the *meta*-substituted compounds (**82** and **83**). Satisfactory results could be achieved in the series of the *para*-substituted amines, where both the malic acid derivative (**75**) and, in two cases, the succinic acid derivative (**76**) could be used for resolution. The most successful resolving agent (**76**) of the parent compound **18** gave crystalline diastereoisomeric salts in five cases among the seven model compounds. Only the two *para*-halogenated amines (**19** and **20**) failed to crystallize with that reagent, but they could be resolved with another member (**75**) of that family of resolving agents.

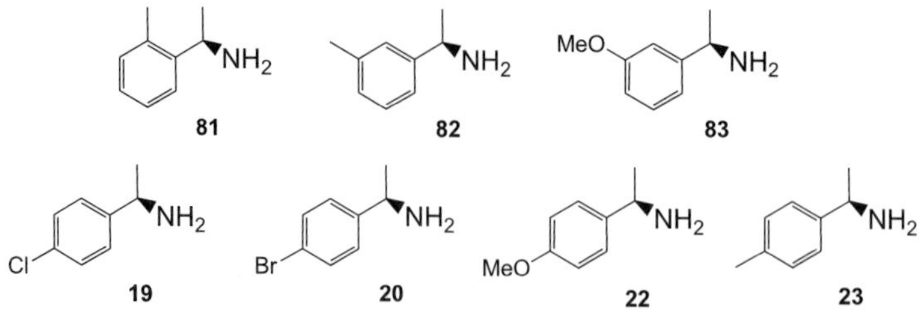

Scheme 9 Ring-substituted 1-phenylethylamines resolved with dicarboxylic acid amide derivatives of **18**

Table 6 Results of the resolutions of substituted 1-phenylethylamines with the amide type derivatives of (R)-18

Racemate	Resolving agent	ee(configuration) (%)	Yield (%)	S-factor
81	(R)-74	46 (S)	98	0.45
81	(R)-76	76 (S)	19 [a]	0.14
82	(R)-76	54 (S)	99	0.53
83	(R)-76	51 (S)	100	0.51
23	(R)-75	55 (S)	100	0.55
22	(R)-75	32 (S)	99	0.32
19	(R)-75	48 (S)	99	0.48
20	(R)-75	49 (S)	100	0.49
23	(R)-76	36 (S)	100	0.36
22	(R)-76	45 (S)	97	0.47

[a] Without recrystallization of the diastereoisomeric salt. In the other cases, the salts were recrystallized from ethyl acetate or acetone

5.2
Effects of Achiral Additives on the Results of Resolutions

Single crystal X-ray diffraction measurements on (S)-18.(R)-74, (R)-18.(R)-74 salts [31] and the (S)-23.(R)-76 salt [32] confirmed that the acid amide moieties are involved in the intermolecular hydrogen bond systems of the crystals. Starting from this fact, we supposed that addition of urea (an imitation of the acid amide structure) or urea derivatives to the reaction mixtures would strongly influence the crystallization properties of the diastereoisomeric salts. According to this concept, racemic 1-phenylethylamine (18) was reacted with a half equivalent of the four homologous dicarboxylic acid amide derivatives (74–77) in acetone, in the presence of a half equivalent of urea, mono- and dimethylureas, and thiourea, respectively. Surprisingly, these achiral additives completely inhibited crystallization of the diastereoisomeric salts of compounds 74, 75, and 76 but the (S)-18.(R)-77 salt crystallized in a much purer form than without additives (Table 7, lines 1 and 2). It has to be mentioned that the precipitated crude diastereoisomeric salt contained a half equivalent of urea, but this additive disappeared from the salt during recrystallizations.

These experimental data led us to conclude that simple achiral molecules, having similar structure to the characteristic part of the chiral reagent, may be suitable inhibitors of the crystallization of the more soluble diastereoisomeric salt. The S-factors of such resolutions did not change too much, but the enantiomeric excess in the crystalline diastereoisomeric salts increased significantly.

Table 7 Results of the resolutions of amine **18** with its glutaric acid amide derivative (**77**) in the presence of urea derivatives

Racemate	Resolving agent	Urea derivative	ee (%)	Yield [a] (%)	S-factor
18	(R)-77	–	77	58	0.35
18	(R)-77	H₂N–C(O)–NH₂	99	42	0.41
18	(R)-77	H₂N–C(S)–NH₂	100	35	0.35
18	(R)-77	H₂N–C(O)–NHCH₃	100	36	0.36
18	(R)-77	CH₃NH–C(O)–NHCH₃	100	33	0.33

[a] The diastereoisomeric salts were recrystallized twice, in each case

5.3
Solvent-Free Resolutions with Homologous Series of Resolving Agents

The solvent of diastereoisomeric salt formation and crystallization processes may radically influence the outcome of resolutions. In order to eliminate such influencing parameters, solvent-free enantiomer separations of bases **18, 20**, and **23** were performed with a series of "derivative resolving agents" [33]. Technically these experiments are simple: one should mix the (oily) racemate with a half equivalent of the resolving agent and warm it carefully until the solid reagent dissolves in the melt (oil). Then it should be cooled down to room temperature, treated with hexane and the precipitated diastereoisomeric salt filtered off. The results of such experiments are collected in Table 8. Even though the efficiencies of these resolutions were slightly smaller than the same resolutions in solution systems, the tendencies were the same: quasi-racemate formation dominated in all cases.

The solvent-free resolution method was also extended to the diastereoisomeric salt-forming reactions of racemic **18, 20**, and **23** with different two-component mixtures of derivative resolving agents ((R)-**74–77**, (R)-**84**, and (R)-**85**, Scheme 10).

The selected solid resolving agents were mixed in a mortar and a half equivalent amounts of these mixtures were added to each of the oily racemates **18, 20**, and **23**. Then the experiments were continued according to the above-mentioned recipe. The oily filtrate contained the optically active free enantiomer ((R)-**18**, (R)-**20**, or (R)-**23**) while the (S) isomers of the bases were liberated from the quasi-racemate-type diastereoisomeric salts. Table 9 contains the results of resolutions carried out with the single resolving agents of

Table 8 Solvent-free resolution of **18** with its optically active derivatives (**74–77**)

Racemate	Resolving agent	ee (configuration) (%)	S-factor [a]
18	(R)-**74**	26	0.26
18	(R)-**75**	27	0.27
18	(R)-**76**	28	0.28
18	(R)-**77**	38	0.38
23	(R)-**76**	29	0.29
20	(R)-**75**	33	0.33

[a] All of the diastereoisomeric salts were isolated quantitatively from the reaction mixture by its trituation with hexane and filtration

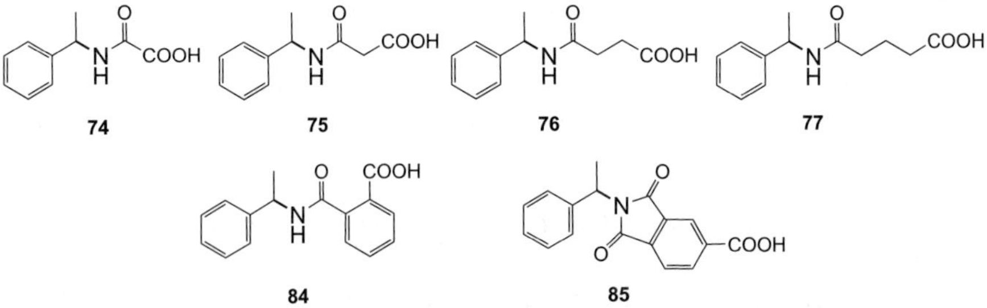

Scheme 10 Series of structurally similar resolving agents for solvent-free resolutions

the homologous series and with the two-component (**A** and **B**) mixtures of them. Compositions of these mixtures were changed from A/B = 0.9/0.1 to A/B = 0.1/0.9 stepwise but Table 9 contains only those compositions where a minimum or maximum level of enantiomer separations could be detected. Since we worked without any solvent, practically 100% of the diastereoisomeric salts were in the crystalline phase in each cases. Therefore, Table 9 contains only the enantiomeric excess of the products liberated from the salts.

In most cases we observed a decrease of the ee values compared to the results achieved with one member of the pairs of resolving agents. In one case, however, significant synergistic effect was observed. Compound **18** could be resolved with the glutaric acid derivative (**77**) providing 38% ee of the diastereoisomeric salt, but compound **84** alone was unsuitable for enantiomer discrimination. However, resolution with a mixture of **77** and **84**, with this latter compound in excess (molar ratio **77**/**84** = 0.2/0.8), resulted in the amine enantiomer in 49% ee. In other words, addition of a small amount of a suitable resolving agent (**77**) to the unsuitable one (**84**) initiated even a better enantiomer discrimination by the latter compound than by the first compound alone [33].

Table 9 Solvent free resolutions carried out with single resolving agents (given in column A or B) and with the two-component mixtures (A and B)

Racemate	Resolving agent A	Resolving agent B	A/B	ee [a] (%)
18	(R)-74	(R)-76	1/0	26
18	(R)-74	(R)-76	0.6/0.4	10
18	(R)-74	(R)-76	0/1	28
18	(R)-77	(R)-84	1/0	38
18	(R)-77	(R)-84	0.2/0.8	49
18	(R)-77	(R)-84	0/1	– [b]
18	(R)-77	(R)-85	1/0	38
18	(R)-77	(R)-85	0/1	0 [c]
23	(R)-76	(R)-74	1/0	29
23	(R)-76	(R)-74	0.4/0.6	7
20	(R)-76	(R)-74	0/1	– [b]
20	(R)-75	(R)-76	1/0	33
20	(R)-75	(R)-76	0.4/0.6	25
20	(R)-75	(R)-76	0.6/0.4	3
20	(R)-75	(R)-76	0/1	0 [c]

[a] All of the diastereoisomeric salts were isolated quantitatively from the reaction mixture by its trituation with hexane and filtration
[b] No crystallisation occured
[c] The diastereoisomeric salt contained the racemic mixture of the amine

6
Resolution of the Dicarboxylic Acid Monoester Derivatives of 1-(Substituted Phenyl)ethyl Alcohols

Change of the amide group into an ester function in the series of 1-phenylethylamine dicarboxylic acid monoamides let us study the resolution of structurally similar racemic monoester derivatives. This formal change of functions could easily be realized by esterification of racemic 1-phenylethylalcohol and its substituted derivatives with the corresponding achiral dicarboxylic acid anhydrides. Then, the synthesized monoesters (Scheme 11) were subjected to diastereoisomeric salt formation reactions with (S)-1-phenylethylamine. In order to get comparable results, the experiments were carried out under the same conditions: each racemic compound was dissolved in ethyl acetate and reacted with a half equivalent of resolving agent. (The only exception was the resolution of compound 87 under thermodynamic control. In this case a molar equivalent of the resolving agent was used.) The precipitated homochiral diastereoisomeric salts were filtered off after 96 h crystallization time and worked up in the usual way. Data on the resolution of monoesters (87–92, Scheme 11) are collected in Table 10.

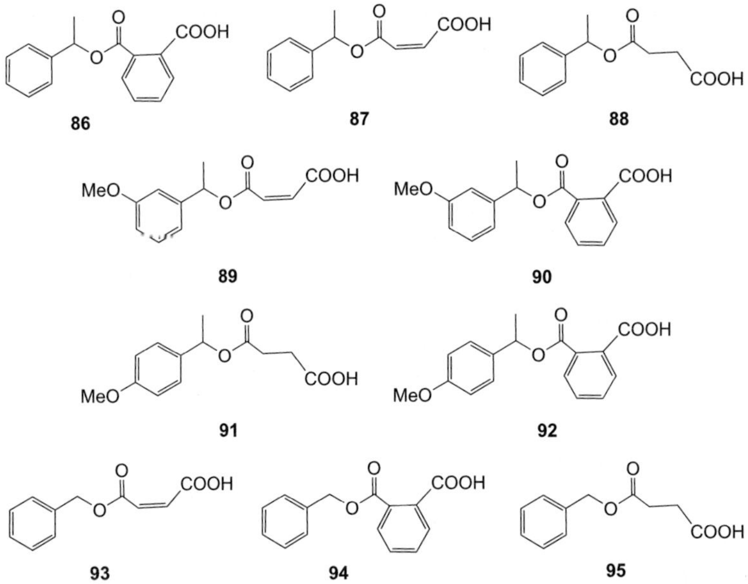

Scheme 11 Dicarboxylic acid monoester derivatives of 1-phenylethyl alcohol (**86–92**) and their achiral analogs (**93–95**)

Table 10 Resolution of structurally similar monoesters **86–92** with (*S*)-**18**

Racemate	Resolving agent	ee (%)	Yield (%)	S-factor
87	(*S*)-**18**	14 (98) [a]	17	0.06 (0.16) [a]
86	(*S*)-**18**	98	45	0.40
88	(*S*)-**18**	56	93	0.52
89	(*S*)-**18**	4	89	0.04
90	(*S*)-**18**	99	38	0.38
91	(*S*)-**18**	21	84	0.18
92	(*S*)-**18**	98	33	0.31

[a] Results in parentheses were observed under kinetic control (filtration after 1 h). In other cases the crystallization time was 96 h

It turned out that only two succinic acid monoesters (**88** and **91**) could be resolved with (*S*)-**18** among the 1-(substituted phenyl)ethyl esters of the saturated dicarboxylic acids (monoesters of oxalic, malonic, succinic, and glutaric acids). On the other hand, satisfactory enantiomer separations of the racemic monoesters of maleic and phtalic acids (**87** and **86, 90, 92**) could be achieved, but almost racemic salt crystallized from the reaction of compounds **89** and (*S*)-**18** (Table 10). Even though the resolutions were not optimized, one can realize that satisfactory results could be achieved with the phtaloyl monoesters (**86, 90,** and **92**). Therefore, these derivatives should be treated as lead molecules in the development of efficient enantiomer separation processes.

In a more general aspect, we can conclude that the change of functional groups (e.g., amide to ester having different hydrogen bond-forming abilities) may cause radical change in the resolution of structurally similar compounds. However, enantiomer separation could be negotiated on the basis of several trials with structurally similar derivatives of the racemates and a resolving agent. In other words, the concept of "derivatives of racemates as resolving agents" could also be applied to the opposite direction of chiral discrimination.

6.1
Resolution of Dicarboxylic Acid Monoesters in the Presence of Another Similar Racemic or Achiral Monoester

The effect of mixing structurally similar racemates (the reverse Dutch resolution) on the efficiency of their resolutions was tested using different mixtures of **86** and **92**. The experiments were carried out in ethyl acetate with a half equivalent of (*S*)-**18** (calculated to the whole amount of the mixture of the racemates). The data are collected in Table 11. Interestingly enough, (*S*)-**86**.(*S*)-**18** and (*S*)-**92**.(*S*)-**18** salts crystallized in 93% and 98% enantiomeric excess, respectively, when neat **86** or **92** were reacted with (*S*)-**18** (Table 11). However, no crystallization occurred when the composition of the racemic mixture was in the range **86**/**92** = 0.9/0.1–0.6/0.4 (therefore these experiments are missing from Table 11). When the relative molar amount of compound **92** achieved 50% in the racemic mixture, a diastereoisomeric salt crystallized containing the two acids in 20% and 86% ee. A further 20% increase in the amount of **92** in the racemic mixture dramatically improved the ee value of **86** to 80% in the precipitated diastereoisomeric salt.

The results were slightly different when the structurally similar but achiral monoesters **93–95** were mixed with the corresponding racemates in different ratios. The enantiomeric excess had a maximum value at **87**/**93** = 0.5/0.5 molar ratio but gradually decreased when compound **86** was mixed with its achiral relative (**94**). Attempts to achieve resolution of **88** failed when the racemate alone or any mixtures of **88** and **95** were reacted with the resolving agent (*S*)-**18**. Surprisingly enough, this tendency stopped at a composition of **88**/**95** = 0.1/0.9. Starting from such a mixture, the heterochiral diastereosiomeric salt ((*R*)-**88**.(*S*)-**18**) crystallized in 21% enantiomeric excess (Table 11).

The above examples of reverse Dutch resolution demonstrated that in these type of mixtures the components can work as strong inhibitors of the resolution. However, in special cases, achiral additives (having similar structure to the racemate) may help to accomplish partial resolution or to

Table 11 Resolution of monoesters **86–88** with (S)-**18** in the presence of another similar racemic (**92**) or achiral monoester (**93–95**)

Racemate A	Racemate or achiral monoester B	A/B	ee of A/B isolated from the crystalline salt (%)	S-factor
86	92	1/0	93/0	0.400
86	92	0.5/0.5	20/86	0.06/0.26 [a]
86	92	0.3/0.7	80/80	0.28/0.28 [a]
86	92	0/1	0/98	0.310
86	94	1/0	76	0.412
86	94	0.6/0.4	64	0.378
86	94	0.3/0.7	10	0.029
86	94	0.1/0.9	7	0.032
87	93	1/0	85	0.587
87	93	0.5/0.5	94	0.431
87	93	0.1/0.9	79	0.368
88	95	1/0	– [b]	– [b]
88	95	0.1/0.9	– 21 [c]	– 0.045

[a] Approximate efficiencies of the separations of the components **A** and **B**
[b] No crystallization occured
[c] An excess of the opposite enantiomer (related to the previous experiments) was found in the precipitated heterochiral salt

improve the optical purity of the crystallizing diastereoisomeric salt. Unfortunately, these effects cannot be predicted on the basis of our present knowledge.

6.2
Resolution of Structurally Similar Dicarboxylic Acid Monoesters with Mixtures of Optically Active and Achiral Amines

In order to test the effect of partial substitution of (S)-**18** with a structurally similar but achiral benzylamine (**96**), series of resolutions were accomplished with racemic **86** and **87**. These model compounds represent typical examples of a highly selective and of a partial resolution process. All experiments were carried out with a half equivalent of the mixture of amines ((S)-**18** + **96**, mixed in different ratios). The results are collected in Table 12. In the case of compound **86** pure (S)-**18** formed crystalline diastereomeric salt with (S)-**86** (98% ee) and the enantiomer excess of the salt changed slightly when the relative amount of benzylamine (**96**) was increased to 60% within the amine mixture. Further addition of the achiral compound instead of the resolving agent dramatically decreased the optical purity of the crystalline salt. In addition, the S-factor decreased gradually.

Table 12 Resolution of compounds **86** and **87** with (S)-**18** in the presence of benzylamine (**96**)

Racemate	Resolving agent A	Achiral amine B	A/B	ee of A in the crystalline salt (%)	S-factor
86	18	96	1/0	98	0.834
86	18	96	0.9/0.1	78	0.570
86	18	96	0.7/0.3	98	0.521
86	18	96	0.4/0.6	85	0.286
86	18	96	0.3/0.7	58	0.066
87	18	96	1/0	14	0.135
87	18	96	0.8/0.2	20	0.137
87	18	96	0.5/0.5	4	0.030
87	18	96	0.4/0.6	−10[a]	−0.050
87	18	96	0.2/0.8	0[b]	0[b]

[a] An excess of the opposite enantiomer (related to the previous experiments) was found in the precipitated heterochiral salt
[b] The diastereoisomeric salt contained the racemic mixture of the acid

Efficiency of the really poor partial resolution of compound **87** could be improved by using an 80/20 mixture of (S)-**18** and benzylamine (**96**) instead of neat resolving agent. The enantiomeric purity of the diastereoisomeric salt fell to zero when the resolving agent mixture contained more than 20% of achiral base. Surprisingly, the heterochiral diastereoisomeric salt was found in the crystalline phase in slight excess when a 40/60 mixture of (S)-**18** and benzylamine (**96**) was added to the racemic monoester **87**. This finding is in accordance with the above-mentioned observation on the partial resolution of **88** via crystallization of the heterochiral diastereoisomeric salt ((R)-**88**.(S)-**18**) in the presence of a large excess of **96** (Table 12). Furthermore, this side-reagent dependent change of chirality of the crystalline diastereoisomeric salt is in accordance with the findings of Sakai and coworkers [34]. In their recent article they reported that the configuration of α-amino-ε-caprolactam (**97**) in the diastereoisomeric salt with N-tosyl-(S)-phenylalanine (**98**) could be changed by the use of appropriate solvents (Scheme 12). In solvents having

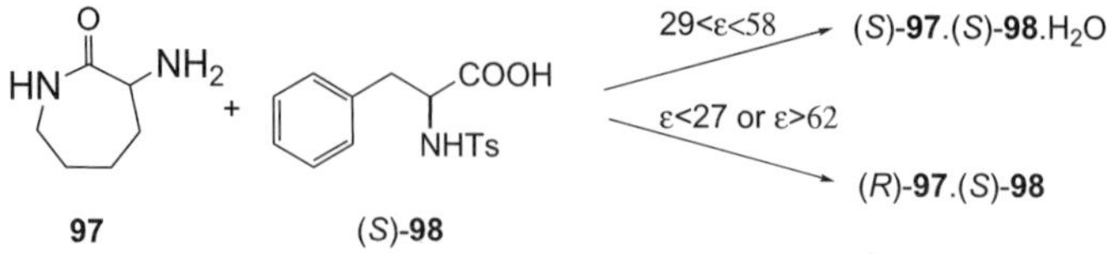

Scheme 12 Solvent-dependent crystallization of the hetero- or homochiral diastereoisomeric salt during resolution of **97**

medium dielectric constant ($29 < \varepsilon < 58$) the diastereoisomeric salt of (S)-**97** precipitated. The other isomer ((R)-**97**) became the major component in the crystalline salt when the solvent was less or more polar (i.e., $\varepsilon < 27$ and $\varepsilon > 62$, respectively) than in the previous case.

7
Parallel Kinetic Resolution with a Quasi-racemate-Type Resolving Agent Mixture

Resolution via diastereoisomeric salt formation is usually treated as a thermodynamically controlled process. In some cases, however, kinetics of crystallization may determine the enantiomeric excess [35] or even the configuration [28] of the enantiomer in the precipitated salt, similarly to the resolution of monoester **94** with its relative (S)-**18** under kinetic control (Table 12). According to this fact, the positive effects of using structurally similar compounds in chiral recognition processes cannot be limited to the resolutions via diastereoisomeric salt formation. Recently, Davies and coworkers [36] reported the parallel kinetic resolution of methyl 5-alkylcyclopentene-1-carboxylates (**99**) with a quasi-racemic mixture of (S)-1-phenylethylamine ((S)-**100**) and (R)-1-(3,4-dimethoxyphenyl)ethylamine ((R)-**101**). The combination of kinetic resolution and enantioselective addition provided the two products **102** and **103** in 97% ee and in 93% de, respectively (Scheme 13).

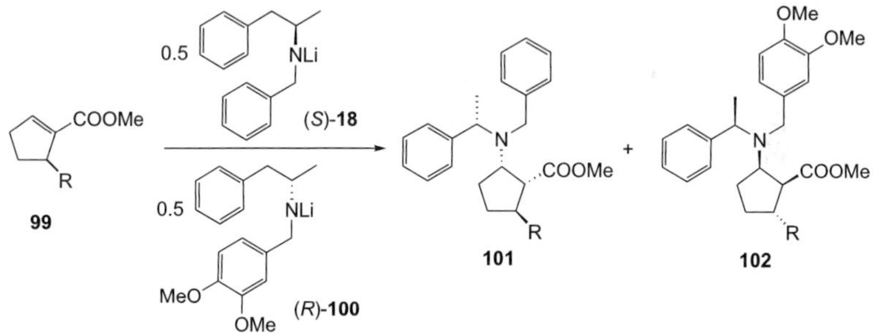

Scheme 13 Parallel kinetic resolution of **99** with the quasi-racemic mixture of (S)-**18** and (R)-**100**

8
Conclusions and Outlook

The reviewed examples demonstrated the scope and limitations of the application of structurally similar compounds in resolution processes. Limitations of this approach come from the fact that we could not find strict rules for definition of sufficient structural similarity. Numerous nonlinear effects were observed during resolutions via diastereoisomeric salt formation. One part of the phenomena could be explained by the catalytic effect of small amounts of additives as nucleation inhibitors. On the other hand, supramolecular assemblies may be formed in the supersaturated solutions before crystallization, and these supramolecules of the resolving agents and the mirror image isomers can be treated as models of receptor–drug complexes in biological systems. It is well known, for example, that the high affinity of morphine and derivatives to the opioid receptors is due to the structural similarity of the morphine skeleton to the tyrosine part of the enkephalines. Numerous modifications could be accomplished without diminishing the analgetic effect of the compound. At the same time, change of the *N*-methyl group in morphine to an allyl group resulted in an antagonist. We suppose that the derivative resolving agents and structurally similar additives work in a similar way. When the deviation from the perfectly complementary structure of the reaction partners (enantiomer and resolving agent) is caused by a group situated far from the stability-determining second order bond-forming centers, the compound is sufficiently similar to its pair. In any other case, the derivative is only partially similar to the target molecule in the diastereoisomeric salt-forming reaction. The positive effect of urea on the resolution of **18** (Table 7) demonstrated that even small molecules are able to efficiently replace the important amide group of the resolving agent in the hydrogen bonding system.

The "derivative resolving agent" approach can shorten the time of selection of good resolving agents and may help to avoid application of expensive and/or toxic reagents. The solvent-free version also offers an environmentally benign solution to enantiomer separation.

Numerous diastereoisomeric salt formation resolutions using an optically active derivative of the racemate have already been tested on an industrial scale [1]. These methods, in combination with second order asymmetric transformation or dynamic kinetic resolution, offer efficient and economic alternatives to asymmetric synthesis for producing optically pure enantiomers.

Acknowledgements Financial support from the Hungarian Scientific Research Found is gratefully acknowledged (OTKA T 048362).

References

1. Kozma D (ed) (2001) Optical resolutions via diastereoisomeric salt formation. CRC, Boca Raton, FL
2. Fogassy E, Nógrádi M, Pálovics E, Schindler J (2005) Synthesis 10:1555
3. Lopata A, Faigl F, Fogassy E, Darvas F (1984) J Chem Res (S) 14:322–324

4. Vries TR, Wynberg H, van Echten E, Koek J, ten Hoeve W, Kellog RM, Broxterman QB, Minnaard A, Kaptein B, van der Sluis S, Hulshof LA, Koostira J (1998) Angew Chem Int Ed 37:2349
5. Jacques J, Collet A, Wilen SH (1981) Enantiomers, racemates and resolutions. Wiley, New York
6. Fogassy E, Kozma D (1995) Tetrahedron Lett 36:5069
7. Kozma D, Madarász Z, Ács M, Fogassy E (1995) Chirality 7:381
8. Kozma D, Fogassy E (1997) Enantiomer 2:51
9. Kozma D, Simon H, Kassai Cs, Madarász Z, Fogassy E (2001) Chirality 13:29
10. Arsenijevic VA (1957) CR Acad Sci 245:317
11. Neumann P (1983) Optical resolution procedures for chemical compounds, vol 1–3. Optical resolution information center, New York
12. Fogassy E, Lopata A, Faigl F, Darvas F, Ács M, Töke L (1980) Tetrahedron Lett 21:647
13. Pallavicini M, Valoti E, Villa L, Piccolo O (1996) Tetrahedron: Asym 7:1117
14. Pallavicini M, Valoti E, Villa L, Piccolo O (1997) Tetrahedron: Asym 8:1069
15. Gizur T, Péteri I, Harsányi K, Fogassy E (1996) Tetrahedron: Asym 7:1589
16. Dai X, Wong A, Virgil SC (1998) J Org Chem 63:2597
17. Takaya H, Mashima K, Koyano K, Yagi M, Kumobayashi H, Taketomi T, Akutagawa S, Noyori R (1986) J Org Chem 51:629
18. Hamada Y, Matsuura F, Oku M, Hatano K, Shioiri T (1997) Tetrahedron Lett 38:8961
19. Matteoli U, Beghetto V, Schiavon C, Scrivanti A, Menchi G (1997) Tetrahedron: Asym 8:1403
20. Benincori T, Gladiali S, Rizzo S, Sannicolo F (2001) J Org Chem 66:5940
21. Dupart de Paule S, Jeulin S, Ratovelomanana-Vidal V, Genet JP, Champion N, Dellis P (2003) Tetrahedron Lett 44:823
22. Dupart de Paule S, Jeulin S, Ratovelomanana-Vidal V, Genet JP, Champion N, Dellis P (2003) Eur J Org Chem 10:1931
23. Faigl F, Kozma D (2004) In: Toda F (ed) Enantiomer separation, fundamentals and practical methods. Kluwer, Dordrecht, p 73
24. Loiodice F, Longo A, Bianco P, Tortorella V (1995) Tetrahedron: Asym 6:1001
25. Dalmolen J, Tiemersma-Wegam TD, Niuwenhuijzen JW, van der Sluis M, van Echten E, Vries TR, Kaptein B, Broxterman QB, Kellog R (2005) Chem Eur J 11:5619
26. Markovits I, Egri G, Fogassy E (2002) Chirality 14:674
27. Bálint J, Egri G, Vass G, Schindler J, Gajáry A, Friesz A, Fogassy E (2000) Tetrahedron: Asym 11:809
28. Fogassy E, Faigl F, Ács M, Grofcsik A (1981) J Chem Res (S) 11:346
29. Bálint J, Egri G, Kiss V, Gajáry A, Juvancz Z, Fogassy E (2001) Tetrahedron: Asym 12:3435
30. Drauz K, Jahn W, Schwarm M (1995) Chem Eur J 1:538
31. Bálint J, Egri G, Czugler M, Schindler J, Kiss V, Juvancz Z, Fogassy E (2001) Tetrahedron: Asym 12:1511
32. Bálint J, Schindler J, Egri G, Hanusz M, Marthi K, Juvancz Z, Fogassy E (2004) Tetrahedron: Asym 15:3401
33. Schindler J, Egressy M, Bálint J, Hell Z, Fogassy E (2005) Chirality 17:565
34. Sakai K, Sakurai R, Nohira H, Tanaka R, Hirayama N (2004) Tetrahedron: Asym 15:3495
35. Faigl F, Turner A, Farkas F, Proszenyák Á, Valacchi M, Mordini A (2004) Arkivoc 7:53
36. Davies SG, Garner AC, Long MJC, Morrison RM, Roberts PM, Savory ED, Smith AD, Sweet MJ, Withey JM (2005) Org Biomol Chem 3:2762

Dutch Resolution of Racemates and the Roles of Solid Solution Formation and Nucleation Inhibition

Richard M. Kellogg[1] (✉) · Bernard Kaptein[2] · Ton R. Vries[1]

[1] Syncom BV, Kadijk 3, 9747 Groningen AT, The Netherlands
r.m.kellogg@syncom.nl

[2] DSM Research, Life Sciences—Advanced Synthesis & Catalysis, P.O. Box 18, 6160 Geleen MD, The Netherlands

1	Introduction	160
1.1	Basic Aspects of Resolutions	162
2	Aggregation in Supersaturated Solutions	163
2.1	Nucleation, Crystal Growth and Tailor-Made Additives	168
2.2	Theory of Nucleation	173
3	Dutch Resolution	175
3.1	Basic Principles	175
3.2	Reverse (Reciprocal) Dutch Resolution	178
3.3	Solid Solution Behavior in Dutch Resolution	182
3.4	Nucleation Inhibition in Dutch Resolution	184
4	Conclusions	192
	References	194

Abstract An overview is given of the principles of Pasteur resolutions via separation of diastereomeric salts. Thereafter, primary nucleation processes of (chiral) organic compounds in supersaturated solution are considered followed by crystal growth in the presence of tailor-made additives. A representative example of a Dutch Resolution is presented, the concept of families of resolving agents is defined and examples are given. The phenomenon of reversed Dutch Resolution, resolution of mixtures of families of racemates, is illustrated. The roles of both solid solution formation and nucleation inhibition in Dutch Resolution are discussed. The work is concluded with the results of a broadly based search for nucleation inhibitors for phenylethyl amine as resolving agent. This search can serve as a model for the discovery of nucleation inhibitors for other resolving agents. The specific role of bifunctional family members of resolving agents as possible nucleation inhibitors is also discussed.

Keywords Nucleation inhibition · Resolving agents · Supersaturation · Families · Aggregation · Chirality · Enantiomeric enrichment

Abbreviations
Abu   2-aminobutyric acid
Asn   asparagine

Asp	aspartic acid
β	degree of supersaturation $C_{initial}/C_{equilibrium}$
Cys	cysteine
de	diastereomeric excess
DSC	differential scanning calorimetry
ee	enantiomeric excess
glc	gas liquid chromatography
Glu	glutamic acid
Ile	isoleucine
His	histidine
Hpg	4-hydroxyphenylglycine
hplc	high performance liquid chromatography
Leu	leucine
Lys	lysine
NMR	nuclear magnetic resonance
Nva	norvaline
Orn	ornithine
Phe	phenylalanine
Phg	phenylglycine
SEM	scanning electron microscopy
STM	scanning tunneling microscopy
Ser	serine
TEM	tunneling electron microscopy
Thr	threonine
Trp	tryptophane
Tyr	tyrosine
Val	valine
XRPD	X-ray powder diffractogram
S factor	Fogassi value: 2 · yield · de in diastereomeric salt resolution

1
Introduction

Dutch Resolution is a modification of the Pasteur technique to resolve racemates into their enantiomers by formation of diastereomeric salts, which can be separated by crystallization [1–10]. Selective formation of the desired enantiomer by enantioselective synthesis, stoichiometric or catalytic, is the complementary/competitive technology that bypasses racemate separation [11]. We note also the recent interest in the special but relatively rare occurrence of spontaneous asymmetric synthesis [12–14].

The classical approach devised by Pasteur for the separation of enantiomers in a racemate was published in 1854 and has not changed through the years [1–3]. Pasteur's original resolution of tartaric acid 1 with the aid of a stoichiometric quantity of quinotoxin, a degradation product of quinine, is shown in Scheme 1.

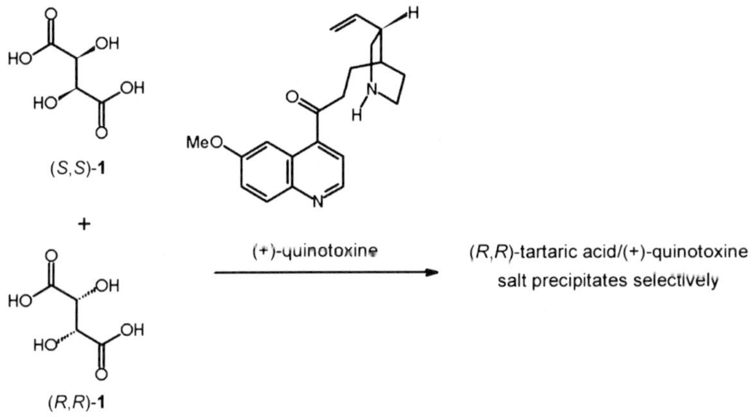

Dutch Resolution of Racemates

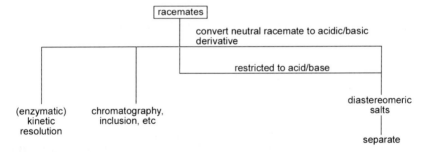

Scheme 1 Original Pasteur experiment for the resolution of racemic tartaric acid

The concept, interaction of an enantiomerically pure base (acid) with a racemic acid (base) by means of proton transfer to form two diastereomeric salts separable by solubility differences, is deceptively simple.

Many organic compounds are, of course, neither especially acidic nor basic and do not lend themselves to this methodology. A rough division on the basis of properties is given in Scheme 2.

Scheme 2 Classification of various methods to separate racemates

To circumvent this complication it may be possible in a multistep synthesis to resolve an acidic or basic intermediate. Neutral compounds may in some cases be converted reversibly to acidic or basic derivatives. Neutral compounds as well as acids and bases may also sometimes be resolved by covalent bonding with an enantiomerically pure material followed by separation of the diastereomers by, for example, crystallization or chromatography followed by release of the chiral auxiliary. Resolution of neutral compounds via inclusion compounds is on occasion possible and practical [15]. On an analytical scale in the majority of cases separation of racemates is possible with chiral hplc and/or glc. Racemates can also often be separated on preparative scale by chiral chromatography and simulated moving bed chromatography can be applied on bulk scale [16, 17]. Each of these techniques is a separate technology, none of which will be discussed here. The applicability of each of these methods as compared to (catalytic) asymmetric synthesis is determined by the nature of the compounds, scale and costs.

Modern drug development is dependent on a rapid and dependable supply of pure enantiomers the structural diversity of which can be enormous. Resolution of racemates obtained as intermediates is a much-used technology to accomplish this. In the early stages of drug development this is often carried out at the gram scale and time is of the essence. A resolution procedure for commercial application may be far different from that used in the laboratory for rapid resolution of material that has never been resolved before. Dutch Resolution will, as is discussed in this overview, turn out to be more suited for rapid, small-scale resolutions.

1.1
Basic Aspects of Resolutions

For early phase pharma applications both enantiomers of the racemate are often required. In a resolution via diastereomeric salts an obvious rapid way to obtain both enantiomers of the racemate is to carry out separate resolutions with each enantiomer of the resolving agent. For resolving agents like mandelic acid and tartaric acid this is feasible because both enantiomers are available in quantity and not excessively expensive. However, for resolving agents like brucine, (+)-camphor sulfonic acid, quinine, quinidine, etc only a single enantiomer is readily available. Sometimes a structurally different resolving agent will lead to salts the least soluble of which will contain the desired other enantiomer of the racemate. Development of new resolving agents (see further) has done much to widen the scope of diastereomeric resolutions. Recently, it has been shown that in some cases by variation of the dielectric constant of the solvent a single resolving agent can be used to derive either enantiomer of the racemate [18, 19]. It is not clear, however, how general this methodology is.

As noted by Jacques, Collet and Wilen [4] "Indeed, what is wrong with the traditional way of carrying out resolutions is a lack of real understanding ... Such understanding is equally necessary ... for the study of systems enriched in one enantiomer produced by other methods" [21]. This lack of understanding applies particularly dramatically to the nucleation processes in solution that lead to subsequent crystallization. Common practical problems are that the resolving agent used must readily form crystalline salts with the racemates, that the solubilities of these salts differ appreciably, i.e. that the eutectic composition should be well removed from 50 : 50, that the chiral centers in the racemate not be too far removed from the acidic or basic centers and that crystallization of the less soluble diastereomer in reasonable diastereomeric excess occurs within an experimentally acceptable time. It has been estimated that around 75% of attempted classical diastereomeric resolutions fail [4]!

These are intrinsic points of the resolution itself. In addition one has to be able to determine fairly quickly how successful a resolution has been. In general this entails access to chiral hplc or glc facilities. Successful performance of resolutions on a regular basis requires experience in the choice of resolving agents and conditions (solvents and temperature, for example), excellent experimental skill in the form of people with experience and suitable analytical capacities readily at hand. Considerable effort is necessary to set up working conditions where all these criteria are met simultaneously.

There are more matters of concern. The salts often do not precipitate in an enantiomerically pure form but rather are only enriched and have to be purified further usually by additional crystallization. Precipitation of a diastereomeric salt can only start once supersaturation has been achieved, that is to say that the concentration considerably exceeds the solubility. The concentrations and solvent or solvent combination can have a profound effect as noted by Sakai [18]. The reader is again referred to Jacques, Collet and Wilen [4] for a detailed and quantitative analysis of many of these factors as well as modified methods of diastereomeric resolutions such as the use of sub-stoichiometric quantities of resolving agents or the Pope and Peachy approach in which part of the resolving agent is replaced by an achiral (mineral) acid or base.

2
Aggregation in Supersaturated Solutions

Crystallization starts with nucleation. This process must be discussed briefly as a preliminary to Dutch Resolution. Nucleation occurs in the area of supersaturation. Many factors affect nucleation: the degree of supersaturation, the rate of cooling and possible presence of "seeds" just to mention a few possibilities. As nuclei transform into crystals a complicated dynamic situation arises in which crystal growth competes with nucleation. The kinetics and mechanisms of primary nucleation probably often differ from those of secondary nucleation that occurs as nuclei split off from growing crystals [22, 23].

Resolutions take place from "impure" solutions in which two diastereomeric salts are present. Rationalization and control of the process require understanding of what happens at the molecular scale in both phases. The mechanisms of crystal growth are reasonably well understood, but the nature of the supramolecular structures in supersaturated solution that lead to nucleation and then to crystallization are usually unknown. Models for prediction of the effects of stereochemistry on these structures are essentially unavailable. Aggregation in solution is obviously highly dependent on the solution itself: for example, aggregation principles in aqueous solution are fundamentally different from those in nonaqueous solvents. Nucleation and crystallization are not the same processes.

An incomplete list of possibilities, in addition to or prior to desired crystallization, of the many things that can happen in the area of supersaturation includes phase separation, liquid crystal formation, micelle formation, vesicle formation, gel formation and chemical reaction of the components. Supersaturation (metastable) zones are the temperature zones between the dissolution temperature and the (lower) temperature at which precipitation (the experimentally observable consequence of nucleation) takes place. The width of such zones depends, among other things, on the concentration and rate of cooling.

The classical approach to explanation of nucleation is to consider that molecules on the surface of a small nucleus will have a positive free energy relative to those in the interior, which will be stabilized through aggregation and which will have a negative free energy relative to those in solution. For a spherical nucleus the surface effect is dependent on the square of the radius of the growing nucleus whereas the bulk effect is dependent on the third power of the radius. In classical theory based on formation of spherical nuclei the critical nucleation radius r_c is that at which overall stabilization is realized. Somewhere on the path from critical nucleation radius to crystal growth an ordered packing will begin after which growth will proceed on crystal surfaces (Fig. 1) [24]. Basic mathematical aspects of the theory of nucleation will be discussed in Sect. 2.2.

Unfortunately, the assumption of formation of ideal spherical nuclei is probably in most cases not applicable and can be badly misleading especially when the individual molecules are chiral. Free energy minimization during nucleation can be (and is) achieved via other shapes. Weissbuch, Lahav and Leiserowitz [25] cogently argue on the basis of various experimental observations that there is much evidence that nuclei are not spherical and that the shape and internal structure of the critical nucleus should be treated as a variable in theoretical approaches. Chirality, particularly in the reso-

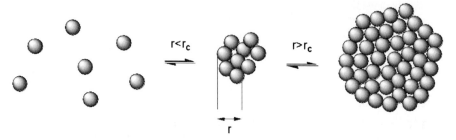

Fig. 1 Simplified model of aggregation in supersaturated solution

lution of diastereomeric salts, will also be expressed in some fashion during nucleation. Molecular chirality should be recognized at the early stage of supramolecular aggregates that lead to nucleation and not only at the stage of crystallization.

There are no spectroscopic methods to observe directly in solution the structure and composition of nanometer-sized aggregates/nuclei. However, clues (in addition to those given in [25]) as to how aggregation can work are available via other sources. We know much about the ordered arrangements of liquid crystals [26]. The aggregation involved in micelle formation has been studied in great depth [27]. Vesicle formation has been the subject of extensive study [28]. These noncrystalline structures are illustrative of what can happen in solution. The techniques of tunneling electron microscopy (TEM) and scanning electron microscopy (SEM) provide surprisingly detailed information at the level of individual molecules for cases where the aggregates can be trapped and immobilized on a surface.

Consider the aggregation behavior of some organic salts, which might be reasonable models for the salts formed in diastereomeric salt resolutions. Ionic interactions are in general much stronger than noncovalent interactions that bind neutral compounds together. An instructive example is provided by salt 2 as shown in Fig. 2 [29, 30].

This pure enantiomer of tartaric acid provided with a gemini ammonium cation does not crystallize but rather gelates solvent $CHCl_3$ quite effectively. Neither the racemate nor the meso compound acts as a gelator nor do they crystallize. Examination of dried and frozen pure enantiomer by transmission electron microscopy (TEM) revealed long, twisted ribbon fibers, left handed for (R,R) and right handed for (S,S) tartrates. On use of a gemini ammonium salt with a C16 chain it was observed that an enantiomerically enriched composition did not undergo phase separation (e.g. a preliminary

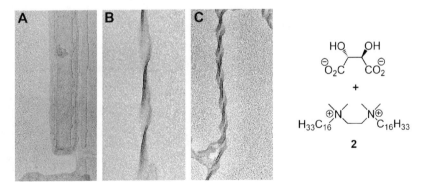

Fig. 2 Cyro-TEM images of the twisted ribbons formed at **A** 0% (racemate); **B** 50% and **C** 100% enantiomeric excess of tartrate salt 1. Permission to use this figure has been kindly given by Nature Publishing Group

to conglomerate formation) but rather formed helical ribbons with a longer pitch. The lower the enantiomeric purity the greater the pitch indicating that there is a continuous composition change of the enantiomers (resembles solid solution behavior in crystalline compounds).

The corresponding D-glucarate and L-malate salts do not form ribbons but rather amphiphilic bilayers. If the aliphatic chains of the ammonium head groups are shorter micelle formation takes place.

It is reasonable to suggest that the crystallization of the tartrates has been derailed by gel formation. The detailed information at the molecular level provided by TEM is indicative of the potential formation of a racemic compound on crystallization rather than a conglomerate. In other words molecular chirality is being expressed at the level of formation of chiral supramolecular aggregates.

The aggregation behavior in aqueous solution exhibited by a chiral nonracemic glycerol derived phosphate 4 is strikingly similar (Fig. 3) [31–33].

The TEM photographs of frozen, immobilized material show for structurally related 3 (Scheme 3) only plate-like aggregates wherein the molecular chirality is not clearly expressed whereas for 4 a long, right-hand coiled helix is formed. These helixes subsequently assemble into superhelixes in which molecular chirality is again clearly expressed at the supramolecular level.

Compound 5 is liquid crystalline and uncharged [34]. With the aid of scanning tunneling microscopy (STM) imaging on a conducting graphite surface the immobilized individual molecules of 5 can be recognized (Fig. 4).

The (S) enantiomer in the unit cell stacks tilted to the right whereas the (R) enantiomer stacks with a tilt to the left. The racemate shows areas with

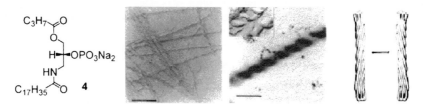

Fig. 3 Illustration of the formation of right-handed helical fibers upon self-assembly of phospholipid 4. Permission to use this figure has been kindly given by the American Chemical Society

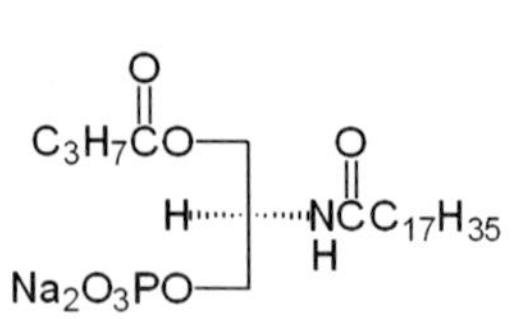

Scheme 3 Phospholipid 2 forms only plate-like aggregates rather than helices

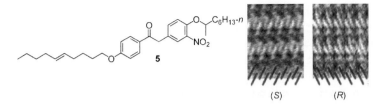

Fig. 4 Enantiomorphous STM images obtained from (S) and (R) enantiomorphs of 5 in which two dimensional chiral monolayer systems can be seen. Permission to use this figure has been kindly given by Wiley

a right-hand tilt and other areas with a left-hand tilt. Clearly racemic 5 exhibits conglomerate behavior and has undergone spontaneous symmetry breaking on deposition. Molecular chirality is expressed immediately in the supramolecular structures observed. Why 5 exhibits conglomerate behavior is unknown.

Regardless of the great contrasts in structures of the molecules as well as conditions (two rather than three-dimensional pictures are obtained) it is a common phenomenon that aggregation observed by STM, TEM and related techniques occurs not by the formation of spherical nuclei but rather by association into rope-like structures, which continue to grow in size. These aggregates might be *rough* models for the structures of the nuclei that lead to crystallization. Helix formation is a common manner to express molecular chirality at the supramolecular level [35–37].

Low molecular weight (< 500) organic gelators have been studied extensively and their behavior well illustrates some of the above points. Such molecules have a pronounced tendency to gel organic solvents rather than to crystallize. Low molecular weight hydrogelators are also common. The aggregation processes are known to proceed along the lines given in Fig. 5. Long fibers form and these, if entangled, lead to gels rather than crystals [38–40].

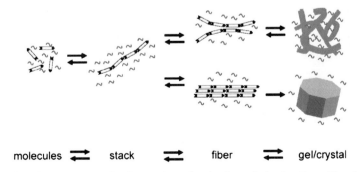

Fig. 5 Model of aggregation of anisotropic molecules into chains leading either finally to crystals or gels

The coils formed lead to gelation of organic solvents. The arrangement is not a thermodynamic minimum; in due course (weeks in general) crystallization can set in.

2.1
Nucleation, Crystal Growth and Tailor-Made Additives

For a successful resolution nucleation must lead to crystallization. Classical resolutions can be very time consuming owing to difficulties in obtaining a crystalline diastereomeric salt. Often this difficulty is explained by the low (50% diastereomeric) purity of the least soluble salt in solution. The interplay of crystallization, together with primary and secondary (occurs as crystallization takes place) nucleation, is profoundly complex [22]. Nucleation and crystal growth inhibition by the more soluble diastereomer could play an important role.

"Family" behavior in the inhibition of crystal growth in the special case of conglomerates was recognized early by Lahav and colleagues, who refer to the effects of "tailor-made" additives [25]. This work will be surveyed briefly as a preliminary to further discussion of Dutch Resolution.

Not much is known about the inhibition of crystal growth of diastereomeric salts. To our knowledge in the literature only one example of addition of a tailor-made additive in a classical resolution has been described [41–43]. Sakai et al. investigated the effect of the various "dimeric" stereoisomers **8a–c** of the racemate to be resolved as habit modifiers in the resolution of (±)-1-phenylethylamine **6a** with (R)-(-)-mandelic acid **7** as illustrated in Scheme 4.

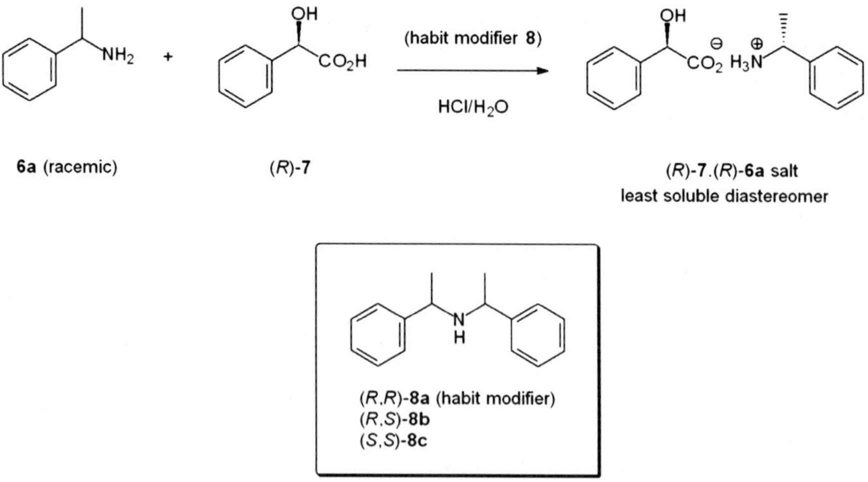

Scheme 4 The effect of habit modifiers in the resolution of phenethylamine **6a** by mandelic acid **7**

It was shown that the (R,R)-dimer **8a** at concentrations as low as 0.007 mol % had a significant impact on the morphology, and changed the crystal shape of the least soluble (R)-**6a**·(R)-**7** diastereomeric salt from needle-like to hexagonal plates and thereby improved the isolation of this salt on centrifugation. Meso **8b** was reported to be a much poorer habit modifier, and addition of (S,S)-**8c** had no effect. The efficient habit modification caused by (R,R)-**8a** is explained by a strong double chiral recognition at the surface of the {011} plane of the growing (R,R)-salt, thus preventing the incorporation of (R)-**6a** ammonium ions.

On the other hand, enantioselective nucleation/growth inhibition has frequently been described for the direct resolution of racemates by preferential crystallization. A prerequisite for the direct resolution of racemates is conglomerate behavior, i.e. the spontaneous crystallization of the enantiomers as mirror image crystals (enantiomorphs). This requirement resembles eutectic behavior as a precondition for an efficient resolution via diastereomeric salt formation. Principles derived from enantioselective nucleation and crystal growth inhibition of conglomerates can, with care, be extrapolated to resolution via diastereomeric salt formation. A compilation of organic compounds known to show conglomerate behavior is given in the standard book of Jacques, Collet and Wilen [4]. It is estimated that approximately 5–10% of all racemates fulfil this requirement.

The role of "tailor-made" additives that act as enantioselective nucleation/growth inhibitors in the direct resolution of conglomerates has been extensively studied especially for amino acids, which exhibit the required conglomerate behavior. In 1982 Lahav et al. described the inhibitor-induced resolution of threonine (Thr), glutamic acid (Glu·HCl), asparagine (Asn·H$_2$O), p-hydroxyphenylglycine (Hpg·pTsOH) and histidine (as the metastable conglomerate His·HCl·H$_2$O) amongst other conglomerates. Direct resolution could be achieved by addition of small amounts of a different but structurally related enantiomerically pure amino acid [44]. For example, the crystallization of (RS)-Thr in the presence of 1–2 mol % of the (S)-enantiomer of Asn, Glu, Asp, His, Lys, Phe or (R)-Cys resulted in the preferential crystallization of the (R)-enantiomer of threonine in 40–95% enantiomeric excess. An overview of the results for the direct resolution of other amino acids and derivatives by tailor-made additives is given in Table 1.

On the basis of these and other examples Lahav defined the "rule of reversal": stereoselective adsorption of an enantiopure "tailor-made" additive S′ at the surface of the growing crystals of the enantiomer with the same absolute configuration (S) results in a drastic decrease in their rate of growth and thus allows preferential (faster) crystallization of the opposite (R)-enantiomer as illustrated in Scheme 5.

Preferential adsorption of the additive (0.5–3 mol %) on the crystal of the analogous absolute configuration is sufficient to inhibit the further growth of this enantiomorph. In the adsorption mechanism the additive preferably

Table 1 Resolution of amino acid conglomerates using tailor-made additives

Conglomerate	Chiral additive	Crystalline enantiomer	Refs.
Glu	(S)-Asp, (S)-Leu	(R)-Glu	[45]
Asp	(S)-Glu	(R)-Asp	[45]
Glu	(S)-Glu(γ-Me)	(R)-Glu	[46]
Cu(Asp)$_2$	(S)-Glu, (S)-Ala, (S)-Ile	(R)-Cu(Asp)$_2$	[47–49]
	(R)-Glu, (R)-Ala	(S)-Cu(Asp)$_2$	
Thr	(S)-Glu, (S)-Gln, (S)-Asn,	(R)-Thr	[44, 49, 51–53]
	(S)-Phe, (S)-His, (S)-Lys,		
	(S)-Asp, (R)-Cys, (S)-Pro,		
	(S)-Ala, (S)-PAL[a] (S)-PMAL[b]	(S)-Thr	
	(R)-Glu, (R)-Asn, (R)-PAL,		
	(R)-PMAL		
Asn·H$_2$O	(S)-Glu, (S)-Gln, (S)-Asp,	(R)-Asn	[44, 49, 51]
	(S)-Ser, (S)-Lys, (S)-Orn,		
	(S)-His, (S)-PAL, (S)-PMAL	(S)-Asn	
	(R)-Glu, (R)-Asp, (R)-PAL,		
	(R)-PMAL		
Glu·HCl	(S)-Lys, (S)-Orn, (S)-His,	(R)-Glu	[44, 51]
	(S)-Thr, (S)-Tyr, (S)-Leu,		
	(R)-Cys, (S)-PAL, (S)-PMAL	(S)-Glu	
	(R)-PAL, (R)-PMAL		
p-Hpg·pTSA	(S)-Phg, (S)-Tyr, (S)-Phg(p-Me),	(R)-p-Hpg	[44]
	(S)-Phe, (S)-DOPA,(S)-(αMe)DOPA		
His·HCl·H$_2$O	(S)-Trp, (S)-Phe, (S)-PAP[c]	(R)-His	[44, 51, 54]
	(R)-PAP	(S)-His	
Met·HCl·H$_2$O	(R)-PAL, (R)-PMAL	(S)-Met	[50]
Cys·HCl	(R)-PAL, (R)-PMAL	(R)-Cys	[51]
Ile·HCl	(S)-PAL, (S)-PMAL, (S)-Phe	(R)-Ile	[51, 55]
Val·HCl	(R)-PAL, (R)-PMAL	(S)-Val	[51, 55]
	(S)-Phe	(R)-Val	
Leu·HCl	(S)-PAL, (S)-PMAL, (S)-Phe	(R)-Leu	[51, 55]
	(R)-PAL, (R)-PMAL	(S)-Leu	
Ac-Nva·NH$_4$[d]	(S)-Ac-Ala·NH$_4$ salt	(S)-Ac-Nva·NH$_4$	[56]
Ac-Abu·NH$_4$[e]	(S)-Ac-Ala·NH$_4$ salt	(S)-Ac-Abu·NH$_4$	[56]

polymeric additives:
[a] (S)-PAL: poly-(N$^\varepsilon$-acryloyl-(S)-lysine)
[b] (S)-PMAL: poly-(N$^\varepsilon$-methacryloyl-(S)-lysine)
[c] (S)-PAP: poly-(p-(acrylamido)-(S)-phenylalanine)
[d] Nva = norvaline [CH$_3$CH$_2$CH$_2$CH(NH$_2$)CO$_2$H]
[e] Abu = 2-aminobutyric acid [CH$_3$CH$_2$CH(NH$_2$)CO$_2$H]

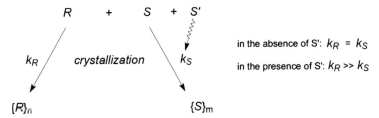

Scheme 5 The effect of tailor-made additives S' in the resolution of a conglomerate

binds to a crystal with the same absolute configuration, but only at those surfaces where the (structurally different) side chain of the adsorbate emerges from the crystal. The growing face of the crystal is thereby changed and growth is slowed [57, 58]. The adsorption to specific crystal faces may result in inhibition of the crystal growth but can also result in a change of the morphology of the crystals as is represented in Fig. 6.

In this manner the change of the crystal habit of one of the enantiomorphs of asparagine on addition of small amounts of (R)-cysteine could be used as the basis for separation by sieving the (R)- and (S)-crystals of asparagine based on their crystal size [59].

Polymers grafted with amino acids as tailor-made additives seem in general to be more efficient than monomeric additives. This is likely due to adsorption to multiple binding sites (polydentate) in addition to the shielding of the crystal surface by the polymer backbone [58]. Polymeric additives also have been successfully applied to prevent lamellar twinning in the cases of methionine, cysteine and valine [60].

In addition to crystal growth inhibition, the addition of polymeric tailor-made additives also can retard the dissolution of the enantiomorphous crystal, as was shown for the dissolution of (R)-methionine·HCl monohydrate

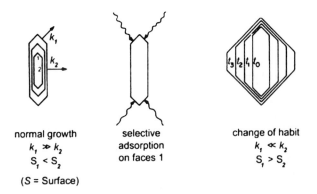

Fig. 6 Specific absorption on fast growing faces of a needle-like crystal and the corresponding morphology changes

crystals in the presence of the polymeric additive poly-(N^ε-methacryloyl-(R)-lysine) [50].

The use of polymeric tailor-made additives is not limited to the similarity principle of the conglomerate and the additive as described above for amino acids. For example, block polymers of polyethyleneglycol and polyethylenimine grafted with various chiral groups R* have been applied in the direct resolution of calcium tartrate and sodium ammonium tartrate (Pasteur's salt) as shown in Scheme 6 [61].

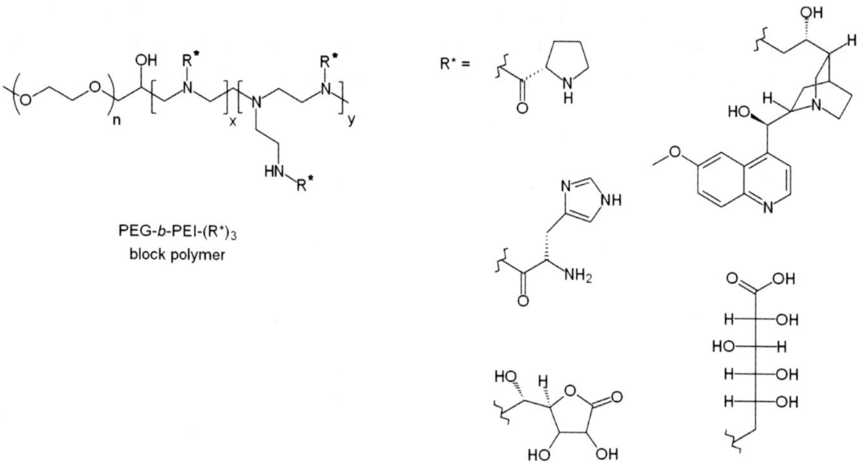

Scheme 6 Examples of polymeric tailor-made additives applied in the resolution of calcium tartrate and sodium ammonium tartrate

There is clearly a delicate balance between adsorption resulting in growth inhibition and incorporation of the inhibitor in the growing crystal, as Lahav et al. observed for the crystallization on Asn·H_2O with Asp as the additive [49]. In this case a true solid solution was formed with incorporation of 10–12% of Asp. Others have found similar co-crystallization results [62, 63].

In contrast to nucleation or crystal growth inhibition, Shiraiwa et al. have described the direct resolution of three different conglomerate forming N-acetyl amino acid ammonium salts on addition of (S)-N-acetyl-alanine ammonium salt as the result of an increased solubility of one of the enantiomers of the conglomerate. In this case large amounts of additives are used, up to 100 mol % based on the racemate [56]. Remarkably enough, this resulted in crystallization of the enantiomer of the conglomerate having the same absolute stereochemistry as the additive.

For a detailed and excellent discussion of (enantioselective) crystal nucleation the reader is referred to a review by Weissbuch, Lahav and Leiserowitz [25].

2.2 Theory of Nucleation

In order to describe the physical effects of enantioselective nucleation inhibition it is necessary to understand the basics of the nucleation theory. Primary nucleation can be considered as the continuous formation and disappearance of nanoscopically small molecular clusters—usually on the order of a hundred or so molecules—that can grow to the new crystalline phase once the critical nucleus size has been exceeded (Fig. 1).

In classical nucleation theory the overall free energy difference ΔG between a solid spherical nucleus and the solution is equal to the sum of surface excess free energy ΔG_S and the volume excess free energy ΔG_V (Eq. 1)

$$\Delta G = \Delta G_S + \Delta G_V = 4\pi r^2 \gamma + 4/3 \pi r^3 \Delta G_{Vm}, \tag{1}$$

where the nucleus radius is r, interfacial energy is γ and volume excess free energy is ΔG_{Vm} per molecule with molecular volume V_m (Fig. 7).

The maximum in ΔG ($\delta \Delta G / \delta r = 0$) corresponds to the critical nucleus size r_c, which is given by (Eq. 2).

$$r_c = -2\gamma / \Delta G_{Vm}. \tag{2}$$

For nonelectrolytes the Thomson–Gibbs expression describes the relationship between particle size and solubility (Eq. 3)

$$\ln S = 2\gamma V_m / k_B T r, \tag{3}$$

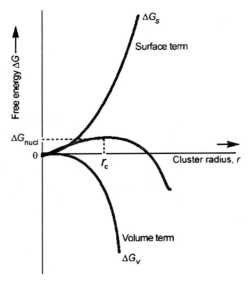

Fig. 7 Excess free energy of nucleation ΔG_{nucl} as a function of the cluster radius, r Eq. 1

where S is the supersaturation ratio, the ratio of the actual activity and the equilibrium activity of the molecules in solution (or for ionic compounds the ratio between the product of the actual activities of all the ions involved and the product of their equilibrium activities). The temperature is T, k_B is the Boltzmann constant and r the radius.

Combining Eqs. 2 and 3 with Eq. 1 for the critical nucleus size, the free energy excess ΔG_{nucl} for the critical nucleus size may be defined as (Eq. 4)

$$\Delta G_{nucl} = 16\pi\gamma^3 V_m^2 / 3(k_B T \ln S)^2 \,. \tag{4}$$

In a recent paper by Kashchiev and Van Rosmalen classical nucleation theory and the current ideas with regard to (one-component) nucleation in solution of nonionic as well as ionic compounds have been reviewed [64]. The role and effect of nucleation inhibitors on the rate of homogeneous as well as heterogeneous primary nucleation can be understood by modification of classical nucleation theory, as recently has been described by Kondipudi and Crook [65]. In classical nucleation theory the rate of (primary) nucleation J is expressed by the Arrhenius reaction velocity Eq. 5

$$J = J_0 \exp[-\Delta G_{nucl}/k_B T] \,, \tag{5}$$

and on substitution in Eq. 4 for ΔG_{nucl} there is obtained (Eq. 6)

$$J = J_0 \exp[-16\pi\gamma^3 V_m^2 / 3k_B^3 T^3 (\ln S)^2] \,. \tag{6}$$

Nucleation inhibitors can alter the supersaturation ratio S, the interfacial energy γ and the molecular volume V_m. This may be expressed mathematically (Eq. 7) by describing ΔG_{nucl} as a function of nucleation inhibitor mole fraction X_{inh} using a power series in X_{inh} of which at low inhibitor concentrations only the first term is taken into account.

$$\begin{aligned}\Delta G_{nucl} &= \Delta G^0_{nucl}(1 + A^1_{inh}X_{inh} + A^2_{inh}(X_{inh})^2 + \ldots) \\ &\approx \Delta G^0_{nucl}(1 + A_{inh}X_{inh}) \,.\end{aligned} \tag{7}$$

ΔG^0_{nucl} is the Gibbs energy in the absence of nucleation inhibitors and the efficacy of the nucleation inhibitor is expressed by the coefficient A_{inh}. Efficient nucleation inhibitors have high A_{inh} values and inhibit nucleation even at low concentrations. The effects of nucleation inhibitors on secondary nucleation and crystal growth may be described as surface phenomena.

Using Langmuir adsorption theory, Kubota and Mullin [67] described the rate of crystal growth G as a function of the adsorption factor θ_{inh}, the fraction of the occupied sites on the surface and effectiveness factor α_{inh} (Eq. 8)

$$G = G_0(1 - \alpha_{inh}\theta_{inh}) \,. \tag{8}$$

Assuming the Langmuir isotherm to apply in crystal growth, θ_{inh} can be expressed as Eq. 9

$$\theta_{inh} = K_{inh}X_{inh}/(1 + K_{inh}X_{inh}) \,, \tag{9}$$

where K_{inh} is the Langmuir constant and X_{inh} the inhibitor concentration. In this way the effectiveness of a nucleation inhibitor on the rate of crystal growth is determined by the Langmuir equilibrium constant K_{inh} for adsorption–desorption and the effectiveness factor α_{inh}. For very effective nucleation inhibitors, $\alpha_{inh} \gg 1$, the growth rate is decreased significantly even at low inhibitor concentrations (X_{inh}) or poor absorption (low Langmuir constant K_{inh}) when only part of the surface is occupied by inhibitor. Kubato and Mulin successfully applied this theory to describe the reduced crystal growth rates as a function of impurity concentration [66, 67]. The rate of secondary nucleation can be treated in an analogous fashion, albeit with different values for α_{inh} and K_{inh} [68].

Using the nucleation theory above for conglomerate crystallization, Kondepudi has been able to model enantioselectivity of the direct resolution of glutamic acid, using (S)-lysine as an enantioselective nucleation inhibitor [65].

3
Dutch Resolution

3.1
Basic Principles

The homochiral set of cyclic phosphoric acids **9a–c** shown in Scheme 7 (see further) served as the starting point for many of the studies [69]. These phosphoric acids can be resolved readily. The absolute configurations of these cyclic acids have been established either from crystallographic structure determinations on salts with amines of known configuration and/or chemical correlation.

The individual phosphoric acids were demonstrated to be excellent resolving agents for a variety of amines [69]. These structurally similar homochiral acids form a family in the same sense that α-amino acids are a family.

As already emphasized diastereomeric resolutions, especially of materials that have never been resolved before, often are time consuming owing to failure to form crystals, slow rates of crystallization and poor diastereomer excesses in the isolated first salts. In an attempt to improve the efficiencies of resolutions and reduce the time necessary for scanning Vries had attempted resolutions with equimolar combinations of various resolving agents used simultaneously in the hope that a crystalline salt with the best resolving agent would precipitate out spontaneously. Although this combinatorial approach with structurally unrelated resolving agents was not especially successful it did become clear with further experimentation that structurally related resolving agents could have a synergistic effect.

As is well illustrated with the example shown in Scheme 7 it was discovered that an equimolar mixture of the family **9a–c**, "P mix", was a superior

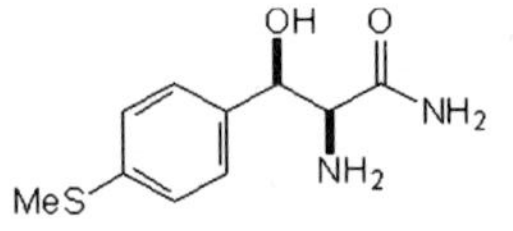

Scheme 7 An example of a Dutch Resolution experiment

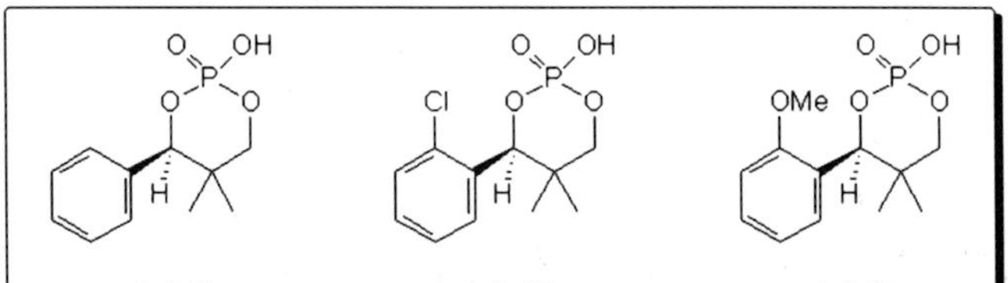

Entry	Resolving Agent	Yield (%)	ee (%)	S-Factor	mix-ratio salt
1	(−)-9a	47	52 (2R,3S)	0.49	−
2	(−)-9b	55	17 (2R,3S)	0.19	−
3	(−)-9c	41	67 (2S,3R)	0.55	−
4	(−)-P-Mix (1:1:1)	25	99 (2S,3R)	0.49	9a:9b:9c (12:35:53)

resolving agent for amines. Diastereomeric excesses of the first salts were generally higher and crystallization occurred readily [70]. The resolution of the thiamphenicol precursor **10** (examined in detail at DSM Research) is a classical example (Scheme 7). Each of the cyclic phosphoric acids **9a–c** is a resolving agent for **10** (entries 1–3) and certainly with **9a** and **9c** the enantiomeric excess of **10** obtained in the first salt is acceptable (note that opposite enantiomers of **10** are obtained). The overall efficiencies are given in terms of the Fogassy S-factor where $S = 2 \cdot \text{yield} \cdot de$ [71]. Because the yield of a perfect resolution can never be more than 50% S-factors hence vary between 0 (no resolution) and 1 (perfect resolution). In the Dutch Resolution experiment (entry 4, Scheme 7) the enantiomeric excess of isolated **10** is essentially perfect. However, the yield decreases. This is a commonly observed phenomenon. The salt obtained forms a mixed crystal in which the cyclic phosphoric acids are present in nonstoichiometric ratio. This is also common to Dutch Resolution.

Other families were quickly found, some examples of which are shown in Scheme 8 together with their trivial names [72]. Some examples of nonproprietary compounds resolved by means of Dutch Resolution are shown in Scheme 9. Other examples are available from the literature [70, 72].

The families used for resolution consist usually of equimolar mixtures of three members and are stereochemically homogeneous, that is to say they have analogous absolute configurations (care is necessary in comparison of absolute configurations on the basis of Cahn–Ingold–Prelog nomenclature

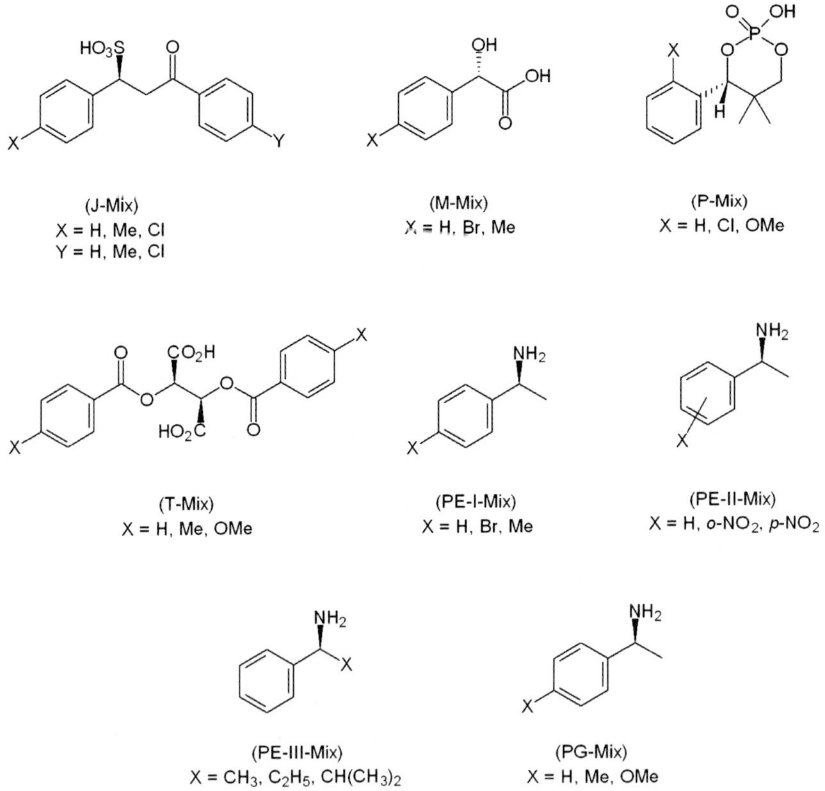

(J-Mix)
X = H, Me, Cl
Y = H, Me, Cl

(M-Mix)
X = H, Br, Me

(P-Mix)
X = H, Cl, OMe

(T-Mix)
X = H, Me, OMe

(PE-I-Mix)
X = H, Br, Me

(PE-II-Mix)
X = H, o-NO_2, p-NO_2

(PE-III-Mix)
X = CH_3, C_2H_5, $CH(CH_3)_2$

(PG-Mix)
X = H, Me, OMe

Scheme 8 Examples of Dutch Resolution mixes

owing to the effects of substitution on priorities). Resolutions are carried out exactly as an experiment with a single resolving agent would be performed. The racemate is dissolved in an appropriate solvent, often an alcohol, and then an equivalent amount of the family of resolving agents (equimolar ratios) is added. The solution is cooled—we prefer in small-scale experiments to use a temperature-programmed cooling bath—until solid precipitates. After a short period, two to three hours generally, the solid is collected and submitted for analysis. From NMR measurements the composition of salts can usually be determined. Chiral hplc is usually used to determine the diastereomeric ratio (enantiomeric ratio if the salt is neutralized prior to determination; the choice is dependent on the racemate).

On the basis of more than a 1000 resolutions of new compounds since this discovery, we now know that:

- Dutch Resolution using acidic or basic families of resolving agents usually leads to the formation of solid salts usually considerably more quickly than in a resolution with a single family member.

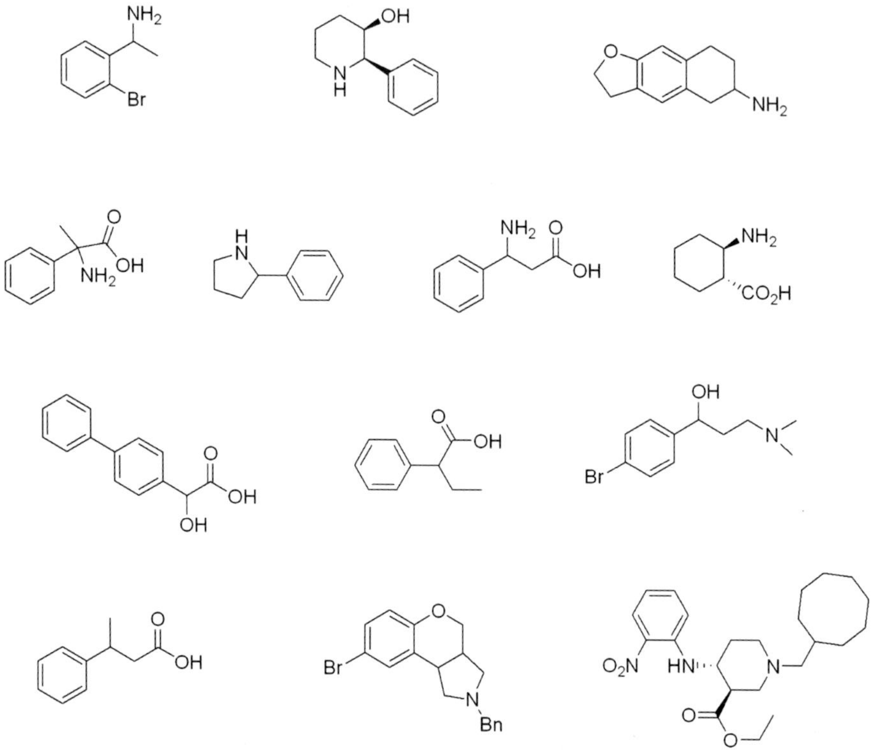

Scheme 9 Some examples of amines, carboxylic acids and amino acids that have been resolved by means of Dutch Resolution

- The salts formed consist of mixed crystals and contain a nonstoichiometric ratio of resolving agents; they are solid solutions in the resolving agents (see further).
- The diastereomeric excesses of the first salts are in general higher than when single resolving agents are used.

In many, if not all, cases the resolutions are kinetically controlled.

3.2
Reverse (Reciprocal) Dutch Resolution

A variation on Dutch Resolution deserves special mention particularly for the insight it provides. This is the simultaneous resolution of racemates. This is nicely illustrated with the sulfonic acids **11a–c** prepared as a new family, J-mix, by the addition of sodium bisulfite to chalcones [72]. Resolution of **11a–c** can be carried out simultaneously using 1,2-aminoalcohol **12** as shown in Scheme 10.

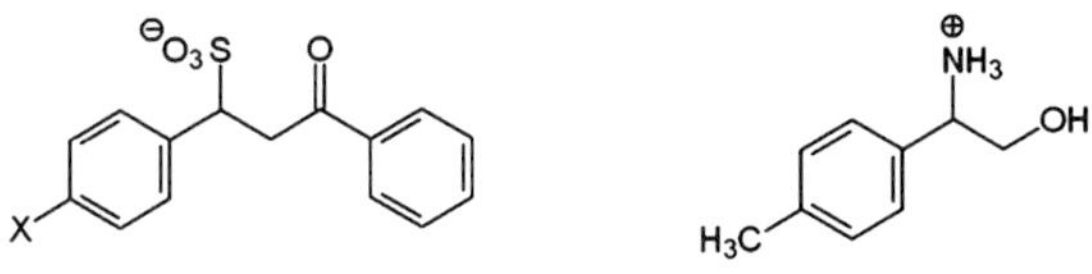

Scheme 10 Reverse Dutch Resolution of chalconesulfonic acids

Racemate 11	e.e. H/Me/OMe	Ratio H:Me:OMe	e.e. after recryst.	Ratio after recryst.
Me	67		99	
H/Me	86/79	1.3:1	99+/99+	2.7:1
H/Me/OMe	79/19/27	5:1:5	98/90/96	10:1:4

This simultaneous resolution proceeds in remarkably high diastereomeric excess. Although separation problems attend such resolutions, the phenomenon has important ramifications. One could even speculate whether this could be a mechanism for propagation of single chirality in pre-biotic chemistry.

A more extensively studied example of the same effect is provided by the resolution of 4-hydroxy and 4-fluorophenyl glycines **13b** and **13c** (Scheme 11) [73].

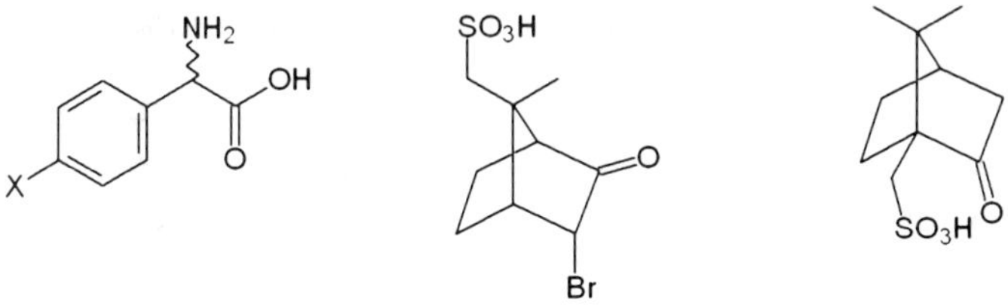

a) X = H   DL-phenylglycine
b) X = OH   DL-4-hydroxyphenylglycine
c) X = F   DL-4-fluorophenylglycine

Scheme 11 The reverse Dutch Resolution of phenylglycine **13a** and its 4-hydroxy (**13b**) and 4-fluoro (**13c**) derivatives

Resolution of **13b**, used in the production of semi-synthetic antibiotics, is possible with (+)-3-bromocamphor-8-sulfonic acid **14**, which, however, is quite expensive. However, resolution with cheaper (+)-10-camphorsulfonic acid **15**, which is used for commercial resolution of phenylglycine **13a**, fails. Resolution of **13c**, of interest in conjunction with synthesis of Substance P analogs, also fails with (+)-**15**. However, resolution of either **13b** or **13c** with

(+)-**15** acid can be accomplished readily as a mixture with racemic **13a** (or D-(−)-**13a**). For example, resolution of a 75 : 25 mixture of **13a** and **13b** in 1 N HCl solution with (+)-**15** gave in 87 : 13 ratio D-(−)-**13a** and D-(−)-**13b** both in 94% ee (100% ee after one recrystallization). Half quantities of D-(−)-**13a** can be used in place of the racemate.

A crystal structure of mixed crystals (26 : 74 ratio) of resolved D-(−)-**13c**/D-(−)-**13a** with (+)-**15** has been determined (Fig. 8a,b)

As anticipated the absolute configurations are clearly analogous and the **13a** and **13c** molecules are randomly distributed in the unit cells. Both from crystallographic observations and differential scanning calorimetry (DSC) results it is clear that a *solid solution* has formed in which the small structural differences (para H, F and OH in **13**) in the enantiomers are not recognized.

Similar observations have been made with racemic alaninol **16**, which steadfastly resists attempts at resolution including use of Dutch Resolution families (Scheme 12) [74–76]. For example, resolution experiments using (R)-(−)-mandelic acid **7** or the M-mix (Scheme 8) only resulted in crystallization of an equimolar mixture of diastereomeric salts, i.e. crystallization of racemic **16**. From the resolution experiments with (R)-(−)-**7** an eutectic composition of 0.53 in favor of (S)-(+)-**16** was found, indicative of an unfavorable resolution.

Whereas resolution of **16** was not possible, resolution of 2-amino-1-butanol **17a** with (R)-(−)-**7** results in a very efficient crystallization of the (R)·(R)-salt. The cause of these differing results is clear from the solubilities of the diastereomerically pure salts as given in Table 2.

We therefore turned our attention to the reverse Dutch Resolution approach for the resolution of alaninol. Addition of the "family member" (RS)- or (R)-(−)-2-amino-1-butanol **17a** to racemic **16** followed by addition of an equivalent of (R)-(−)-**7** leads to the formation of mixed crystals and simultaneous successful resolution of **16**. For example, the resolution of 0.75 equiv. of

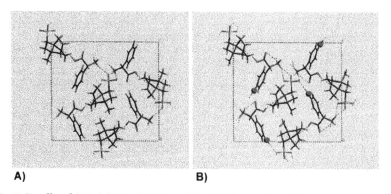

Fig. 8 Unit cells of **A** D-(−)-**13a** (+)-**15** and **B** mixed crystal D-(−)-**13a**/D-(−)-**13c** (+)-**15**. Permission to use this figure has been kindly given by Elsevier Publishing Company

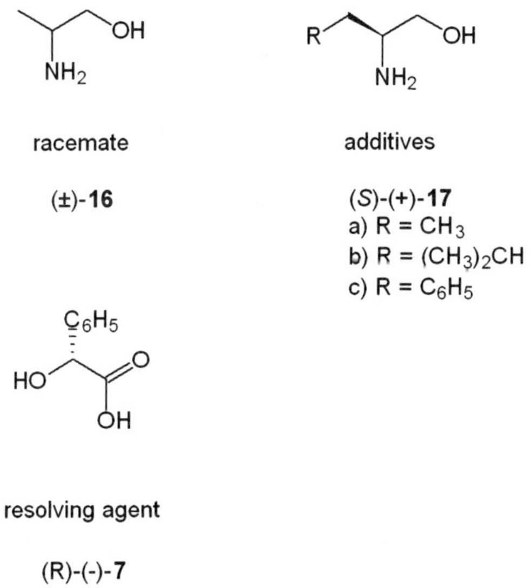

Scheme 12 The reverse Dutch Resolution of alaninol 16 by mandelic acid 7

Table 2 Solubilities of the diastereomeric salts of 16 and 17a with (R)-(−)-7 in iPrOH/water 19 : 1 at 25 °C

Diastereomer	Solubility (mg/g)
(S)-(+)-alaninol(16)·(R)-(−)-mandelic acid(7)	107
(R)-(−)-16·(R)-(−)-7	64
(S)-(+)-aminobutanol (17a)·(R)-(−)-7	> 800
(R)-(−)-17a·(R)-(−)-7	21

(RS)-16 with 0.25 equiv. of (R)-17a and 1.0 equiv. of (R)-(−)-7 gave a mixed crystal salt in 40% yield containing (R)-(−)-16 (ee 92%) and (R)-(−)-17a in a ratio of 57 : 43 (34% resolution yield of (R)-(−)-16). As observed before, the mixed crystals formed in this Dutch resolution experiment form a solid solution, as was shown by comparison of the X-ray powder diffractograms (XRPD) of the mixed crystals with the pure salts. On the other hand, when the more soluble (S)-(+)-17 salt was added to the crystallization of racemic alaninol 16 with (R)-(−)-7 from iPrOH/water 19 : 1 almost chemically and diastereomerically pure (R)-(−)-16 · (R)-(−)-7 salt was obtained in 26% yield (ee 94%(R)-16; 16 : 17a ratio 98 : 2). We will return to this phenomenon in the discussion of nucleation inhibition (see further).

Judged from the foregoing a single resolving agent can be used to resolve simultaneously two or more "family" related racemates provided that a solid solution (mixed crystals) can be formed on precipitation. This is a key observation.

3.3
Solid Solution Behavior in Dutch Resolution

The mixed crystals observed in either classical or reverse (reciprocal) Dutch Resolutions are solid solutions. Criteria are necessary to predict whether solid solution behavior is possible and the relation thereof to the outcome of Dutch Resolution. Kitagorodsky has emphasized that the mutual solubility necessary for formation of mixed crystals is highly dependent on geometrical similarities [77]. One can intuitively expect such similarities in "family members" although a precise definition of what constitutes a "family" is difficult to give. At a minimum one would expect similar chemical characteristics (α-amino acids, structurally similar cyclic phosphoric acids like **9a–c**, etc.) and the same sense of absolute configuration (homochirality).

Solid solution behavior in Dutch Resolution has been investigated by use of molecular modeling techniques [78]. The resolution of (1R,2S)-ephedrine **18** (Scheme 13) by the cyclic phosphoric acids **9a–c** (Scheme 7) is used as a model.

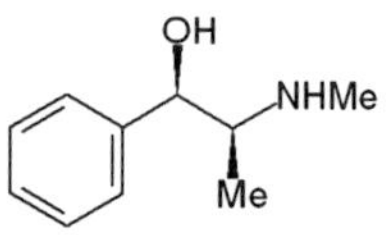

Scheme 13 (1R,2S)-(–)-Ephedrine

With the aid of molecular modeling procedures using the crystal structures of the pure diastereomeric salts as starting points mixed salts can be generated by isomorphous replacement and the energy compared to that of pure salts. From the results solid solution formation was predicted for the diastereomeric salts of **18** with binary combinations of the cyclic phosphoric acids **9a–c**. However, full or partial solid solution behavior was not predicted for all combinations. For example, calculations for both diastereomeric salts of **18** with **9a** and **9c** led to prediction of complete eutectic behavior. Both full and partial solid solution behavior as well as eutectic behavior with the various binary combinations of phosphoric acids have indeed been observed as confirmed by construction from experiment of the appropriate phase diagrams.

In the case of solid solution behavior the influence of variation in the composition of the phosphoric acids (the resolving agent) can have considerable effect on the solubility ratio for the two diastereomeric salts. From this specific example [78] it is difficult to predict whether solid solution behavior will have a similar role in other Dutch Resolution experiments. On the other hand, even in the case that solid solutions are not observed, a kinetic effect may play an important role (see further). The reader is referred to the original paper for extended discussion of the approach [78].

This same system of **9a–c** with **18** has been investigated experimentally with regard to its behavior in nucleation and crystal growth [79]. A comment is necessary. The acids **9a–c** may be resolved with **18**. The same acids can, as expected, be used to resolve racemic **18**. However, one should note that these resolution using **9a** actually should fail. For **9a** (phencyphos) the solubilities of the two diastereomeric salts with **18** are virtually identical. For kinetic reasons the (–)-**9a** salt with **18** apparently crystallizes out faster. It is not clear why this is so. Dutch Resolution of racemic **18** with P-mix (**9a–c**) also proceeds extremely well and gives after one recrystallization resolved **18** in 98% d.e. as a 1 : 1.2 : 1.1 salt with **9a, 9b** and **9c** [70].

Optical microscopy has been used to follow crystal growth for this system in solution. With (+)-**9a** alone in supersaturated solution ($\beta = 1.5$–2.0) where $\beta = C_{initial}/C_{equilibrium}$ at 25 °C, the salt with (–)-**18** provides either elongated or blocky crystals or flat ribbons. Both (+)-**9b** and (+)-**9c** provide also blocky crystals. However, crystallization of a mixture of (+)-**9a** and (+)-**9b** together with (–)-**18** provided helical ribbons, which proved to be a solid solution with a 65 : 35 ratio of **9a** and **9b**. Helical ribbons were also observed with mixtures of the salts of (–)-**18** with (+)-**9b** and (+)-**9c**. The effects observed are shown in Fig. 9a–c as well as Fig. 10.

Nucleation in the cases of the pure salts of (–)-**18** with (+)-**9b** and (+)-**9c** was quite difficult whereas nucleation from the mixtures proceeded rapidly and easily. This is a hallmark of Dutch Resolution, namely the quite ready formation of crystals. A reasonable supposition is that the formation of chiral supramolecular structures in the supersaturated region is a forerunner to crystallization.

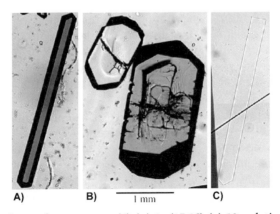

Fig. 9 Crystal shapes from a pure (S)-(+)-**9a**·(1R,2S)-(–)-**18** solution: **A** elongated; **B** blocky crystals and **C** ribbons. Permission to use this figure has been kindly given by Elsevier Publishing Company

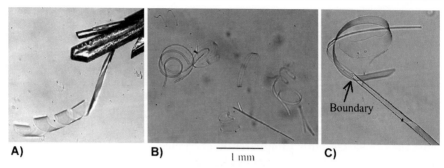

Fig. 10 Growth from a 1:1 (S)-(+)-**9a**·(1R,2S)-(−)-**18** and (R)-(+)-**9b**·(1R,2S)-(−)-**18** two compound solution: **A** crystals and wide helical ribbons; **B** thin helical ribbons and **C** structure or thickness boundary in a ribbon. Note that although the Cahn–Prelog–Ingold descriptors of **9a** and **9b** are opposite, the compounds actually have the same relative absolute configurations. Permission to use this figure has been kindly given by Elsevier Publishing Company

3.4
Nucleation Inhibition in Dutch Resolution

Data obtained from many Dutch Resolution experiments coupled with experimental observation were carefully re-analyzed. In many cases one of the (usually) three components of the family was not, or only slightly, incorporated, although the resolution proceeded better with all components of the family present. Consideration of the data reported in [70] led to the conclusion that when PE-II mix (a mixture of phenylethylamine **6a** with the o- and p-nitrated derivatives **6b,c**) is used the nitrated compounds are often poorly or not at all incorporated into the precipitated salts. This lack of incorporation is visually striking owing to the fact that the nitrated derivatives are yellow. The first salts obtained are in general colorless as a consequence of *the absence* of the nitrated derivatives. Also in various resolutions using **9a–c**, P mix, the o-methoxy **9c** and o-chloro **9b** derivatives are also often poorly or not at all incorporated. Similar trends are seen with T-mix although there is considerable variation in which tartaric acid is not incorporated.

Owing to the practical consideration that use of an equimolar mixture of resolving agents leads to additional complications in work-up, especially at the larger scale, attempts were made to develop procedures whereby additives rather than equimolar amounts were used. The observations mentioned above led to the investigation of the nitrated phenylethylamines **6b,c** to see whether Dutch Resolution results could also be obtained with these materials as *additives* [80]. The resolution of mandelic acid **7** by enantiomerically pure phenylethylamine **6a** (Scheme 14, see also Scheme 4) was chosen as the test system. On use of 10 mol percent of a mixture of nitrated **6b,c** with

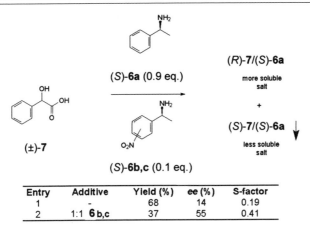

Entry	Additive	Yield (%)	ee (%)	S-factor
1	-	68	14	0.19
2	1:1 **6 b,c**	37	55	0.41

Scheme 14 Nucleation inhibition in the resolution of mandelic acid 7 by phenethylamine 6a

phenylethylamine **6a** as the resolving agent it was immediately observed in the laboratory, simply by feel of the flask with the hand and later confirmed by quantitative measurements (see further), that the precipitation of solid material occurred at a significantly lower temperature than with **6a** alone. The enantiomeric excess was improved considerably (Scheme 14) and this is reflected in the S-factors. Care has been taken in all experiments to ensure that the error in reproducibility of measurements is not greater than 5%. No nitro derivative **6b,c** was found in the isolated salt.

With the aid of turbidity measurements the effects of the nitro additives **6b,c** on each diastereomeric salt could be measured individually. By means of these measurements the dissolution temperature can be determined (full transmission) whereas the cut off of transmission marks the formation of the first crystals, or the nucleation temperature. The results of the measurements are given in Fig. 11 wherein the darkened horizontal bars indicate the width of the observed metastable zones.

In the presence of additives **6b,c** the metastable zone width for the least soluble diastereomer of mandelic acid 7 with phenylethylamine **6a** is widened slightly as the result of a slightly lower nucleation temperature. On the other hand the additive has an enormous effect on the *more soluble* diastereomer and widens the metastable zone width drastically. The dissolution temperature in this case is also lowered. The major effect, however, is that of the lowering of the nucleation temperature of the more soluble diastereomer. Put in qualitative terms the more soluble diastereomer is held in solution longer allowing the less soluble diastereomer to precipitate.

Analysis of the isolated first resolution salt of **6a** with 7 was carried out and no nitro additives **6b,c** could be detected. This behavior is clearly that of a *nucleation inhibitor*. Another role has been discovered for a family member, namely that of nucleation inhibitor.

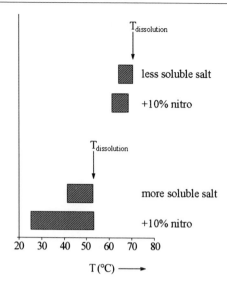

Fig. 11 Width of metastable zones of the less and more soluble 6a·7 salt with/without nitro-additive 6b/c, as derived from turbidity measurements

The observation by Sakai et al. that "dimeric" phenylethylamine derivative (R,R)-8a is a habit modifier of the least soluble diastereomer of 6a with 7 (Scheme 4) has already been mentioned [41–43]. However, (see further) we do not observe that 8a acts as a nucleation inhibitor.

We do observe nucleation inhibition by various family members of resolving agents in other experiments. The 2-chloro and 2-methoxy substituted phosphoric acids 9a and 9c are often also poorly incorporated in Dutch Resolution experiments although they have a positive effect. This was examined quantitatively for the resolution of 19 as shown in Scheme 15.

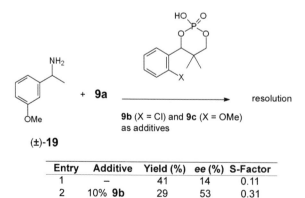

Entry	Additive	Yield (%)	ee (%)	S-Factor
1	–	41	14	0.11
2	10% 9b	29	53	0.31
3	10% 9c	33	28	0.18

Scheme 15 Nucleation inhibition in the resolution of 19 by cyclic phosphoric acid 9a

Again the additives are not incorporated and the diastereomeric excesses improve appreciably although yields go down some. The lower yields are a general observation and in line with what is expected in the case of nucleation inhibition.

Nitrated phosphoric acid **9d** also was observed to act as a very efficient nucleation inhibitor in the resolution of **20** as shown in Scheme 16.

Additional information was obtained from deeper study of the reversed Dutch Resolution of alaninol **16** discussed previously (Scheme 12). From turbidity measurements it was shown that the nucleation temperature for the crystallization of the (S)-(+)-**16** salt with (R)-(−)-**7** was lowered on addition of 10 mol % of the (S)-(+)-**17a** salt with (R)-(−)-**7**, resulting in a larger metastable zone width, whereas the same additive had no effect on the metastable zone width for the diastereomeric (R)-(−)-**16**·(R)-(−)-**7** salt. The more bulky (S)-(+)-leucinol **17b** or (S)-(+)-phenylalaninol **17c** salts proved to be even better nucleation inhibitors of the (S)-(+)-**16**·(R)-(−)-**7** salt (Fig. 12).

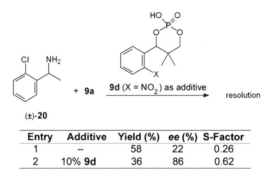

Entry	Additive	Yield (%)	ee (%)	S-Factor
1	–	58	22	0.26
2	10% **9d**	36	86	0.62

Scheme 16 Nucleation inhibition in the resolution of **20** by cyclic phosphoric acid **9a**

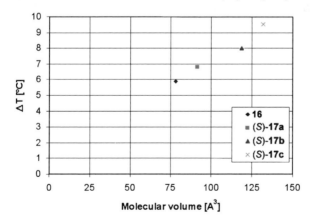

Fig. 12 The metastable zone widths (ΔT) for the (S)-(+)-**16**·(R)-(−)-**7** salt and the corresponding zone widths found on addition of 10 mol % (S)-(+)-**17a**, **17b** and **17c** salts plotted versus the molecular volumes (as calculated from the modeling program Cerius) of **16** and **17a–c**

For this limited selection of amino alcohols tested as nucleation inhibitors a linear correlation is observed between the molecular volume and the metastable zone width (for the more soluble salt). Clearly both the size of the substituent and the absolute configuration influence the efficacy of the nucleation inhibition, where larger substituents result in an increased value for A_{inh} in Eq. 7.

Nucleation inhibition has been studied further with the system of ephedrine **18** with cyclic phosphoric acids **9a** and **9c** [81]. As previously mentioned resolution of **9a** by **18** is in principle not possible because the system is eutectic with an eutectic composition of almost exactly 50 : 50. This system lends itself perfectly to study of additives. Acid **9c** is found to be a strong nucleation inhibitor. In particular the metastable zone width for the diastereomer (–)-**9a**·(–)-**18** is widened dramatically on addition of 10 mol % (–)-**9c** (absolute configuration analogous to that of (–)-**9a**). In other words a "family member" of the racemate rather than of the resolving agent can also act as a potent nucleation inhibitor.

A more broadly based study was carried out both to establish the generality of nucleation inhibition and to allow identification of potential nucleation inhibitors. There are no specific guidelines for identification of "family members" that engage in solid solution behavior as opposed to those that act as nucleation inhibitors. The task would be easier if one understood the relationship, if any, between solid solution behavior and nucleation inhibition (see further).

The following approach has been used for one specific case to identify nucleation inhibitors. The PE-II "family" of compounds structurally related to phenylethylamine **6a** as resolving agent has been quantitatively investigated in the resolution of mandelic acid **7** (Scheme 14). We decided to use this model system as the basis for further study whereby the structural diversity of the "family" related to **6a** was varied as far as synthetic accessibility and/or commercial availability allowed. Structural variation was achieved by (a) modification of the aryl portion by substitution and/or (b) structural modification of the side chain of **6a**.

A major source of potential inhibitors was a readily accessible enantioselective synthetic scheme for phenylbutylamine **21** and many derivatives thereof [82, 83]. Phenylbutylamine **21** differs structurally from **6a** in the length of the aliphatic side chain rather than by substitution in the aromatic ring. We were extremely encouraged to find that this compound is also a good nucleation inhibitor (Scheme 17). On the basis of optimization experiments it was found that 6% additive gave in general the maximum effect.

The inhibitor is not incorporated into the first salts and turbidity measurements provided a picture entirely analogous to that of Fig. 11 [84]. Again the inhibitor has little effect on the width of the metastable zone width of the least soluble diastereomer whereas it has a profound widening effect on the metastable zone width for the more soluble diastereomer. In other words the

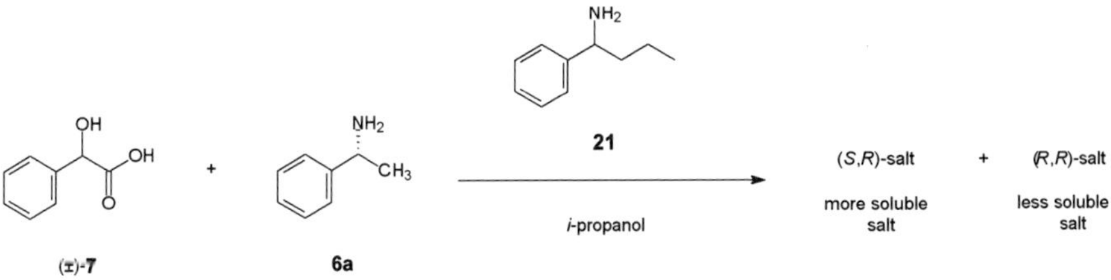

Dutch Resolution of Racemates

Scheme 17 Compound 21 as nucleation inhibitor for the resolution of mandelic acid 7 by phenethylamine 6a

Resolving Agent	Additive	Yield (%)	ee (%)	S-Factor
(R)-6a	–	68	14	0.19
(R)-6a	6% (R)-21	60	42	0.50
(R)-6a	6% (±)-21	62	35	0.43
(R)-6a	6% (S)-21	61	30	0.37
(R)-21	–	No salts		

more soluble diastereomer is held in solution longer and this is exactly what is required for an efficient resolution.

However, not every derivative of **21** is a nucleation inhibitor. Fogassi S-factors [71] for the resolution of racemic **7** by (R)-**6a** in the presence of 6% enantiomerically pure (R)-ring substituted derivatives are summarized in the bar graph of Fig. 13. The bar farthest right (94% resolving agent) can best be used for comparison purposes: 6% additive is used and this S-factor (0.24) corresponds to the concentrations used in the inhibition experiments. It is clear that the ring-substituted derivatives shown in Fig. 13, with the exception of m-nitro-substituted derivative **44**, **21** itself, the α-napthhyl derivative **43** and 2-hydroxy-substituted **42** are at best only mild inhibitors. Compounds like the 2-bromo derivative **24**, 3-bromo derivative **23**, 3-fluoro derivative **25**, and especially the 3-chloro derivative **22** have a profoundly negative effect on the S-factor for the resolution. The origin of this effect is not well understood, but may be the result of a more efficient nucleation inhibition of the less soluble diastereomer or from nucleation catalysis of the more soluble diastereomer.

Modification of the structure of the aliphatic side chain as summarized in the bar graph of Fig. 14 leads to greater insight. In this case not every inhibitor was available in enantiomerically pure form. In some cases (see graph) it was possible to test both the enantiomerically pure (R) inhibitors (analogous to resolving agent (R)-**6a**), the (S) inhibitor (in principle the "wrong" absolute configuration) and the racemate. A good example is **21** itself. The "wrong" (S) absolute configuration (S-factor 0.37) is still better than the blank result of S-factor 0.24. The racemate, S-factor 0.43, compares well with the (R)-enantiomer, S-factor 0.50. Mild branching in the side chain, for example **60**, clearly leads in general to inhibitors about as effective as **21** itself. Cyclization

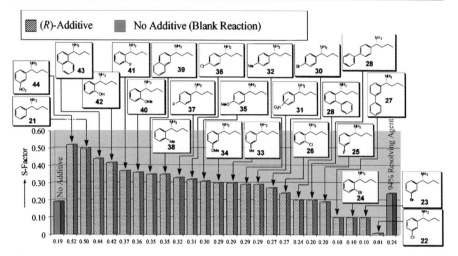

Fig. 13 S-Factors for the resolution of mandelic acid by (R)-phenethyl amine in the presence of potential (R) nucleation inhibitors **21–44**

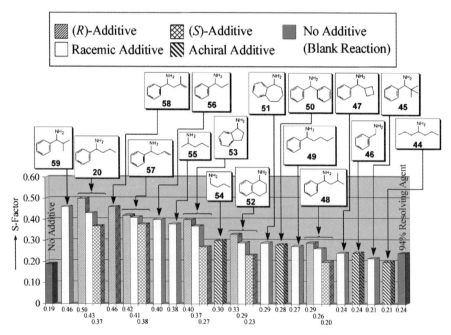

Fig. 14 S-Factors for the resolution of mandelic acid **7** by (R)-phenethylamine **6a** in the presence of potential nucleation inhibitors **44–59**. The inhibitors are in many cases racemic (see codes)

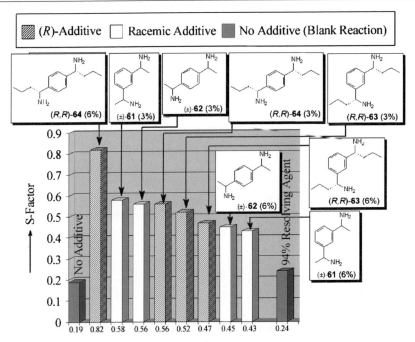

Fig. 15 S-Factors for the resolution of mandelic acid 7 by (R)-phenethylamine 6a in the presence of bifunctional nucleation inhibitors 61–64

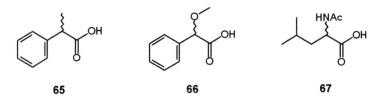

Entry	Racemate	Additive (mol %)	Yield (%)	de (%)	S-Factor
1	65	–	60	7	0.08
2	65	(±)-61 (3%)	36	59	0.43
3	65	(±)-62 (3%)	54	52	0.56
4	65	(R,R)-63 (6%)	18	97	0.35
5	66	–	65	1	0.02
6	66	(±)-61 (6%)	16[i]	96	0.30
7	67	–	58	13	0.15
8	67	(±)-62 (3%)	35	75	0.53
9	67	(R,R)-63 (3%)	44	48	0.43
10	67	(R,R)-64 (6%)	56	32	0.36

[i] Salt isolated after 6 days.

Fig. 16 S-Factors for the resolution of 65–67 by (R)-phenethylamine 6a in the presence of bifunctional nucleation inhibitors 61–64

(compounds **52–54**) is not very effective, however, and achiral derivatives **45** and **47** have essentially no effect, as one would intuitively expect.

The observation that racemates approach the effectiveness of the "correct" absolute configurations of the inhibitors is important. This indicates that in screening processes that racemates, often far more readily available, may safely be used in tests for nucleation inhibitory effects.

However, by far the profoundest effects are obtained with bifunctional analogs of phenylethylamine **61–64** as shown in Fig. 15 for the resolution of mandelic acid **7**.

These same derivatives work remarkably well also for other resolutions such as those of **65–67**. Nonoptimalized results are summarized in Fig. 16. The effects on the obtained diastereomeric excesses can be dramatic. For example, the first resolution salts of **65** and **66** are obtained almost diastereomerically pure.

4
Conclusions

Simple explanations and applications of "Occam's Razor" type of arguments to Dutch Resolution in particular and resolutions in general should be regarded with skepticism. Complicated phenomena often have complicated explanations. Family behavior and its relation to both nucleation inhibition and solid solution behavior are key issues in Dutch Resolution. The tailor-made additives of Lahav et al. are, of course, family members of the conglomerates that undergo crystallization. The effects of these additives on the growing crystal surface have been carefully defined by this group and Lahav and colleagues have clearly pointed out much of what is not known in the complex process of nucleation and crystallization [25].

Probably the first example of family behavior was noted by Barton and Kirby who found more than forty years ago that compounds with structures similar to the material undergoing crystallization can influence the crystallization [85–87]. In the early nineties Brock, Schweizer and Dunitz pointed out that the crystallization of a single enantiomer may well be kinetically inhibited by the presence of its mirror-image molecule [88]. A theory has been developed and tested for the effect of chiral impurities on the crystallization of a conglomerate (the goal, of course, is to be able to direct crystallization of racemates towards a conglomerate) [89, 90]. Arguments have been advanced recently based on analysis of α-amino acids that the racemic crystal becomes in general thermodynamically more stable as the "degree of chirality" of the molecule increases [91].

Additives can stimulate rather than inhibit nucleation. Although the example involves nonchiral molecules the freezing of water can be aided, rather than retarded, by certain long-chain alcohols as additives. Detailed models at the molecular level have been developed to understand this [92, 93]. One certainly wonders whether nucleation catalysts for diastereomeric salt resolutions also could be found. The behavior of "negative inhibitors" like 22–25 (Fig. 13), which induce significantly poor resolutions, certainly deserve further investigation as possible nucleation catalysts.

It has been observed that circular stirring in one direction can induce both nucleation and spontaneous resolution [94, 95]. This phenomenon, in which the chirality of the crystals formed is random, probably arises from secondary nucleation effects caused by the stirring and is most likely unrelated to Dutch Resolution.

Crystal growth and the effects thereon of chirality, including that of diastereomeric salt pairs, is complicated and does not readily lend itself to calculational approaches [96] although from the study of well-designed crystalline systems some principles of chiral recognition may be derived [97]. Saigo et al. have recently presented a combination of calculational work supported by crystallographic data on O-substituted phenylphosphonothioic acids as resolving agents for amines. A molecular cluster that leads to prism-type crystals seems to be essential for high degrees of chiral recognition [98]. Whether these results can be extrapolated to other systems is not known.

In Dutch Resolution there seems to be two major effects, namely formation of solid solutions and nucleation inhibition. The major effects observed in nucleation inhibition are with the *more soluble* rather than *less soluble* diastereomer. It may well be that "design" should be directed more towards what *does not* precipitate rather than the salt that does precipitate even though this seems counterintuitive.

In practical terms a classical Dutch resolution wherein several family members of the resolving agent are used in equimolar quantities can be very effective in the laboratory. Larger-scale application is in general less attractive because of the problems associated with recovery of a mixture of resolving agents. However, addition of a small amount of an inhibitor, the crux of nucleation inhibition experiments, to a single resolving agent is far more attractive from a technological standpoint. To all appearances a successful inhibitor forms a solid solution with *the more soluble* diastereomer and barely affects the less soluble diastereomer. For that reason the inhibitor is generally not, or only barely found, in the precipitated salt. Recovery of the (single) resolving agent becomes a far easier task.

But how does one find inhibitors [99]? We obviously need to screen a number of acidic and basic resolving agents to answer that question completely. Such work is in progress. We believe that inhibitors are best found among the family members of a resolving agent. It will probably be necessary to screen several possible inhibitors to find an optimum. There is no guarantee that a potent inhibitor for one racemate will be equally effective for another.

On the basis of results now in hand we suspect that polyfunctional compounds may in general be quite effective inhibitors. Probably the inhibition we observe occurs to a great extent in the primary nucleation process. We observe, for example, that relatively large amounts (3–6 mol % in our experience) are required whereas nucleation inhibition of crystal growth normally proceeds with far lower concentrations. We have also found that "habit modifiers" like **8a**, known to inhibit the growth of specific crystal faces of the salt of phenylethylamine **6a** with mandelic acid **7**, have little, if any, effect when used as nucleation inhibitors under the conditions described in foregoing sections (Scheme 4).

The understanding of diastereomeric resolutions in general as well as the particular case of Dutch Resolution is far from complete. Some of the factors involved are now apparent. Our hope is also that the reader will be convinced that diastereomeric resolutions are practical and, owing to these new insights, represent also an area of scientific challenge and opportunity.

Acknowledgements The work on nucleation inhibition in the Syncom group has been carried out by Dr. J. Nieuwenhuijzen and Dr. J. Dalmolen and is described in their Ph.D theses (University of Groningen). We are grateful to many people for their thoughts and suggestions and mention in particular Prof. E. Vlieg and Dr. J. Loh (Radboud University Nijmegen), Dr. Q.B. Broxterman (DSM Research), Dr. K. Pouwer (Syncom), Mr. M. Leeman (Syncom) and Ms. G. Brasile (Syncom).

References

1. Pasteur L (1853) CR Acad Sci 37:162
2. Pasteur L (1853) Ann Chim (Paris) 3:38–437
3. Woodward RB, Doering WE (1945) J Am Chem Soc 67:860
4. Jacques J, Collet A, Wilen SH (1994) Enantiomers, Racemates, and Resolutions. Krieger Publ. Co., Malabar, Florida
5. Kozma D (ed) (2002) CRC Handbook of Optical Resolutions via Diastereomeric Salt Formation. CRC Press, New York
6. Eliel EL, Wilen SH (1994) Stereochemistry of Organic Compounds. Wiley, New York
7. Eliel EL, Wilen SH, Doyle MP (2001) Basic Organic Stereochemistry. Wiley, New York
8. Wilen SH (1971) Resolving Agents and Resolutions in Organic Chemistry, vol 6. Topics in Stereochemistry. Wiley, New York
9. Collet A, Brienne M-J, Jacques J (1980) Chem Rev 80:215
10. Marchand P, Lefèbvre L, Querniard F, Cardinaël P, Perez G, Counioux JJ, Coquerel G (2004) Tetrahedron: Asymmetry 15:2455
11. Ojima I (ed) (1993) Catalytic Asymmetric Synthesis. Wiley, New York
12. Gridnev ID, Serafimov JM, Quiney H, Brown JM (2003) Org Biomol Chem 1:3811
13. Soai K, Shibata T, Morioka H, Choi K (1995) Nature (London) 378:767
14. Mislow K (2003) Collect Czech Chem Commun 68:849
15. Toda F (ed) (2004) Enantiomer Separation: Fundamentals and Practical Methods. Kluwer, Dordrecht
16. Lehoucq S, Verheve D, VandeWouwer A, Cavoy E (2000) AICHE J 46:247
17. Kaspereit J (2005) J Chromat A 1092:43
18. Sakai K, Sakurai R, Nohira H, Tanaka R, Hirayama N (2004) Tetrahedron: Asymmetry 15:3495

19. Tamura R, Fujimoto D, Lepp Z, Misaki K, Miura H, Takahashi H, Ushio T, Nakai T, Hirotsu K (2002) J Am Chem Soc 124:13139
20. Borghese A, Libert V, Zhang T, Alt CA (2004) Organ Proc Res Dev 8:532
21. Jacques J, Collet A, Wilen SH (1994) Enantiomers, Racemates, and Resolutions. Krieger Publ. Co., Malabar, Florida, p vi
22. Mullin JW (1997) Crystallization, 3rd ed. Butterworth Heinemann, Oxford
23. Davey R, Garside J (2000) From Molecules to Crystallizers. An Introduction to Crystallization, Oxford University Press, Oxford
24. Boistelle R, Astier JP (1988) J Crystal Growth 90:14
25. Weissbuch I, Lahav M, Leiserowitz L (2003) Cryst Growth Des 3:125
26. Kelker H, Hatz R (1980) Handbook of Liquid Crystals. Verlag Chemie, Weinheim
27. Gelbart WM, Ben-Shaul A, Roux D (eds) (1994) Micelles, Membranes, Microemulsions, and Monolayers. Springer, New York
28. Rosoff M (ed) (1996) Vesicles. Dekker, New York
29. Oda R, Huc I, Candau SJ (1998) Angew Chem Int Ed Engl 37:2689
30. Oda R, Huc I, Schmutz M, Candau SJ, MacKintosh FC (1999) Nature 399:566
31. Cornelissen JJLM, Rowan AE, Nolte RJM, Sommerdijk NAJM (2001) Chem Rev 101:4039
32. Spector MS, Selinger JV, Singh A, Rodriguez JM, Price RR, Schnur JM (1998) Langmuir 3493
33. Singh A, Burke TG, Calvert JM, Georger JH, Herendeen B, Price RR, Schoen PE, Yager P (1988) Chem Phys Lipid 47:135
34. Stevens F, Dyer DJ, Walba DM (1996) Angew Chem Int Ed Engl 35:900
35. Prins LJ, Huskens J, de Jong F, Timmerman P, Reinhoudt DN (1999) Nature 398:498
36. Claussen RC, Rabatic BR, Stupp SI (2003) J Am Chem Soc 125:12680
37. Chung DS, Benedek GB, Konikoff FM, Donovan JM (1993) Proc Natl Acad Sci USA 90:11341
38. de Loos M (2005) PhD Thesis, University of Groningen
39. Schoonbeek FS, van Esch JH, Hulst R, Kellogg RM, Feringa BL (2000) Chem Eur J 6:2633
40. van Esch J, De Feyter S, Kellogg RM, De Schryver F, Feringa BL (1997) Chem Eur J 3:1238
41. Sakai K, Maekawa Y, Saigo K, Sukegawa M, Murakami H, Nohira H (1992) Bull Chem Soc Jpn 65:1747
42. Sakai K, Yoshida S, Hashimoto Y, Kinbara K, Saigo K, Nohira H (1998) Enantiomer 3:23
43. Sakai K (1999) J Synth Org Chem Jpn 57:458
44. Addadi L, Weinstein S, Gati E, Wiessbuch I, Lahav M (1982) J Am Chem Soc 104:4610
45. Purvis JL (1959) US Patent 2 790 001
46. Fike HL (1960) US Patent 2 937 200
47. Harada K (1965) Nature 205:590
48. Harada K, Tso W (1972) Bull Chem Soc Jpn 45:2859
49. Addadi L, van Mil J, Lahav M (1981) J Am Chem Soc 103:1249
50. Zbaida D, Lahav M, Drauz K, Knaup G, Kottenhahn M (2000) Tetrahedron 56:6645
51. Zbaida D, Lahav M, Drauz K, Knaup G, Kottenhahn M (2001) PCT Pat Appl WO 01/58835
52. Shiraiwa T, Yamauchi M, Yamamoto Y, Kurokawa H (1990) Bull Chem Soc Jpn 63:3296
53. Shiraiwa T, Kubo M, Fukuda K, Kurokawa H (1999) Biosci Biotechnol Biochem 63:2212
54. Weissbuch I, Zbaida D, Addadi L, Leiserowitz L, Lahav M (1987) J Am Chem Soc 109:1869

55. Shiraiwa T, Ikawa A, Sakaguchi K, Kurokawa H (1984) Chem Lett 113
56. Shiraiwa T, Yamauchi M, Yamauchi T, Yamane T, Nagata M, Kurokawa H (1991) Bull Chem Soc Jpn 64:1057
57. Addadi L, Berkovitch-Yellin Z, Weissbuch I, van Mil J, Shimon LJW, Lahav M, Leiserowitz L (1985) Angew Chem Int Ed Engl 24:466
58. Wiessbuch I, Lahav M, Leiserowitz L (2001) Adv Crystal Growth Res 381
59. Yokota M, Doki N, Hatakeyama T, Sasaki S, Kubota N (2004) J Chem Engin Jpn 37:1284
60. Berfeld M, Zbaida D, Leiserowitz L, Lahav M (1999) Adv Mater 11:328
61. Mastai Y, Sedlák M, Cölfen H, Antonietti M (2002) Chem Eur J 8:2429
62. Garcia C, Collet A (1992) Tetrahedron: Asymmetry 3:361
63. Kojo S, Uchino H, Yoshimura M, Tanaka K (2004) Chem Commun 2146
64. Kashchiev D, van Rosmalen GM (2003) Crystal Res Technol 38:555
65. Kondipudi DK, Crook KE (2005) Cryst Growth Des 5:2173
66. Kubota N, Yakoto M, Mullin JW (1997) J Crystal Growth 182:87
67. Kubota N, Mullin JW (1995) J Crystal Growth 152:203
68. Randolph AD, Larson MD (1988) In: Theory of Particulate Processes. 2nd edn. Academic Press, San Diego
69. ten Hoeve W, Wynberg H (1985) J Org Chem 50:4508
70. Vries T, Wynberg H, van Echten E, Koek J, ten Hoeve W, Kellogg RM, Broxterman QB, Minnaard A, Kaptein B, van der Sluis S, Hulshof L, Kooistra J (1998) Angew Chem Int Ed Engl 37:2349
71. Fogassy E, Lopata A, Faigl F, Darvas F, Ács M, Toke L (1980) Tetrahedron Lett 21:647
72. Kellogg RM, Nieuwenhuijzen JW, Pouwer K, Vries TR, Broxterman QB, Grimbergen RFP, Kaptein B, La Crois RM, de Wever E, Zwaagstra K, van der Laan AC (2003) Synthesis 10:2003
73. Kaptein B, Elsenberg H, Grimbergen RFP, Broxterman QB, Hulshof LA, Pouwer KL, Vries TR (2000) Tetrahedron: Asymmetry 11:1343
74. Kaptein B, Vries TR, Nieuwenhuijzen JW, Kellogg RM, Grimbergen RFP, Broxterman QB (2006) In: Ager D (ed) New Developments in Crystallization-Induced Resolution. CRC Taylor & Francis, Boca Raton, Fl, p 97–122
75. Stoll A, Peyer J, Hofmann A (1943) Helv Chim Acta 26:929
76. Den Hollander CW, Leimgruber W, Mohacsi E (1972) US patent 3 682 925 (to Hoffman-La Roche Inc.)
77. Kitagorodsky AI (1984) Mixed Crystals. Springer, Berlin
78. Gervais C, Grimbergen RFP, Markovits I, Ariaans GJA, Kaptein B, Bruggink A, Broxterman QB (2004) J Am Chem Soc 126:655
79. Loh JSC, van Enckevort WJP, Vlieg E (2004) J Crystal Growth 265:604
80. Nieuwenhuijzen JW, Grimbergen RFP, Koopman C, Kellogg RM, Vries TR, Pouwer K, van Echten E, Kaptein B. Hulshof LA, Broxterman QB (2002) Angew Chem Int Ed 41:4281
81. Loh JSC, van Enckevort WJP, Vlieg E, Gervais C, Grimbergen RFP, Kaptein B (2006) Cryst Growth Des 6:861
82. van der Sluis M, Dalmolen J, de Lange B, Kaptein B, Kellogg RM, Broxterman QB (2001) Org Lett 3:3943
83. Dalmolen J, van der Sluis M, Nieuwenhuijzen JW, Meetsma A, de Lange B, Kaptein B, Kellogg RM, Broxterman QB (2004) Eur J Org Chem 1544
84. Dalmolen J, Tiemersma-Wegman TD, Nieuwenhuijzen JW, van der Sluis M, van Echten E, Vries TR, Kaptein B, Broxterman QB, Kellogg RM (2005) Chem Eur J 11:5619

85. Barton DHR, Kirby GW (1962) J Chem Soc 806
86. Shieh W-C, Carlson JA (1994) J Org Chem 59:5463
87. Chaplin DA, Johnson NB, Paul JM, Potter GA (1998) Tetrahedron Lett 39:6777
88. Brock CP, Schweizer WB, Dunitz JD (1991) J Am Chem Soc 113:9811
89. Kondipudi DK, Crook KE (2005) Cryst Growth Des 5:2173
90. Kubota N, Yokota M, Mullin JW (2000) J Cryst Growth 212:480
91. Huang J, Yu L (2006) J Am Chem Soc 128:1873
92. Gavish M, Popovitz-Biro R, Lahav M, Leiserowitz L (1990) Science 250:973
93. Popovitz-Biro R, Wang JL, Majewski J, Shavit E, Leiserowitz L, Lahav M (1994) J Am Chem Soc 116:1179
94. Kondepudi DK, Kaufman RJ, Singh N (1990) Science 250:975
95. McBride JM, Carter RL (1991) Angew Chem Int Ed Engl 30:293
96. Karamertzanis PG, Price SL (2005) J Phys Chem B 109:17134
97. Custelcean R, Ward MD (2005) Cryst Growth Des 5:2277
98. Kobayashi Y, Morisawa F, Saigo K (2006) J Org Chem 71:606
99. Kinbara K (2005) Synlett 5:732

…

New Resolution Technologies Controlled by Chiral Discrimination Mechanisms

Kenichi Sakai[1,2] (✉) · Rumiko Sakurai[1] · Hiroyuki Nohira[3]

[1]R&D Division, Yamakawa Chemical Industry Co., Ltd., Kitaibaraki, 319-1541 Ibaraki, Japan
kenichi_sakai@tfc.toray.co.jp

[2]Present address:
Specialty Chemicals Research Laboratory, Toray Fine Chemicals Co., Ltd., Minato-ku, 455-8502 Nagoya, Japan

[3]Department of Applied Chemistry, Faculty of Engineering, Saitama University, Shimo-Okubo, Sakura, 338-8570 Saitama, Japan

1	Introduction	201
2	Chiral Purity Improvement by Crystal Habit Modification with a Tailored Inhibitor	201
2.1	Introduction	201
2.2	Causes of Lower Chiral Purity of the Diastereomeric Salt	202
2.3	Effective Chiral Additive	203
2.4	Crystal Habit Modification Mechanism	205
2.4.1	Determination of Crystal Growth and Inhibition Directions	205
2.4.2	Crystal Structure of (R)-PEA:(R)-MA Salt	206
2.4.3	Crystal Habit Modification Mechanism	207
2.5	Application of Crystal Habit Modification to Industrial Resolution Process	208
3	A New Approach for Finding a Suitable Resolving Agent: Space Filler Concept	208
3.1	Introduction	208
3.2	Molecular Size and Resolution Results	209
3.3	Space Filler Concept	212
3.4	Resolution of MMT	213
3.5	Space Filler Concept for Finding Optimum Resolving Agent	215
4	Chirality Control by Dielectrically Controlled Resolution with a Single Resolving Agent	216
4.1	Introduction	216
4.2	DCR Phenomenon	217
4.2.1	Resolution of (RS)-ACL with (S)-TPA	217
4.2.2	Resolution of (RS)-PTE with (S)-MA	219
4.2.3	Resolution of (RS)-MPRD with (R,R)-TA	224
4.2.4	Resolution of (RS)-CHEA with (S)-MA	226

4.3	Strategic Resolution Process Utilizing the DCR Phenomenon	227
4.3.1	ACL–TPA System	227
4.3.2	PTE–MA Resolution System	228
5	Conclusion and Prospects	228
References		229

Abstract It is well known that optical resolution via crystallization is still a useful and practical method for obtaining enantiomerically pure compounds for both laboratory experiment and industrial production, although it is a classical technique discovered nearly 160 years ago. In particular, diastereomeric salt formation using a resolving agent (diastereomer method) has been well applied in various fields such as the pharmaceutical, agrochemical, and liquid crystal industries. Despite these affluent reported and patented examples, unfortunately no concrete theory to determine an optimum resolution condition has been devised, and only empirical procedures seem to provide a unique path in process development. In this chapter, three novel approaches for optical resolution via diastereomeric salt formation are presented: (1) chiral purity improvement by crystal habit modification with a tailored chiral additive; (2) a new approach for finding a suitable resolving agent based on the space filler concept; and (3) chirality control by dielectrically controlled resolution.

Keywords Optical resolution · Crystal habit modification · Space filler · Dielectrically controlled resolution (DCR) · Diastereomeric salt formation

Abbreviations

ACL	α-Amino-ε-caprolactam
bisPEA	Bis(1-phenylethyl)amine
CHEA	1-Cyclohexylethylamine
DCR	Dielectrically controlled resolution
de	Diastereomeric excess
DMT	3-(Dimethylamino)-1-(2-thienyl)propan-1-ol
E	Resolution efficiency
ee	Enantiomeric excess
ε	Solvent dielectric constant
MA	Mandelic acid
ML	Molecular length
MMT	3-(Methylamino)-1-(2-thienyl)propan-1-ol
MPRD	2-Methylpyrrolidine
MTBE	*tert*-Butyl methyl ether
PEA	1-Phenylethylamine
PTE	1-Phenyl-2-(4-methylphenyl)ethylamine
TA	Tartaric acid
TPA	*N*-Tosyl-(*S*)-phenylalanine

1
Introduction

To date, many new and attractive techniques for the production of enantiopure compounds have been devised. Among these known chiral technologies, optical resolution via diastereomeric salt formation is still a useful key technology to obtain enantiopure or diastereopure compounds in industrial-scale production; nevertheless, it is old-fashioned technology discovered nearly 160 years ago by Pasteur [1, 2]. In fact, at the present time, it has been estimated that more than half of the chiral drugs in the current pharmaceutical market are produced by the diastereomeric salt formation method using enantiopure resolving agent because of ease of operation and wide applicability [3]. Currently, chiral key intermediates for important chiral drugs such as indinavir (antiviral) [4–8], sertraline (antidepressant) [9–14], orlistat (antiobsessional) [15–17], and duloxetine (antidepressant) [18–21] are widely known to be efficiently produced by this method on an industrial scale. It is believed at present that several tens of thousands of examples have been reported so far.

Upon resolution process development, finding an optimum resolving agent is the first concern for the chemist. Although various attempts have been made by numerous researchers, a useful concrete theory to find a satisfactory resolving agent has not yet been proposed [22, 23]. Therefore, it is unavoidable to carry out "trial and error" laboratory experiments.

We have been attempting to establish a simple and convenient method to find optimum optical resolution conditions including resolving agent, and we found three new technologies useful for the industrial-scale production of enantiopure compounds. This chapter deals with our current research results focused on control of the chiral molecular recognition mechanism.

2
Chiral Purity Improvement by Crystal Habit Modification with a Tailored Inhibitor

2.1
Introduction

In general, the first purpose of research on resolution in the laboratory is to find an optimum resolving agent for the target substrate and its combination with an optimum solvent. Thus, it is rarely concerned with the less-soluble diastereomeric salt obtained as a product of the resolution reaction between acidic and basic molecules. Practically, however, when resolution conditions optimized in the laboratory are applied at an industrial scale, unexpected scale-up issues are often observed: the solid–liquid separability of the salt from the mother liquor is a typical example arising from the salt crystal shape. Unsuccessful separation remarkably affects the chiral purity of the salt as a result of contamination with residual mother liquor containing undesired enantiomer. During process development of the resolution of (RS)-1-phenylethylamine (PEA) with enantiopure mandelic acid (MA), we encountered such a low salt separability that gave unexpectedly lower chiral purity compared with that of laboratory data. This scale-up issue has been solved by controlling the crystal habit modification of the salt to give an easily separable shape with a chiral additive derived from its resolution system without any deterioration in industrial operation.

2.2
Causes of Lower Chiral Purity of the Diastereomeric Salt

Solid–liquid separation is the most influential step for reproducing the chiral purity of the salt in optical resolution and for operational efficiency on a large scale, such as industrial production. In the case of the salt being thin plates, in general, solid–liquid separation by centrifuge becomes very difficult, because the crystals become arranged perpendicularly to the centrifugal force, resulting in contamination with a considerable amount of the mother liquor (Fig. 1). The optical purity of the obtained diastereomers therefore becomes lower than that expected from the actual diastereomeric composite. In such a case, it is desired to change the shape of the crystals into a more easily separable one by some habit modification.

Lahav et al. have extensively studied the habit modification of amino acids by the coexistence of a structurally similar additive [24–26], and applied that habit modification to the determination of their absolute configurations [27]. They proposed a two-step mechanism involving "binding" and "inhibition" for this habit modification [26].

It is widely known that PEA can be resolved with enantiopure MA via diastereomeric salt formation in industrial-scale production [28–31]. How-

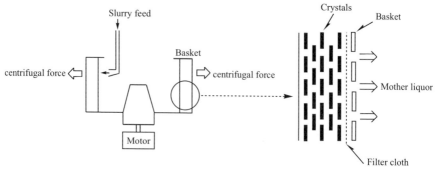

Fig. 1 Crystal arrangement in a centrifuge [34]

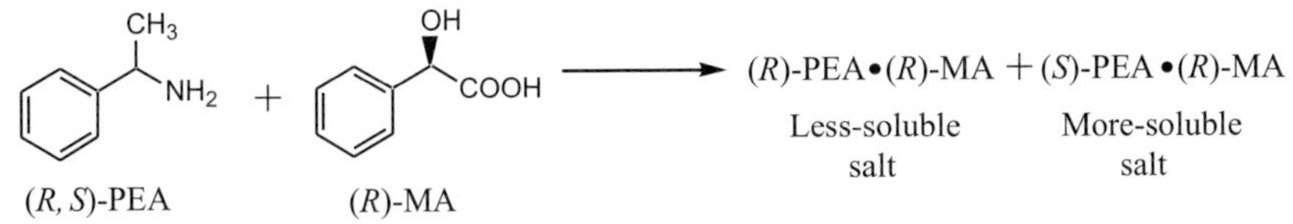

CH₃–CH(NH₂)–Ph + Ph–CH(OH)–COOH ⟶ (R)-PEA•(R)-MA + (S)-PEA•(R)-MA

(R,S)-PEA   (R)-MA   Less-soluble   More-soluble
                                         salt   salt

Scheme 1 Resolution of (RS)-PEA with (R)-MA

ever, the optical purity of the PEA obtained was not stable (97–99% ee), whereas more than 99% ee has been steadily obtained in the laboratory (Scheme 1).

In order to clarify this unexpected fact, the manufacturing conditions were carefully reviewed. As a result, it was found that the crystal shape of the salt was variant (thin long hexagonal and thick hexagonal), and the thin long crystal shape caused difficulties in separating mother liquor containing undesired enantiomer (Fig. 2). Moreover, it was found that long thin hexagonal plates were crystallized when fresh (newly produced) racemic PEA was used, whereas thick hexagonal crystals were obtained when racemized PEA was mainly used. This observation indicates that some impurities in the racemized PEA affected the crystal shape of the salt.

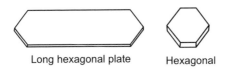

Long hexagonal plate   Hexagonal

Fig. 2 Crystal shapes of less-soluble salt composed of PEA and MA [33]

2.3
Effective Chiral Additive

In order to find a key substance (effective impurity), the resolution process was investigated. As a result, Schiff base and *sec*-amine were found as impurities derived from the racemization step using alkali metal catalyst (Fig. 3). These impurities were chromatographically isolated and they were added individually to the resolution reaction. It was found that only *sec*-amine (bis-PEA) affects crystal habit modification to give thick hexagonal salt crystals. However, *sec*-amine has three stereoisomers, (R,R), (RS), and (S,S), based on two chiral centers. These stereoisomers were individually synthesized and added to the resolution reaction. The test results are summarized in Table 1. It was found that the stereochemistry of bis-PEA dramatically affected the morphological change of the less-soluble salt. Moreover, the presence of a coexisting salt was also influential for realizing morphological change of the salt crystals.

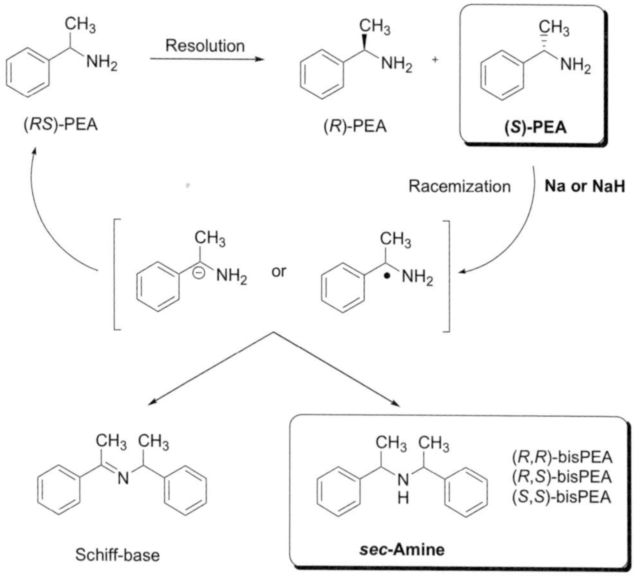

Fig. 3 Resolution process of PEA with enantiopure MA [34]

Table 1 Effective additive for morphological change of (R)-PEA:(R)-MA salt [33]

Entry	Additive	Concentration (mol%) of less-soluble salt	Coexisting salt	Morphological change[a]
1	(R,R)-bisPEA·HCl	0.007	PEA·HCl	A
2	(RS)-bisPEA·HCl	0.29	PEA·HCl	A
3	(S,S)-bisPEA·HCl	1.00	PEA·HCl	U
4	(R,R)-bisPEA·HCl	1.00	None	U
5	(RS)-bisPEA·HCl	1.00	None	U
6	(S,S)-bisPEA·HCl	1.00	None	U
7	(R,R)-bisPEA·HCl	0.05	Benzylamine·HCl	A
8	(R,R)-bisPEA·HCl	0.05	Ethylamine·HCl	A
9	(R,R)-bisPEA·HCl	0.05	NH$_4$Cl	A
10	(R,R)-bisPEA·HCl	0.05	NaCl	A
11	(R,R)-bisPEA·HCl	0.05	AcONH$_4$	A
12	(R)-N-BnPEA·HCl	2.41	PEA·HCl	U
13	Diethylamine·HCl	0.50	PEA·HCl	U

[a] A and U mean the formation of affected hexagonal crystals and unaffected long hexagonal crystals, respectively

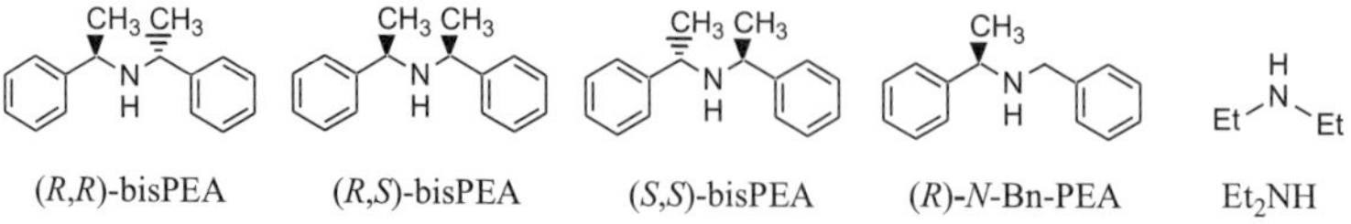

As shown in Table 1, it was found that *sec*-amine (bis(1-phenylethyl)amines; bisPEA), having at least one of the same absolute configurations of PEA in the less-soluble diastereomeric salt, was effective in producing thick crystals. If the target less-soluble salt is (R)-PEA:(R)-MA, the effective concentrations of (R,R)- and (RS)-bisPEA are 0.007 and 0.29 mol %, respectively, to the racemic substrate (RS)-PEA existing in the resolution reaction. None of the other compounds, such as (S,S)-bisPEA, (R)-N-benzyl-1-phenylethylamine, and diethylamine, was effective at all. These results revealed that crystal habit modification with a chiral additive is extremely structure specific and stereospecific to the target substrate.

2.4
Crystal Habit Modification Mechanism

2.4.1
Determination of Crystal Growth and Inhibition Directions

In order to determine the directions of crystal growth and inhibition, the angles of the affected and unaffected crystals were measured microscopically. Model shapes are shown in Fig. 4. The obtuse angle (α) and the acute angle (β) in both affected and unaffected crystals were measured by a protractor using 20 crystals shown in micrographs. It was found that the angles α and β observed in both crystals were perfectly identical ($\alpha = 128°$ and $\beta = 103°$). This means that the longest axis of the unaffected crystal was shortened by inhibition. In other words, the direction of crystal growth inhibition by the additive was determined to be the longest axis of the unaffected crystal. These crystal angles were demonstrated to be identical to the angles calculated by crystal parameters obtained in the X-ray crystallographic analysis indicated in the following Sect. 2.4.2. That is, this inhibition direction by the additive was proved to be along the crystallographic b-axis.

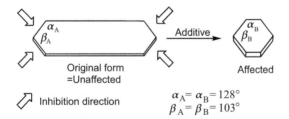

Fig. 4 Schematic crystal morphology of the less-soluble diastereomeric salt in the resolution of (RS)-PEA with enantiopure MA [34]

2.4.2
Crystal Structure of (R)-PEA:(R)-MA Salt

The crystal structure of the less-soluble diastereomeric salt (R)-PEA:(R)-MA obtained in the resolution of (RS)-PEA with (R)-MA from the resolution solvent (water) was determined by X-ray crystallographic analysis [32]. The crystal structure is indicated in Fig. 5.

As shown in Fig. 5, hydrogen-bonding networks were observed only on the plane composed of crystallographic b–c-axes. No hydrogen-bonding connection was observed along the a-axis: this means that the thickness of the crystal is along the a-axis. Molecules of PEA and MA are individually aligned along the b-axis and make single-molecule layers by using hydrogen bonds. Molecules of MA bind to each other by O – H ··· O hydrogen bonds to make rigid layers along the b-axis. The molecule PEA seems to be sandwiched by N – H ··· O and O – H ··· N hydrogen bonds between two layers of MA from the crystallographic + c and – c sides, making a 2_1 column. Accordingly, it is considered that the PEA molecule is unequivocally recognized by a cavity made of these MA layers. Since high chiral purity (R)-PEA (> 99% ee) is exclusively recognized with (R)-MA during resolution, the cavity composed of MA layers must play a major role in recognizing (R)-PEA.

A crystal lattice is shown in Fig. 5. Crystal angles were calculated from the crystal data obtained. The obtuse angle (α) and the acute angle (β)

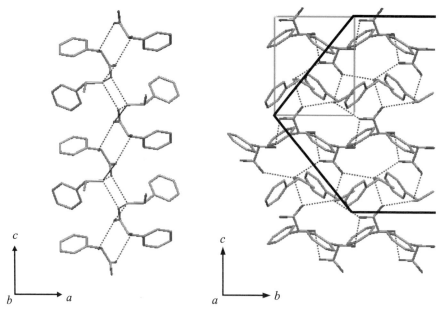

Fig. 5 Crystal shape and structure of (R)-PEA:(R)-MA salt viewed along the a- and b-axes [33, 34]

were $\alpha = 129°$ and $\beta = 101°$, respectively, and they were perfectly identical to those observed by the direct observation of the salt crystals with a protractor (Sect. 2.4.1). These results clearly suggest that the crystal growth direction is along the b-axis. In other words, it is certain that the cavity composed of MA layers along the b-axis recognizes oncoming (R)-PEA along the b-axis. Attached (R)-PEA makes two hydrogen bonds with two molecules of (R)-MA located on the $+c$ and $-c$-axes. Primary amine (R)-PEA attached to the crystal surface can bind an oncoming MA molecule with its residual hydrogen bonding hand. The crystal grows this way.

2.4.3
Crystal Habit Modification Mechanism

The less-soluble diastereomeric salt, (R)-PEA:(R)-MA, is obviously inhibited by a chiral additive along the b-axis which is the fastest growth direction. It is also clear that crystal growth inhibitors are the secondary amines, having at least one of the same absolute configurations of PEA composed of the less-soluble salt. On the basis of these facts, a crystal habit modification mechanism by chiral additive was proposed as shown in Fig. 6.

sec-Amine $((R,R)$-bisPEA$)$ having the same absolute configuration is recognized by the chiral cavity composed of (R)-MA layers as if it is (R)-PEA. Since the *sec*-amine has only two hands for making hydrogen bonds, it has not sufficient ability to bind the oncoming MA molecule for the next layer along the b-axis. Therefore crystal growth along the b-axis is inhibited.

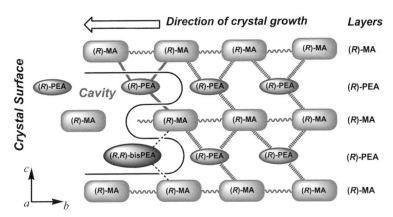

Fig. 6 Molecular recognition of (R)-PEA and (R,R)-bisPEA by (R)-MA in the less-soluble diastereomeric salt [34]

2.5
Application of Crystal Habit Modification to Industrial Resolution Process

Chiral purity improvement of the less-soluble diastereomeric salt was initiated from observation of the facts in plant production. For the purpose, adding a recrystallization step is one of the solutions without doubt. However, an increase in the number of steps usually causes a decrease of efficiency in industrial-scale production. Therefore, an improvement of crystal shape so that it can be easily separated from the mother liquor is a better solution for this issue. On the other hand, racemization of the undesired substrate is one of the essential techniques for realizing the economical process. Racemized substrate is usually recycled to the next batch as a part of the raw material. In the racemization step, it is better to minimize decomposition and/or side reactions. Racemization of PEA by using Na, NaH, $NaNH_2$, or Al_2O_3 – NaOH – K is widely known. However, there was no information on by-products in the racemazation reaction. NaH (60% dispersion) was selected as a racemization catalyst and the reaction conditions (temperature and time) were optimized to obtain enough bisPEA to lead to crystal habit modification (bisPEA content 0.1–3.0%) without an excessive decomposition reaction. Racemized PEA containing a sufficient amount of bisPEA is recycled to the next batch as a part of the raw material (RS)-PEA. Since this new methodology has been applied in the actual production process, the optical purity of enantiopure PEA has been well stabilized. Detailed crystal habit modification mechanisms of the phenomenon [33, 34] and its application to other compounds [35] are reported in the literature.

3
A New Approach for Finding a Suitable Resolving Agent: Space Filler Concept

3.1
Introduction

In general, the first research target for optical resolution is without doubt to find an optimum resolving agent, and various researchers have been trying to establish a quick methodology to find it. For example, Borghese et al. have proposed an efficient, fast screening methodology to find a resolving agent on the basis of the eutectic composition and its solubility of the salt [36]. Dyer et al. have also proposed a new methodology based on thermal analyses of the diastereomeric salts [37]. However, these methodologies need a number of experiments to collect basic physicochemical data, prior to applying them to the target resolution. On the other hand, Kellogg et al. offered a new quick methodology named "Dutch resolution" using a structure-related family of resolving agents in one resolution experiment [38–40]. This method gives a remarkably rapid result in a survey of the suitable resolving agents. However, there is an unavoidable drawback that the target stereoisomer is not always crystallized with a single resolving agent; a multicomponent salt is favorably crystallized in some cases. On the other hand, Saigo et al. have tried to demonstrate the molecular recognition between racemic substrate and resolving agent during salt crystallization by three functions: relative molecular lengths, CH–π interactions, and van der Waals interactions between resolving agent and racemic substrate [41–45]. They successfully designed tailor-made resolving agents for the given target substrates, although these specially tailored resolving agents are usually very costly and had to be given up in industrial-scale production for economic reasons. Regrettably, the number of economical and readily available resolving agents is still limited in the present market. Therefore, we should consider how the readily available resolving agents in hand can be aptly used in various resolution targets.

As described above, various methodologies have been proposed, but no concrete theory to find the optimum resolving agent has yet been proposed. We have investigated a molecular structure relationship between resolving agents and racemic substrates for both successful and unsuccessful resolution results, and devised one facile procedure for finding a suitable resolving agent for the target substrate [46–48]. On the basis of this finding, we contrived a new methodology, the "space filler concept", for realizing the most essential factor for obtaining successful resolution, the closest packing in a crystal lattice of the less-soluble salt. In this section, the original concept of the space filler and its application to one of the pharmaceutical intermediates are presented.

3.2
Molecular Size and Resolution Results

In optical resolution via diastereomeric salt formation, amine and acid molecules recognize each other by various interactions based on their molecular structures and functional groups. During salt crystallization, these molecules arrange in order and pack as tightly as possible by using hydrogen bonds, CH–π interactions, and van der Waals interactions to realize the closest packing in the crystal lattice. In the salt crystal, functional groups of amine and acid molecules are facing each other to form a hydrophilic layer along the 2_1 column constructed with hydrogen-bonding networks, as observed in the literature [41–48]. In order to learn the effect of relative molecular size between racemic substrate and resolving agent, we have investigated a correlation between relative molecular length and resolvability in a series of resolution systems between 1-aryl-1-alkylamines and 2-hydroxycarboxylic acids (Fig. 7).

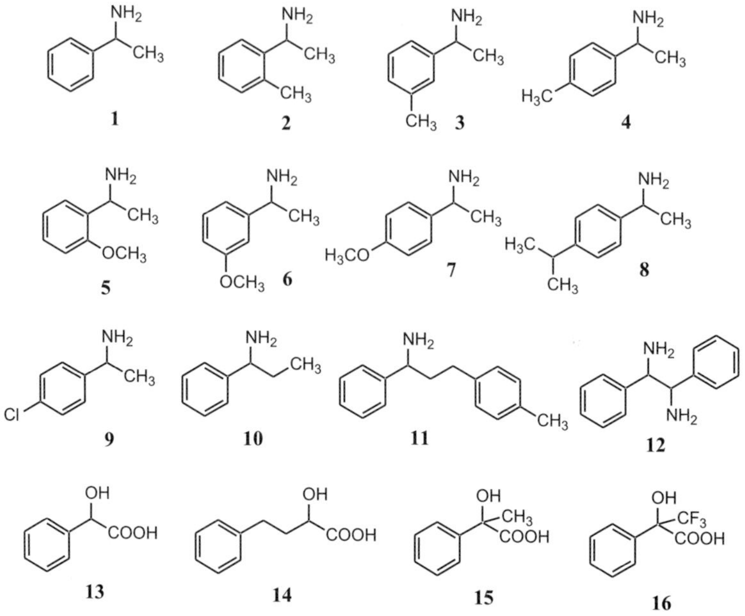

Fig. 7 The 1-aryl-1-alkylamines and 2-hydroxycarboxylic acids studied [48]

Molecular length (ML) is determined by counting the number of heavy atoms along the bond connection from the α-atom to the far side of the molecule (Fig. 8). For example, the ML of mandelic acid (Fig. 8b) is determined to be "5". A molecule with a p-substituent is continuously counted from the end of the phenyl group; in the case of p-methoxy-1-phenylethylamine, the ML is determined to be "7". Some resolution results were evaluated based on the difference of molecular lengths between resolving agent and racemic substrate. The evaluation results are summarized in Table 2.

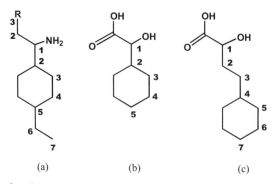

Fig. 8 Counting of molecular lengths [46, 48]

Table 2 Relation between relative molecular length and resolvability [48]

Entry	Racemate	Resolving agent	Resolvability Yield[a] %	de[b] %	Level[c]	Difference of molecular length[d]	Symmetry of p-substituent[e]	Refs.
1	1	13	76	99	A	0	Y	[49]
2	2	13	71	100	A	0	Y	[50]
3	5	13	69	81	B	0	Y	[50]
4	3	13	94	12	F	0	Y	[50]
5	6	13	70	99	A	0	Y	[51]
6	4	13	88	4	F	1	Y	[50]
7	7	13	85	0	F	+2	N	[48]
8	8	13	80	–	B	+2	Y	[56]
9	9	13	34	3	F	+1	Y	[48]
10	10	13	81	92	A	0	Y	[48]
11	11	13	74	97	A	0	Y	[52]
12	12	13	82	100	A	0	Y	[53]
13	4	14	77	85	B	–1	Y	[54, 55]
14	7	14	98	54	C	0	N	[48]
15	13	1	80	75	B	0	Y	[48]
16	14	1	108	–	F	–2	Y	[48]
17	14	4	58	81	B	–1	Y	[57]
18	14	10	Oil		F	–2	Y	[48]
19	15	1	92	–	B	0	Y	[58]
20	16	1	44	–	B	0	Y	[59]

[a] Calculated based on a half amount of racemate
[b] Based on enantiomeric purity of the target amine or acid in the salt
[c] Level of resolution: A ≥ 90% ee, 70≤ B < 90% ee, 50 ≤ C < 70% ee, F < 50% ee or impossible to resolve
[d] Difference of molecular length = Amine – Acid molecule
[e] p-Substituent of amine molecule: symmetrical = methyl, chloro, *iso*-propyl; unsymmetrical = methoxy

In Table 2, the resolvability (degree of resolution of the target enantiomer) is roughly classified into four levels based on the chiral purity of the target substrate in the deposited salt as follows: A ≥ 90% ee, 70 ≤ B < 90% ee, 50 ≤ C < 70% ee, F < 50% ee or not resolved. We have mainly focused on the chiral purity of the salt crystal deposited as the first crop. The difference in molecular lengths is indicated as a subtracted number of "Acid – Amine". If the acid molecule is one heavy atom longer or shorter than the amine, the difference of molecular lengths is determined to be +1 or –1, respectively. The symmetry of the p-substituent of the aryl group is indicated in terms of the two-dimensional symmetry written on paper, and is not three-dimensionally concerned. From the data shown in Table 1, the following six rules were collected. Rules 1–5 are applied to the resolution of a racemic amine with enantiopure acid; Rule 6 is applied to the resolution of a racemic acid with enantiopure amine:

1. If the molecular lengths of acid and amine are equal, optical resolution is expected to be successful (entries 1–5).
2. The *m*-substituent of the aryl group will not affect resolvability up to the methoxy group (entry 2).
3. When the molecular length of the amine is one heavy atom longer than that of the acid, moderate efficiency may be expected (entry 6).
4. If the *p*-substituent is two-dimensionally symmetrical (such as isopropyl), moderate efficiency may be expected (entry 7); nevertheless there is considerable difference of molecular lengths (entries 8–11).
5. If the four rules shown above are inconsistent in the intended resolution system, resolution will be unsuccessful or expected to give lower a efficiency.
6. In the resolution of racemic acid with enantiopure amine, the chiral purity of the acid in the salt is expected to be lower than that of the opposite resolution case (resolution of racemic amine with enantiopure acid; entries 12–17).

The chiral purity of molecule **13** in the salt obtained in the resolution of (*RS*)-**13** with enantiopure **1** (entry 12) is lower than that in the opposite case, the resolution of (*RS*)-**1** with enantiopure **13** (entry 1). This finding can be explained on the basis of the hydrogen-bond ability of the OH group of acid **13**. It is known that racemic **13** and enantiopure **13** have different crystal structures [60, 61]; molecules of racemic **13** arrange much more tightly among (*R*)- and (*S*)-stereoisomers to form a rigid three-dimensional hydrogen-bond network, whereas molecules of enantiopure **13** can only arrange on the two-dimensional sheetlike hydrogen-bond network, and tend to form a chiral layer. Chiral recognition by chiral amine **1** is interrupted by an interaction based on $O-H\cdots O$ hydrogen bonding between **13** molecules, since the $O-H\cdots O$ hydrogen bond is much stronger than the $N-H\cdots O$ bond. As a result, the chiral purity of **13** becomes lower than that of the opposite resolution system, resolution of (*RS*)-**1** with enantiopure **13**.

Although the resolution examples shown here are limited to clearly demonstrate chiral discrimination between racemic substrate and resolving agent, it is apparently revealed that molecular length is a very important factor for rationally choosing a resolving agent.

3.3
Space Filler Concept

In order to demonstrate a role of relative molecular length between racemic substrate and resolving agent during salt crystallization in optical resolution, typical molecular arrangements of the less-soluble diastereomeric salt crys-

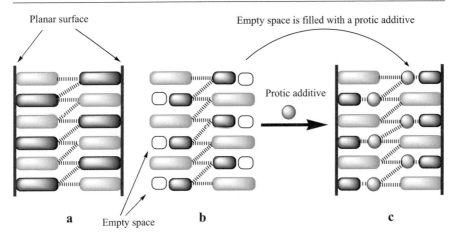

Fig. 9 Typical molecular arrangement in the diastereomeric salt crystal and concept of space filler [70]

tals are schematically depicted in Fig. 9a, on the basis of various crystal structures of the diastereomeric salts in the literature [41–45, 51, 62, 63].

If the molecular lengths of base and acid are the same or closely similar (Fig. 9a), the outer surface of the column becomes smooth and flat (planar surface). In such a case, the salt crystal is usually precipitated with high chiral purity (Fig. 9a), while molecules are packed tightly and the closest packing is realized. On the other hand, if the molecular lengths of base and acid differ from each other (Fig. 9b), the outer surface of the column becomes uneven (nonplanar surface). In such a case, the closest packing will generally not be realized due to empty spaces caused by the difference in molecular length, and the chiral purity of such a salt crystal is usually racemic or lower than that of the other case crystallized with a planar surface, even if it is crystallized. These critical molecular arrangements and molecular packing indicate that the relative length between amine and acid molecules must play a key role in the determination of chiral discrimination during optical resolution. This observation prompted us to devise a new approach for realizing the closest packing by compensating an empty space with a protic additive (Fig. 9c). In other words, the closest packing can be realized if the empty space is effectively filled with an additive of appropriate length, as shown in Fig. 9c. This idea was applied to our challenge of producing a new pharmaceutical intermediate for an antidepressant.

3.4
Resolution of MMT

Duloxetine (LY-248686), (S)-(+)-N-methyl-3-(1-naphthyloxy)-3-(2-thienyl)-propylamine, has been expected to be not only a new potent antidepressant but also a norepinephrine (NE) reuptake inhibitor, a 5-HT (serotonin) reuptake inhibitor, and a new treatment drug for stress urinary incontinence [64]. In order to produce a key enantiopure intermediate for the synthesis of duloxetine, various strategies have been proposed, and optical resolution of the dimethylamine, 3-(dimethylamino)-1-(2-thienyl)propan-1-ol (DMT), with (S)-mandelic acid (MA), a commercially readily available resolving agent, via diastereomeric salt formation was chosen for industrial production [65, 66]. However, because of the critical problems such as low yield and considerable impurities that arose from the final demethylation step, a direct synthesis using the monomethylamine, 3-(methylamino)-1-(2-thienyl)propan-1-ol (MMT), was expected to provide a new intermediate for the production of duloxetine (Fig. 10). However, at that time, resolution of (RS)-MMT had not been fulfilled by anyone, although the molecular structures of DMT and MMT were basically the same based on a sense of the molecular length.

Thus, we at first tried to resolve (RS)-MMT under the identical conditions applied in the resolution of (RS)-DMT by using (S)-MA as a resolving agent (Scheme 2) and methyl tert-butyl ether (MTBE) or MTBE/EtOH

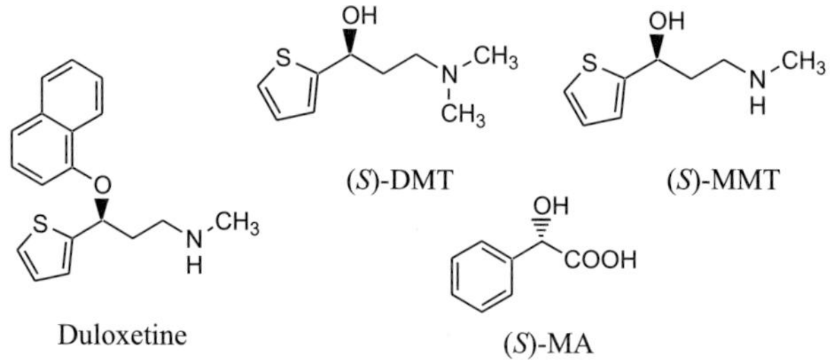

Fig. 10 Duloxetine, its chiral intermediates (DMT and MMT), and resolving agent [68]

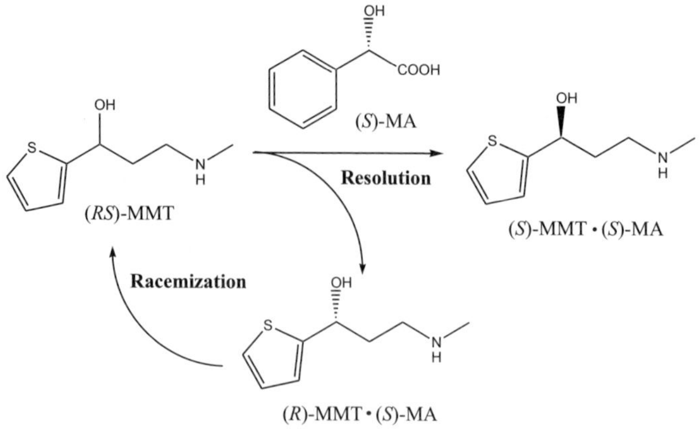

Scheme 2 Resolution of (RS)-MMT with (S)-MA

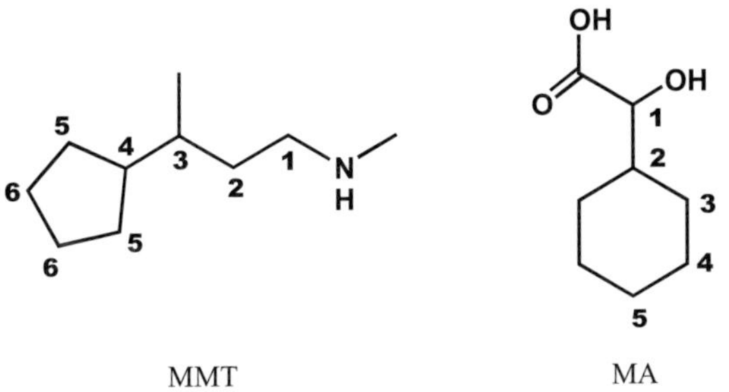

Fig. 11 Molecular lengths based on the space filler concept [46]

as a solvent, which is favorable in the resolution of (RS)-DMT with (S)-MA, but no crystals were obtained at all. Therefore, we decided to apply our new space filler concept to obtain salt crystals by using the same resolving agent, (S)-MA. The molecular lengths of MMT and MA are 6 and 5, respectively (Fig. 11). According to the space filler concept, an appropriate protic molecule to fill a space arising from the difference of molecular lengths is determined to be water molecule having one heavy atom, based on the calculation of MMT(6)–MA(5). Thus, resolution solvents have been optimized using water and water-containing solvents. As a result, we successfully found that n-butanol containing 2 equivalents of water gave the monohydrated salt with the highest resolution efficiency [67–72]. These results suggest that water molecules would play a key role in making the less-soluble salt crystal as a result of the close packing of (S)-MMT, (S)-MA, and water molecules.

3.5
Space Filler Concept for Finding Optimum Resolving Agent

Finding an optimum resolving agent for the given target racemate is the first concern for research on optical resolution via diastereomeric salt formation. Despite various methodological challenges, a rational approach for the prediction of a resolving agent suitable for industrial-scale production has not yet been fully proposed. We investigated a relationship among successful and unsuccessful resolution results and found that the relative molecular lengths of racemic substrate and resolving agent are an important factor for realizing the planar outer surface of the 2_1 column to give successful salt crystals with high chiral purity. Based on the findings, we devised the space filler concept which is targeted to compensate an empty space arising from the difference in molecular lengths (ML) of racemic substrate and resolving agent to give the closest packing of the salt crystals. Namely, if the molecular length of the resolving agent is identical to that of the racemic substance, resolution is expected to be successful. In contrast, if the difference in molecular length between the resolving agent and the racemic substrate is one or more, this difference can be compensated with a protic solvent such as water (ML = 1) or methanol (ML = 2) or other protic solvent such as alcohol (ML > 3). On the basis of this concept, the resolution conditions of (RS)-3-(methylamino)-1-(2-thienyl)propan-1-ol (MMT; ML = 6) with (S)-mandelic acid (MA; ML = 5) were designed and succeeded in obtaining the monohydrated less-soluble diastereomeric salt by filling an empty space (ML = 1) arising from the difference of molecular lengths of MMT and MA, with the water molecule (ML = 1) as a space filler. The present concept is the first rational design procedure for optimizing optical resolution conditions including both resolving agent and solvent. The space filler concept may not be universal for all resolution systems, but it is certain that the number of resolution experiments can be minimized if this concept is considered prior to experiment. Moreover, this methodology may be useful in crystal engineering for tailor-made crystallization of both chiral and/or achiral compounds.

4
Chirality Control by Dielectrically Controlled Resolution with a Single Resolving Agent

4.1
Introduction

The diastereomeric salt formation method using a resolving agent as a chiral selector is one of the most useful methods for obtaining a target stereoisomer from its racemic mixture [22, 23]. Since it is generally assumed that a specific chiral selector acts only on one of the stereoisomers, the opposite stereoisomer must be recognized by the enantiomorph of the chiral selector. It seems that there is no way to change this one-to-one situation from the viewpoint of general resolution chemistry. Recently, however, we have accidentally discovered quite an unusual phenomenon which means we should reconsider the common sense on chiral discrimination. The specific chiral selector not only can recognize both enantiomers individually to be deposited as the less-soluble salt from the different solvent, but also can control the diastereomeric excess (% de) of the less-soluble diastereomeric salt by simply adjusting the dielectric constant (ε) of the solvent used. This phenomenon was named dielectrically controlled resolution (DCR). In this chapter, DCR phenomena observed in four cases of optical resolution of racemic amines with enantiopure acids as resolving agents are presented (Fig. 12). A strategic resolution process based on the DCR phenomenon is also described. Detailed molecular mechanisms of the DCR phenomena observed in the four resolution systems are described in the chapter by Sakai et al. (in this volume) [73].

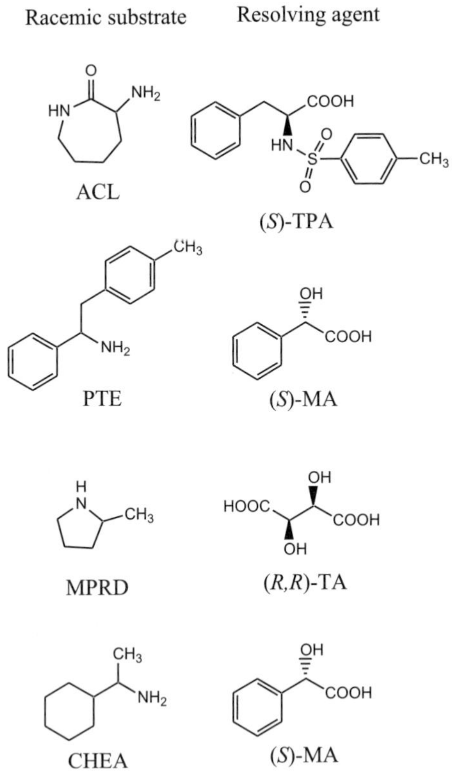

Fig. 12 Resolution system showing DCR phenomenon

4.2
DCR Phenomenon

Resolution experiments were performed under overhead stirring without seeding in order to avoid the occurrence of preferential crystallization with specific seed crystals. An equimolar resolving agent was used to the target racemate and the solvent weight was determined by the solubility of the solid substances at 50 °C in each resolution experiment. Components, chiral purities, and absolute configurations of the salts obtained were determined by elemental, chiral HPLC, and X-ray crystal structure analyses, respectively.

4.2.1
Resolution of (*RS*)-ACL with (*S*)-TPA

The DCR phenomenon was discovered during the resolution process development of (*RS*)-α-amino-ε-caprolactam (ACL) with *N*-tosyl-(*S*)-phenylalanine

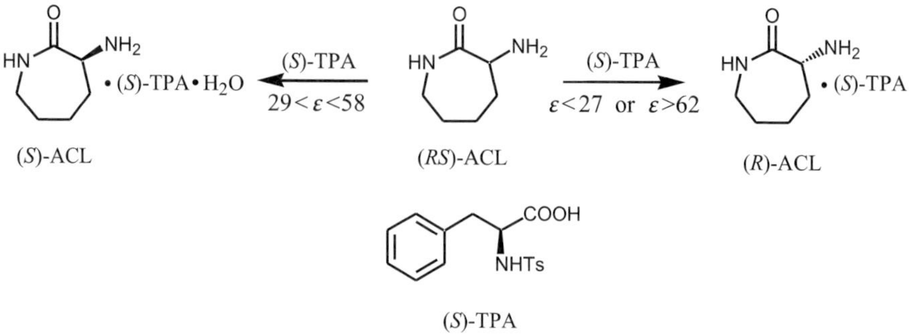

Scheme 3 Resolution of (RS)-ACL with (S)-TPA

(TPA) as a resolving agent (Scheme 3). Surprisingly, it was found that the chirality of the target substrate deposited as the less-soluble diastereomeric salt varied depending on the solvents, as shown in Table 3. These data seemed to indicate that only MeOH was effective for obtaining the (S)-stereoisomer, whereas the (R)-stereoisomer was obtained from water, EtOH, and 2-PrOH, although the chiral purities were lower.

In order to rationalize this unusual phenomenon, various physicochemical properties of solvents were examined. As a result, it was found that a change in the chirality and chiral purity of the salt obtained was aligned with the dielectric constant (ε) of the solvent used. Resolution experiments were ex-

Table 3 Resolution of (RS)-ACL with (S)-TPA in alcohol–water solvents[a] [77]

Entry	Solvent[b]	Solvent weight (vs (RS)-ACL) (w/w)	Yield[c] (%)	Diastereomeric excess (% de)	Resolution efficiency E (%)	Absolute configuration
1	MeOH	10	30	93	56	S
2	60% MeOH	11	9	95	17	S
3	45% MeOH	8	48	3	3	S
4	35% MeOH	6	16	13	4	R
5	10% MeOH	19	37	35	26	R
6	EtOH	32	68	7	10	R
7	90% EtOH	15	60	10	12	S
8	81% EtOH	12	24	99	48	S
9	2-PrOH	50	64	32	41	R
10	89% 2-PrOH	11	59	29	34	R
11	Water	18	30	28	17	R

[a] Resolving agent (S)-TPA/(RS)-ACL = 1.0 (molar ratio)
[b] Mixed solvents are indicated by alcohol contents in weight %
[c] Yield is calculated based on (RS)-ACL

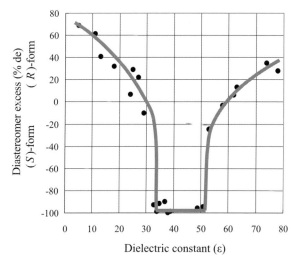

Fig. 13 Resolution of (*RS*)-ACL with (*S*)-TPA only by changing solvent: relation between diastereomeric excess (de %) and dielectric constant of solvent [78]

tended to cover the full range of ε values (5–78). The experimental results are summarized in Table 4. The relation between ε and chirality/chiral purity of the salt is depicted in Fig. 13 [74–78].

As can be seen from Fig. 13, the (*S*)-salt mainly containing (*S*)-ACL was deposited as the less-soluble diastereomeric salt from the solvents with a relatively medium range of dielectric constant ($29 < \varepsilon < 58$), such as 45–100% MeOH, 70–90% EtOH, DMSO, DMF, etc. On the other hand, the (*R*)-salt mainly containing (*R*)-ACL was deposited as a less-soluble diastereomeric salt from the solvents with an outer ε range for (*S*)-ACL: $\varepsilon < 27$ such as EtOH, 85–100% 2-PrOH, and EDC, or $\varepsilon > 62$ such as 30% EtOH, 10–35% MeOH, or water. From the X-ray crystallographic analyses of these crystals, it was found that they were (*S*)-ACL:(*S*)-TPA:H_2O and (*R*)-ACL:(*S*)-TPA, respectively. That is, the water molecule has played a key role in changing the molecular recognition system.

4.2.2
Resolution of (*RS*)-PTE with (*S*)-MA

The DCR phenomenon was also observed in the resolution system of (*RS*)-1-phenyl-2-(4-methylphenyl)ethylamine (PTE) as a substrate to be resolved and (*S*)-MA as a resolving agent (Scheme 4). At first, we examined six sorts of alcohols from C1 to C4 (dielectric constant $\varepsilon = 16$–33) and water ($\varepsilon = 78$) as the resolution solvent, while an equimolar resolving agent (*S*)-2 was used to the racemate (*RS*)-PTE. The resolution results are summarized in Table 5 [79].

Table 4 Resolution of (RS)-ACL with (S)-TPA in various solvents [78]

Entry	Solvent	Dielectric constant (ε)	Solvent weight[a] (vs (RS)-ACL) (w/w)	Yield %[b]	de %	Absolute config. R/S	Resolution efficiency (E)[c]
1	Chloroform	5	7	24	69	R	33
2	EDC	11	6	44	61	R	54
3	MIBK	13	45	63	41	R	52
4	2-PrOH	18	50	65	32	R	42
5	EtOH	24	32	70	7	R	10
6	89% 2-PrOH	25	11	60	29	R	35
7	85% 2-PrOH	27	10	55	22	R	24
8	90% EtOH	29	15	63	10	S	13
9	MeOH	33	10	32	93	S	60
10	81% EtOH	34	12	25	99	S	50
11	95% MeOH	35	16	16	92	S	29
12	DMF	37	27	28	90	S	50
13	74% EtOH	38	14	13	100	S	26
14	1,2-Ethanediol	39	43	38	99	S	75
15	DMSO	49	32	20	96	S	38
16	60% MeOH	51	11	10	95	S	19
17	55% MeOH	53	5	42	25	S	21
18	45% MeOH	58	8	49	3	S	3
19	30% EtOH	62	10	45	6	R	5
20	35% MeOH	63	6	16	13	R	4
21	10% MeOH	74	19	38	35	R	27
22	Water	78	18	31	28	R	17

[a] Solvent volume is a minimum volume to obtain a clear solution at 50 °C except for the case using 74% EtOH
[b] Yield is calculated based on (RS)-ACL
[c] Resolution efficiency (E, %) = yield (%) × diastereomeric excess (% de) × 2/100

As can be seen from Table 5, the chirality of the less-soluble diastereomeric salt crystallized was variable depending on the solvent. Namely, the (R)-PTE:(S)-MA salt was crystallized with relatively higher diastereomeric excess (de %) and resolution efficiency (E) from nonbranched alcohols such as 1-butanol, 1-propanol, and ethanol (Table 5, entries 2, 4, and 5) except from methanol, which is the very basic nonbranched alcohol. The (RS)-PTE:(S)-MA salt was crystallized from branched alcohols such as 2-butanol and 2-propanol, and from methanol (Table 5, entries 1, 3, and 6). On the other hand, to our surprise, the (S)-PTE:(S)-MA:H$_2$O salt was crystallized only from water, although the water weight was unpractical for industrial-scale production (Table 5, entry 7). The resolution results obtained from various alcohols obviously reveal that the molecular recognition between PTE and MA

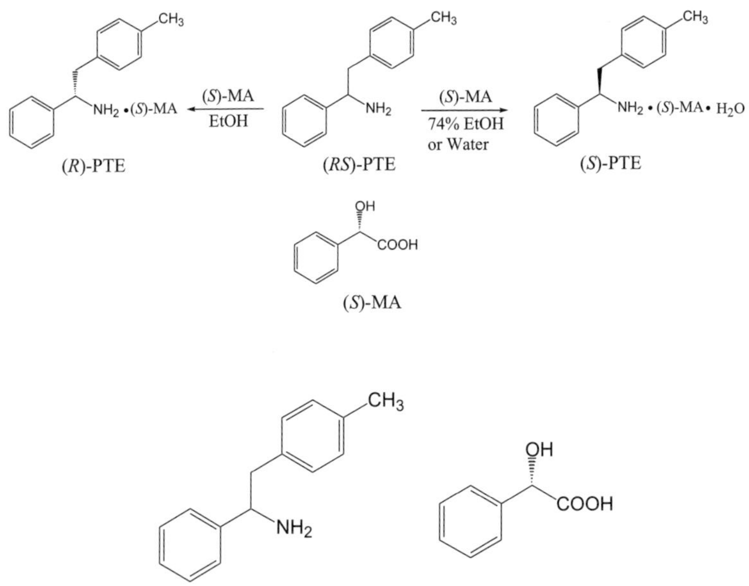

Scheme 4 Resolution of (RS)-PTE with (S)-MA

Table 5 Resolution of (RS)-PTE with (S)-MA in alcohols and water [79]

Entry	Solvent	ε	Solvent weight vs (RS)-PTE[a] (w/w)	de %[b]	R/S	Yield[c]	E[d]
1	2-BuOH	16	47	1	RS	55	4
2	1-BuOH	17	18	92	R	25	46
3	2-PrOH	18	49	1	RS	54	1
4	1-PrOH	22	21	23	R	23	11
5	EtOH	24	12	96	R	16	31
6	MeOH	33	4	0	RS	27	0
7	Water	78	228	98	S	24	47

(S)-MA/(RS)-PTE = 1.0 molar ratio
[a] Solvent weight was determined by the solubility of the solid substances at 50 °C
[b] Diastereomeric excess (% de) = [A − B] ×100/(A + B), where A and B are both diastereomers
[c] Calculated based on (RS)-PTE
[d] Resolution efficiency (E, %) = yield (%) × diastereomeric excess (% de) × 2/100

molecules has been affected by the solvent molecular structures. Although the function of the solvent molecular structure is not yet clear, a crowded solvent association state structured by the branched alcohols may spoil the

Table 6 Resolution of (RS)-PTE with (S)-MA in various alcoholic solvents [79]

Entry	Solvent	Alcohol concentration (%)	ε	Solvent weight vs (RS)-PTE [a] (w/w)	de % [b]	R/S	Yield [c]	E [d]
1	2-BuOH	100	16	47	1	RS	55	2
2		92	22	19	96	S	37	71
3		85	25	16	98	S	34	66
4		75	32	11	96	S	37	71
5		65	38	9	97	S	36	69
6	1-BuOH	100	17	18	92	R	25	46
7		99	18	16	19	R	42	16
8	2-PrOH	100	18	49	1	RS	54	1
9		95	21	20	42	S	45	38
10		82	29	12	95	S	38	72
11		70	36	12	97	S	39	76
12		60	42	13	97	S	37	71
13		45	51	16	65	S	48	62
14		15	69	100	48	S	51	48
15	1-PrOH	100	22	21	23	R	23	11
16		92	26	7	48	S	53	51
17		83	32	6	96	S	34	67
18		75	36	6	97	S	37	71
19		50	50	9	98	S	36	70
20		25	64	28	97	S	40	76
21	EtOH	100	24	12	96	R	16	30
22		95	27	10	28	S	38	21
23		90	29	7	35	S	47	33
24		85	32	8	68	S	41	56
25		74	38	10	99	S	34	67
26		60	46	11	66	S	44	58
27		50	51	13	48	S	59	50
28		30	62	39	36	S	65	36
29		16	69	120	98	S	34	67
30	MeOH	100	33	4	0	RS	27	0
31		86	39	3	18	S	56	20
32		70	47	6	39	S	56	43
33		44	58	34	68	S	44	60
34		20	69	100	95	S	34	65
35	Water	0	78	228	98	S	24	47

(S)-MA/(RS)-PTE = 1.0 molar ratio
[a] Solvent weight was determined by the solubility of the solid substances at 50 °C
[b] Diastereomeric excess (% de) = [A - B] ×100/(A + B), where A and B are both diastereomers
[c] Calculated based on (RS)-PTE
[d] Resolution efficiency (E, %) = yield (%) × diastereomeric excess (% de) × 2/100

chiral discrimination ability of the (S)-MA molecule to PTE molecules. In other words, the solvent association state structured by the nonbranched alcohol with less steric hindrance may enhance the discrimination ability of (S)-MA. The result from methanol may be exceptional because of its specific physicochemical properties compared with those of other alcohols. Moreover, the fact that the opposite chirality (S)-PTE was crystallized only from water indicates that the water molecule plays a distinctive role in the chiral discrimination of (S)-MA to (S)-PTE, as observed in the ACL–TPA resolution system. These facts unequivocally suggest that the chirality and chiral purity of PTE could be controlled by using a mixed solvent of alcohols and water; it can be expected that the chirality and chiral purity of the target substrate PTE in the salt will be inclined to the (S)-enantiomer side according to the increase in water content. In particular, ethanol and its mixture with water are expected to be successful in obtaining both enantiomers of PTE only by controlling the water content, since ethanol and water gave (R)-PTE and (S)-PTE with relatively higher chiral purities, respectively.

Next, we examined various mixed solvents of alcohol and water as the resolution solvent (ε = 16–78). Each alcohol was added to water as long as it gave a homogeneous clear solution at ambient temperature. Concentrations of 1-butanol and 2-butanol were limited and the maximum concentrations were 99 and 65%, respectively, whereas other alcohols were completely miscible with water at any concentration. The resolution results are summarized in Table 6. In order to efficiently display and compare the resolution results obtained from different solvents, the trends of the chirality changes are depicted by the order of the solvent dielectric constant ε in Fig. 14.

As shown in Fig. 14, all alcoholic solvents except for methanol exhibited the same tendency in a range of ε < ca. 40, and their chiral purities were

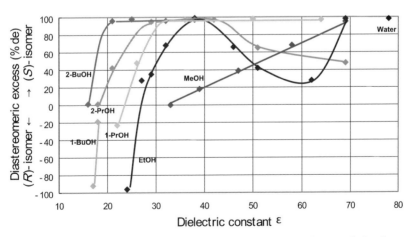

Fig. 14 Resolution of (RS)-PTE with (S)-MA only by changing solvent: relation between diastereomeric excess (de %) and dielectric constant of solvent [79]

rapidly inclined to be (S)-enriched from the (R)-enriched or (RS)-form according to the increase of water content, namely an increase in the solvent ε value. On the other hand, the chiral purities of the salts obtained from aqueous ethanol with $\varepsilon > 40$ were suppressed at 36% de by $\varepsilon = 62$ and retrieved at $\varepsilon = 69$ to afford enantiopure (S)-1 (98% de). Although the reason for this phenomenon is not yet clear, it suggests that the presence of water molecules is not a single factor for chiral discrimination. Namely, it proves that the reaction environment controlled by the solvent dielectric constant plays a very distinctive role in utilizing the water molecule for crystallizing the less-soluble diastereomeric salt (S)-PTE:(S)-MA:H$_2$O during the resolution reaction. From a practical point of view, it is noteworthy that the chirality of the salt has been drastically turned from enantiopure (R)-PTE to enantiopure (S)-MA when water was added to the nonbranched alcohol, ethanol; (R)-PTE (96% de) and (S)-PTE (99% de) were crystallized from ethanol ($\varepsilon = 24$) and 74% ethanol ($\varepsilon = 38$), respectively. On the other hand, when water was added to branched alcohols such as 2-butanol and 2-propanol, the chiral purities of the crystallized salts were improved to give the enantiopure (S)-form from the racemic form at $22 < \varepsilon < 38$ and $29 < \varepsilon < 42$, respectively. These results suggest that we should experiment with changing the reaction environment along the solvent dielectric constant, even if a racemic salt was obtained from pure solvent at the first trial of the resolution experiment. Interestingly, a completely different trend was observed when methanol was used; chiral purity was improved from the (RS)-form to (S)-form linearly according to the increase of water content, and the maximum chiral purity of (S)-PTE (95% de) was obtained at a concentration of 20% ($\varepsilon = 69$, Table 6, entry 34). It is considered that this phenomenon may be related to particular physicochemical properties of methanol.

4.2.3
Resolution of (RS)-MPRD with (R,R)-TA

The DCR phenomenon observed in the resolution system of (RS)-2-methyl-pyrrolidine (MPRD) and (R,R)-tartaric acid (TA) was found in a very sharp range of ε values (Scheme 5). Ethanol–water mixtures were used as solvents. Experimental results are summarized in Table 7. In order to clearly display and compare the resolution results obtained from different ε solvents, the trends of the chirality changes are depicted by the order of the solvent dielectric constant ε in Fig. 15 [80].

As is clearly shown in Fig. 15, the DCR phenomenon was observed in a very narrow range. The (R)-enantiomer was recognized in solvents with $\varepsilon = 24$–25, whereas the (S)-enantiomer was recognized in those with $\varepsilon = 30$–32. The (R)-enantiomer was crystallized as (R)-MPRD:(R,R)-TA and the (S)-enantiomer was crystallized as (S)-MPRD:(R,R)-TA:H$_2$O. Although their chiral purities were not high, this phenomenon suggests that chiral molecu-

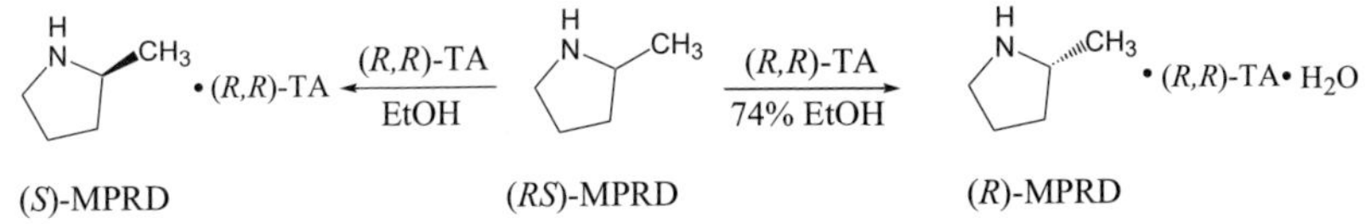

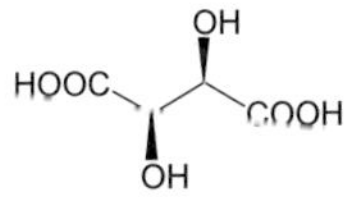

Scheme 5 Resolution of (RS)-MPRD with (R,R)-TA

Table 7 Resolution of (RS)-MPRD with (R,R)-TA from EtOH/H$_2$O solvent [80]

Entry	Solvent	(w/w) vs (RS)-MPRD	Dielectric constant (ε)	de %	R/S	Yield	Resolution efficiency (E)
1	100% EtOH	40	24	7	R	72	11
2	99% EtOH	35	24.4	37	R	64	47
3	97% EtOH	31	25	33	R	62	41
4	95% EtOH	24	26.5	16	R	64	20
5	94% EtOH	21	27	9	R	68	12
6	90% EtOH	17	29	5	R	42	4
7	89% EtOH	16	30	16	S	44	14
8	86% EtOH	12	32	19	S	33	12
9	83% EtOH	10	33	8	S	51	8
10	79% EtOH	8	35	12	S	54	13
11	74% EtOH	4	36	12	S	50	12

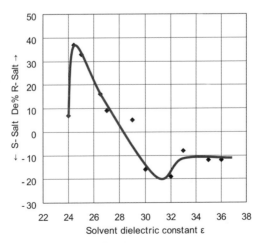

Fig. 15 Resolution of (RS)-MPRD with (S)-TA only by changing solvent: relation between diastereomeric excess (de %) and dielectric constant of solvent

lar recognition between (RS)-MPRD and (R,R)-TA was controlled in a very specific reaction environment.

4.2.4
Resolution of (RS)-CHEA with (S)-MA

During resolution process development of (RS)-1-cyclohexylethylamine (CHEA) with (S)-MA, it was found that (R)-2-methoxy-2-phenylacetic acid and (S)-2-methyl-2-phenylacetic acid served as new resolving agents for racemic CHEA [81]. Moreover, the DCR phenomenon was observed when (S)-MA was used as a resolving agent. (S)-CHEA:(S)-MA and (R)-

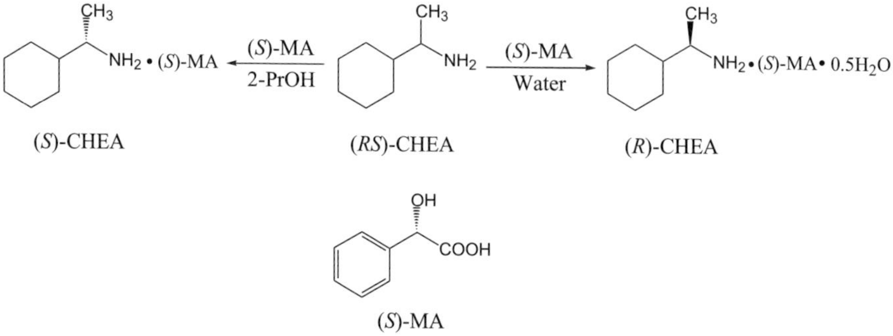

Scheme 6 Resolution of (RS)-CHEA with (S)-MA

Table 8 Resolution of (RS)-CHEA with (S)-MA in various solvents: effect of dielectric constant of the solvent used [82]

Solvent	ε	Solvent weight vs (RS)-CHEA (w/w)	Yield vs (RS)-CHEA (%)	de %	Absolute configuration
Chloroform	5	11.2	Not crystallized		
99% Chloroform	6	11.2	Not crystallized		
2-PrOH	18	6.5	3	24	S
97% 2-PrOH	20	5.0	20	16	S
MeOH	33	2.0	Not crystallized		
81% EtOH	34	2.1	Not crystallized		
88% MeOH	38	2.0	Solidified		
		3.0	Not crystallized		
DMSO	49	2.2	9	30	S
50% EtOH	51	2.5	Not crystallized		
40% MeOH	60	1.0	Solidified		
		2.0	Not crystallized		
Water	78	12.0	7	91	R

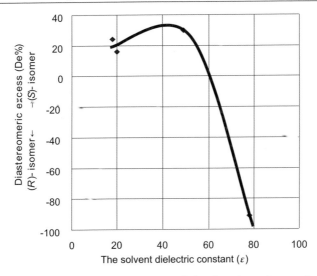

Fig. 16 Resolution of (RS)-CHEA with (S)-MA only by changing solvent: relation between diastereomeric excess (de %) and dielectric constant of solvent [82]

CHEA:(S)-MA:0.5H$_2$O were crystallized from 2-propanol and water, respectively (Scheme 6). The experimental results are summarized in Table 8, and the trends in the chirality changes are depicted by the order of the solvent dielectric constant in Fig. 16 [82].

As clearly shown in Fig. 16, the chirality was drastically changed. The (S)-isomer was recognized at $\varepsilon < 60$ and the (R)-isomer was recognized at $\varepsilon > 60$.

4.3
Strategic Resolution Process Utilizing the DCR Phenomenon

As shown in this chapter, these examples strongly suggest that the DCR phenomenon is not a specific but a general phenomenon that should be observed in various resolution systems. This phenomenon led us to consider continuous resolution by a simple solvent switch, because the resolution of one enantiomer-enriched sample usually gives a much better result than that of the racemic sample as a starting material. The practical resolution process for industrial-scale production can be optimized by utilizing the DCR phenomenon. Typical operation procedures are described below.

4.3.1
ACL–TPA System

A practical continuous resolution process was devised with a solvent switch method [76]. MeOH was the best solvent for the first resolution to obtain

(S)-ACL:(S)-TPA:H$_2$O salt ((S)-salt), although 89% 2-PrOH was suitable for the second resolution to obtain (R)-ACL:(S)-TPA salt ((R)-salt). After the (S)-salt (30%, 93% de, E 56%) was collected by filtration in the first resolution from MeOH, an equimolar mixture of (R)-enriched ACL (40% de) and (S)-TPA was recovered as a condensate by evaporating the mother liquor of the first resolution, and 89% 2-PrOH was added. The mixture was heated to dissolve the condensate and gradually cooled to crystallize the (R)-salt. The (R)-salt was obtained in high resolution efficiency (41%, 93% de, E 75%). The resolution processes were repeated and quite reproducible results were obtained (first resolution: 29–31%, 91–93% de, E 54–56%; second resolution: 41–42%, 91–94% de, E 75–77%).

4.3.2
PTE–MA Resolution System

A practical continuous resolution process based on the DCR phenomenon was also devised with a solvent switch method [79] as applied in the ACL–TPA resolution system. In order to develop the optimum resolution conditions of (RS)-PTE with (S)-MA suitable for industrial-scale production, namely to produce both (R)-PTE and (S)-PTE with high efficiency, ethanol ($\varepsilon = 24$) and 74% ethanol ($\varepsilon = 38$) were selected as resolution solvents, while using a simple combination solvent from a practical viewpoint. Since the resolution efficiency of (R)-PTE from ethanol was relatively lower, the first resolution was designed to produce (S)-PTE from 74% ethanol. Accordingly, the second resolution was designed to produce (R)-PTE by using ethanol from the mother liquor of the first resolution containing (R)-enriched PTE. As a result, the (S)-PTE:(S)-MA:H$_2$O was obtained at an identical efficiency (33% yield, 96% de, E 64%) to the single-resolution result (34%, 99% de, E 67%). The mother liquor of the first resolution was evaporated to dryness to recover the mixture of (R)-enriched PTE (49% de) and equimolar (S)-MA, followed by switching the solvent to ethanol for the second resolution. As intended, the resolution efficiency for the (R)-PTE:(S)-MA salt was significantly improved (23%, 98% de, E 46%) compared with the result observed in the single resolution (16%, 96% de, E 31%).

5
Conclusion and Prospects

The three examples, namely crystal habit modification, space filler concept, and DCR phenomenon, shown in this chapter clearly indicate that we should pay attention to the solvent effects in addition to the usual concerns for chiral discrimination, such as molecular structures of resolving agents and racemic substrates and their hydrogen-bonding abilities. From these research results, we conclude that the solvent is a very important factor as a reaction environment for optimizing the resolution conditions. In other words, if we could illuminate the mechanisms of solvent inclusion and chiral discrimination among molecules participating in the resolution reaction, we would be able to quickly prepare reliable strategies for realizing a cost-effective resolution process.

References

1. Pasteur LM (1848) Compt Rend 26:535
2. Pasteur LM (1848) Ann Chim Phys 24(3):442
3. Rouhi AM (2003) Chem Eng News (May 5) p 46
4. Indinavir (2001) Merck Index 13th edn, No. 4970
5. Murakami H, Tobiyama T, Sakai K (1997) JP Kokai 1997-48,762
6. Murakami H, Satoh S, Tobiyama T, Sakai K, Nohira H (1996) EP 710,652
7. Murakami H, Satoh S, Tobiyama T, Sakai K, Nohira H (1996) USP 5,792,869
8. Murakami H, Satoh S, Tobiyama T, Sakai K, Nohira H (1996) Chem Abstr 125:87209
9. Sertraline (2001) Merck Index 13th edn, No. 8541
10. Quallich GJ (2000) USP 6,593,496
11. Quallich GJ (2000) Chem Abstr 34:41980
12. Mendelovici M, Nidam T, Pilarsky G, Gershon N (2001) USP 6,552,227
13. Mendelovici M, Nidam T, Pilarsky G, Gershon N (2001) Chem Abstr 135:242019
14. Taber PG, Pfisterer DM, Colberg JC (2004) Org Proc Res Dev 8:385
15. Orlistat (2001) Merck Index 13th edn, No. 6935
16. Karpf M, Zutter U (1991) EP 443,449
17. Karpf M, Zutter U (1991) Chem Abstr 115:255980
18. Duloxetine (2001) Merck Index 13th edn, No. 3498
19. Berglund RA (1994) EP 650,965, USP 5,362,886, JP 1995-188,065
20. Berglund RA (1994) Chem Abstr 122:132965
21. Sakai K, Sakurai R, Yuzawa A, Kobayashi Y, Saigo K (2003) Tetrahedron Asymmetr 14:1631
22. Jacques J, Collet A, Wilen SH (1981) Enantiomers, racemates, and resolutions. Wiley, New York, p 251
23. Kozma D (2002) CRC handbook of optical resolution via diastereomeric salt formation. CRC, Boca Raton
24. Addadi L, van Mil J, Lahav M (1981) J Am Chem Soc 103:1249
25. Addadi L, Gati E, Lahav M (1981) J Am Chem Soc 103:1251
26. Addadi L, Berkovitch-Yellin Z, Domb N, Gati E, Lahav M, Leiserowitz L (1982) Nature 296:21
27. Berkovitch-Yellin Z, Addadi L, Idelson M, Leiserowitz L, Lahav M (1982) Nature 296:27
28. Sakai K, Murakami H, Saigo K, Nohira H (1994) JP Appl 1994-1757
29. Sakai K, Murakami H, Saigo K, Nohira H (1994) Chem Abstr 120:298229
30. Murakami H, Sakai K, Tobiyama T (2000) JP Appl 2000-297,066
31. Murakami H, Sakai K, Tobiyama T (2000) Chem Abstr 133:296269
32. Orthorhombic, $P2_12_12_1$, $Z = 4$, $a = 25.581(7)$, $b = 6.867(4)$, $c = 8.348(2)$, $V = 1474.4(9)$, $R = 0.048$, $Rw = 0.051$. Analytical results were identical to data by Brianso (crystal obtained from ether/ethanol solvent system): Brianso M-C (1979) Acta Crystallogr

B35:2751. Polymorphological data were also reported: $P2_1$, $Z = 2$, Hashimoto K, Sumida Y, Terada S, Okamura K (1993) J Mass Spectrom Soc Jpn (Shitsuryo Bunseki) 41:87
33. Sakai K, Maekawa Y, Saigo K, Sukegawa M, Murakami H, Nohira H (1992) Bull Chem Soc Jpn 65:1747
34. Sakai K (1999) J Org Synth Chem Jpn 57:458
35. Sakai K, Yoshida S, Hashimoto Y, Kinbara K, Saigo K, Nohira H (1998) Enantiomer 3:23
36. Borghese A, Libert V, Zhang T, Charles AA (2004) Org Proc Res Dev 8:532
37. Dyer UC, Henderson DA, Mitchell MB (1999) Org Proc Res Dev 3:161
38. Kellogg RM, Nieuwenhuijzen JW, Pouwer K, Vries TR, Broxterman QB, Grimbergen RGP, Kaptein B, La Crois RM, de Wever E, Zwaagstra K, van der Laan AC (2003) Synthesis 10:1626
39. Nieuwenhuijzen JW, Grimbergen RGP, Koopman C, Kellogg RM, Vries T, Pouwer K, van Echten E, Kaptein B, Hulshof LA, Broxterman QB (2002) Angew Chem Int Ed 41:4281
40. Vries T, Wynberg H, van Echten E, Koek J, ten Hoeve W, Kellogg RM, Broxterman QB, Minnaard A, Kaptein B, van der Slues S, Hulshof LA, Kooistra J (1998) Angew Chem Int Ed 37:2349
41. Kinbara K, Sakai K, Hashimoto Y, Nohira H, Saigo K (1996) Tetrahedron Asymmetr 7:1539
42. Kinbara K, Sakai K, Hashimoto Y, Nohira H, Saigo K (1996) J Chem Soc Perkin Trans 2 2615
43. Kinbara K, Hashimoto Y, Sukegawa M, Nohira H, Saigo K (1996) J Am Chem Soc 118:3441
44. Kinbara K, Harada Y, Saigo K (1998) Tetrahedron Asymmetr 9:2219
45. Kinbara K, Saigo K (2003) Top Stereochem 23:207, and references therein
46. Sakai K (2004) CSJ Chem Chem Ind (Kagaku To Kogyo) 5:507
47. Sakai K, Saigo K, Murakami H, Nohira H (1993) Symposium on chiral compounds, Tokyo, 22 Oct 1993
48. Sakai K (1994) PhD dissertation, Saitama University, Japan
49. Kamiya N, Yoshikawa M, Tobiyama T, Nohira H, Fujimura R (1990) JP Appl 1990-2451
50. Kinbara K, Sakai K, Hashimoto Y, Nohira H, Saigo K (1996) Tetrahedron Asymmetr 7:1539
51. Sakai K, Hashimoto Y, Kinbara K, Saigo K, Murakami H, Nohira H (1993) Bull Chem Soc Jpn 66:3414
52. Nohira H (1984) JP 1984-110,656
53. Saigo K, Kubota N, Takebayashi S, Hasegawa M (1986) Bull Chem Soc Jpn 59:931
54. Sakai K, Nohira H, Hashimoto Y, Kinbara K, Saigo K, Murakami H (1993) Symposium on molecular chirality, 20 April 1993, Tokyo
55. Sakai K, Yoshida S, Hashimoto Y, Kinbara K, Saigo K, Nohira H (1998) Enantiomer 3:23
56. Saigo K (1985) JP Appl 1985-104,045
57. Nohira H, Yoshida S (1989) JP Appl 1989-308,244
58. Smith L (1911) J Prakt Chem 84:731
59. Dale JA, Dull DL, Mosher HS (1969) J Org Chem 34:2543
60. Cesario M, Guilhem J (1974) Cryst Struct Commun 127
61. Cesario M, Guilhem J (1974) Cryst Struct Commun 131
62. Sakai K, Maekawa Y, Saigo K, Sukegawa M, Murakami H, Nohira H (1992) Bull Chem Soc Jpn 65:1747

63. Sakai K, Yoshida Y, Hishimoto Y, Kinbara K, Saigo K, Nohira H (1998) Enantiomer 3:23
64. Sorbera LA, Castaner RM, Castaner J (2000) Drug Future 25:907
65. Berglund RA (1995) EP 650,965, US 5,362,886, JP 1995-188,065
66. Berglund RA (1995) Chem Abstr 122:132965
67. Sakai K, Sakurai R, Yuzawa A, Hatahira K (2002) JP Appl 2002-289,068
68. Sakai K, Sakurai R, Yuzawa A, Kobayashi Y, Saigo K (2003) Tetrahedron Asymmetr 14:1631
69. Sakai K (2003) Symposium on molecular chirality 2003, Shizuoka, Japan, 19 Oct 2003, IL-8
70. Sakai K (2004) Japan Process Chemistry winter symposium, Tokyo, 3 Dec 2004
71. Sakurai R, Yuzawa A, Murakami H, Kobayashi Y, Saigo K, Sakai K (2005) Japan Process Chemistry summer symposium 2005, Tokyo, 28 July 2005, P-45
72. Sakai K, Sakurai R, Yuzawa A, Hatahira K (2006) USP Appl 2006/006,3943
73. Sakai K, Sakurai R, Hirayma N (2006) Molecular mechanism of dielectrically controlled resolution (DCR) (in this volume). Springer, Berlin Heidelberg New York
74. Sakai K (2003) Symposium on molecular chirality, Shizuoka, Japan, 10 Oct 2003, IL-8
75. Sakurai R, Sakai K, Yuzawa A, Hirayama N (2003) Symposium on molecular chirality, Shizuoka, Japan, 19 Oct 2003, PA-11
76. Sakai K, Sakurai R, Yuzawa A, Hirayama N (2003) Tetrahedron Asymmetr 14:3713
77. Sakai K, Sakurai R, Yuzawa A, Hatahira K (2003) JP Appl 2003-338,118
78. Sakai K, Sakurai R, Hirayama N (2004) Tetrahedron Asymmetr 15:1073
79. Sakai K, Sakurai R, Nohira H, Tanaka R, Hirayama N (2004) Tetrahedron Asymmetr 15:3495
80. Sakurai R, Yuzawa A, Sakai K, Hirayama N (2006) Cryst Growth Des 6:1606
81. Sakai K, Yokoyama M, Sakurai R, Hirayama N (2006) Tetrahedron Asymmetr 17:1541
82. Sakai K, Sakurai R, Hirayama N (2006) Tetrahedron Asymmetr 17:1812

Molecular Mechanisms of Dielectrically Controlled Resolution (DCR)

Kenichi Sakai[1,2] · Rumiko Sakurai[1,3] · Noriaki Hirayama[4] (✉)

[1]R&D Division, Yamakawa Chemical Industry Co. Ltd., Kitaibaraki, 319-1541 Ibaraki, Japan

[2]*Present address:*
Specialty Chemicals Research Laboratory, Toray Fine Chemicals Co. Ltd., Minatoku, 455-8502 Nagoya, Japan

[3]*Present address:*
Basic Medical Science and Molecular Medicine, Tokai University School of Medicine, Isehara, 259-1193 Kanagawa, Japan

[4]Basic Medical Science and Molecular Medicine, Tokai University School of Medicine, Isehara, 259-1193 Kanagawa, Japan
hirayama@is.icc.u-tokai.ac.jp

1	Introduction	234
2	(RS)-α-amino-ε-Caprolactam: N-tosyl-(S)-Phenylalanine System	235
2.1	X-ray Analysis of the Diastereomeric Salts	236
2.2	Molecular Structures	237
2.3	Crystal Structures	240
2.4	Molecular Recognition	243
3	(RS)-2-Methylpyrrolidine: (R,R)-Tartaric Acid System	245
3.1	X-ray Analysis of the Diastereomeric Salts	246
3.2	Molecular Structures	248
3.3	Crystal Structures	249
3.4	Molecular Recognition	252
4	(RS)-Phenyl-2-p-Tolyl Ethylamine: (S)-Mandelic Acid System	254
4.1	X-ray Analysis of the Diastereomeric Salts	255
4.2	Molecular and Crystal Structures	255
4.3	Molecular Recognition	260
5	(RS)-Cyclohexylethylamine: (S)-Mandelic Acid System	261
5.1	X-ray Analysis of the Diastereomeric Salts	261
5.2	Molecular and Crystal Structures	262
5.3	Molecular Recognition	268
6	Molecular Mechanism of DCR	268
7	Conclusions and Scope	270
	References	271

Abstract It is widely believed that the chiral discrimination process is solely dependent on the stereochemistry of the relevant molecules. However, through systematic studies on several resolution systems with popular chiral selectors, we have discovered a new fact that triggers modification of this prevailing concept of chiral resolution. The studies have demonstrated that one enantiomer of a chiral selector can recognize both enantiomers of a target molecule in different solvent systems with different dielectric constants. The phenomenon was termed dielectrically controlled resolution (DCR). Since DCR was observed in different resolution systems and was not too specific to a particular system, DCR was expected to widely occur in various resolution systems. We have investigated the molecular mechanism underlying this interesting phenomenon based on X-ray analysis of the relevant diastereomeric salts. The disclosed mechanism clearly indicates that a chiral selector can inherently recognize both enantiomers of a target molecule and only the dielectric property of the solvent employed in the resolution process governs the selection of the enantiomer.

Keywords Chiral resolution · Dielectrically controlled resolution · Molecular recognition · Solvent effect · X-ray crystallography

Abbreviations
ACL α-amino-ε-caprolactam
TPA   N-tosyl-phenylalanine
MPRD 2-methylpyrrolidine
TA   tartaric acid
PTE   phenyl-2-p-tolyl ethylamine
MA   mandelic acid
CHEA cyclohexylethylamine

1
Introduction

Diastereomeric salt formation using a resolving agent as a chiral selector is one of the most useful methods for obtaining a target stereoisomer from its racemic mixture [1]. It is widely believed that the chiral discrimination process is solely dependent on the stereochemistry of the relevant molecules. The conventional idea of chiral separation takes it for granted that where a chiral selector of molecule (R)-A is necessary to obtain a less soluble diastereomeric salt with a target molecule of (R)-B, (S)-A is absolutely necessary to separate the enantiomer of the molecule B.

No special attention has been given to the effect of the solvent on the process of discrimination. The role of the solvent is obviously to dissolve both the chiral selector and target molecule, but it is highly probable that the properties of the media influence molecular recognition more than a little. Therefore, we have undertaken a series of experiments to study the effects of solvent on chiral resolution, which is a typical phenomenon of molecular recognition.

In the process of our studies we have recently discovered a typical example of a drastic effect of solvent on chiral resolution [2, 3]. In this particular resolution system, chiral discrimination is controlled by the dielectric properties of the solvents used. Since the dielectric property of the solvent controls chiral resolution, we termed this phenomenon dielectrically controlled resolution (DCR). It was the first discovery of DCR. After this discovery, we studied various resolution systems carefully from the view point of solvent effect and recognized that the DCR phenomena were clearly observed at least in three other resolution systems.

It is very likely that the clues to this novel phenomenon can be found in the crystal structures of the diastereomeric salts. For these four different resolution systems, we have successfully obtained single crystals of the diastereomeric salts from various solvent systems with different dielectric constants. On the basis of the detailed comparison of these crystal structures, we have disclosed a common molecular mechanism underlying DCR. The present study has clearly indicated that DCR is not a special phenomenon but that such solvent effects can be observed in various resolution systems.

In this work, the molecular mechanism of DCR will be discussed in detail.

2
(RS)-α-amino-ε-Caprolactam: N-tosyl-(S)-Phenylalanine System

In the resolution process of (RS)-α-amino-ε-caprolactam(ACL) with N-tosyl-(S)-phenylalanine(TPA), high optical yields for (R)- and (S)-ACL can be attained only by adjusting the dielectric constants (ε) of the solvents employed [2, 3]. The results of a series of resolution experiments obtained for the (RS)-ACL : (S)-TPA system in various solvent systems are given in Table 1. The dielectric constant of the mixed solvent was calculated as a weighted average of the values of the components at 20 °C [4].

(S)-ACL is preferentially selected in the solvents with ε between 27 and 62. In the solvent with ε lower than 27, (R)-ACL is exclusively precipitated. For instance, (S)-ACL is selectively obtained (93%) from a methanol solution ($\varepsilon = 33$) and (R)-ACL (92%) from a 2-propanol : water (89 : 11) solution ($\varepsilon = 25$). The DCR phenomenon of this resolution system was the first ex-

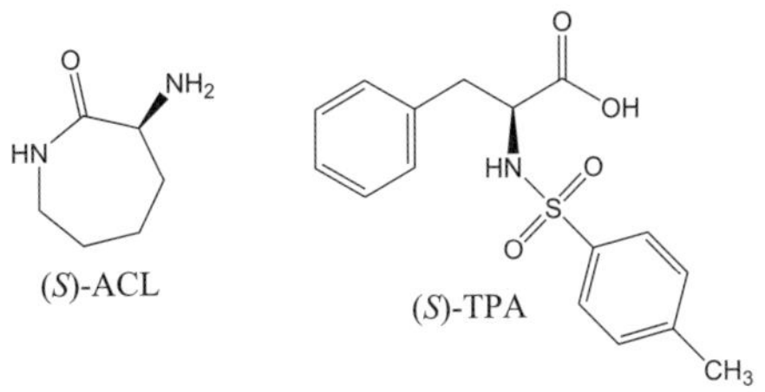

Scheme 1 Chemical structures of ACL and TPA

Table 1 Resolution of (RS)-ACL with (S)-TPA in various solvents

Solvent	ε	Solvent volume[1] (vs (RS)-ACL) (w/w)	Yield (%)[2]	De %	Absolute configuration R/S	Resolution efficiency (E)[3]
chloroform	5	7	24	69	R	33
EDC	11	6	44	61	R	54
MIBK	13	45	63	41	R	52
2-PrOH	18	50	65	32	R	42
EtOH	24	32	70	7	R	10
89% 2-PrOH	25	11	60	29	R	35
85% 2-PrOH	27	10	55	22	R	24
90% EtOH	29	15	63	10	S	13
MeOH	33	10	32	93	S	60
81% EtOH	34	12	25	99	S	50
95% MeOH	35	16	16	92	S	29
DMF	37	27	28	90	S	50
74% EtOH	38	14	13	100	S	26
1,2-ethandiol	39	43	38	99	S	75
DMSO	49	32	20	96	S	38
60% MeOH	51	11	10	95	S	19
55% MeOH	53	5	42	25	S	21
45% MeOH	58	8	49	3	S	3
30% EtOH	62	10	45	6	R	5
35% MeOH	63	6	16	13	R	4
10% MeOH	74	19	38	35	R	27
Water	78	18	31	28	R	17

[1] Solvent volume is a minimum volume to obtain clear solution at 50 °C except for a case using 74% EtOH
[2] Yield is calculated based on a half amount of (RS)-ACL
[3] Resolution efficiency (E, %) = yield (%) × diastereomeric excess (% de) × 2/100

ample of DCR. A molecular mechanism of the phenomenon was disclosed based on X-ray analysis of the diastereomeric salts [5].

2.1
X-ray Analysis of the Diastereomeric Salts

The crystals of the diastereomeric salts between (R)-ACL and (S)-TPA, and (S)-ACL and (S)-TPA are designated as (R)-ACL:(S)-TPA and (S)-ACL:(S)-TPA:W, respectively. W denotes a water molecule in the crystal structure. The details of a procedure to obtain the two diastereomeric salts are described in the literature [5]. The (S)-ACL:(S)-TPA salt was obtained (93%) from a methanol solution ($\varepsilon = 33$), recrystallized in a 74% ethanol solution ($\varepsilon = 38$)

Table 2 Crystallographic data of two diastereomeric salts of the ACL:TPA system

	(R)-ACL:(S)-TPA	(S)-ACL:(S)-TPA:W
formula	$C_{22}H_{29}N_3O_5S$	$C_{22}H_{29}N_3O_5S \cdot H_2O$
space group	$P2_1$, Z = 2	$P2_1$, Z = 2
crystal system	monoclinic	monoclinic
formula weight	447.55	465.56
a (Å)	11.2302(4)	12.723(1)
b (Å)	5.5390(2)	5.3119(3)
c (Å)	17.850(2)	17.908(1)
β (°)	95.444(2)	100.305(3)
V (Å^3)	1105.4(1)	1190.8(2)
Z	2	2
D calc(g cm^{-1})	1.345	1.298
θ max (°) with CuKα	68.2	68.2
observed reflections [$I > 2\sigma(I)$]	2355	2358
$R[I > 2\sigma(I)]$	0.067	0.080
final Δ max/σ	0.000	0.016
$\Delta\rho$ (max; min) (eÅ^{-3})	1.04; –0.58	0.85; –0.68

and single crystals suitable for diffraction experiments were obtained. The (R)-ACL:(S)-TPA salt was obtained (92%) from a 2-propanol-water (89:11) solution ($\varepsilon = 25$), recrystallized from an absolute ethanol solution ($\varepsilon = 24$) and single crystals suitable for diffraction experiments were obtained.

The single crystals of (S)-ACL:(S)-TPA were relatively small and highly fragile. All of the crystals were stable under the laboratory conditions and were mounted on glass fibers and used for data collection. Data were collected on a Rigaku RAPID diffractometer using Cu-Kα radiation and processed with the Process Auto software package [6]. The structures were solved by direct methods [SIR92] [7]. Full-matrix least squares refinement was carried out using observed reflections with $I > 2.00\sigma(I)$. The crystal data, experimental details and structure refinement are summarized in Table 2. Most of the hydrogen atoms were obtained from difference syntheses and refined with the riding model. The absolute configurations were fixed to the known ones and not reconfirmed by the X-ray analyses independently. All of the calculations were performed using the software system CrystalStructure [8].

2.2
Molecular Structures

An ORTEP III [9] drawing of the molecular structure of (S)-TPA in the (R)-ACL:(S)-TPA salt is shown in Fig. 1a together with the atomic numbering

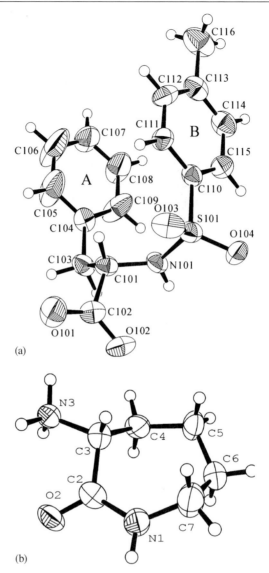

Fig. 1 Molecular structures of (S)-TPA (**a**) and (R)-ACL (**b**) in the (R)-ACL:(S)-TPA crystal. Atomic numbering and labeling of ring systems are also shown

and the labeling of ring systems. Selected geometrical parameters characterizing the conformations in the two different crystal structures are given in Table 3. The chain moieties connecting the two phenyl rings have significantly different conformations as indicated by the torsion angles. The molecule in the (S)-ACL:(S)-TPA crystal takes a more folded conformation. The two rings in the (S)-ACL:(S)-TPA crystal are almost parallel. These results indicate that

(S)-TPA is a relatively flexible molecule and can adapt its conformation to different crystalline environments.

In spite of the conformational differences, the two phenyl rings are located on the same side of the interconnecting chains in these crystal structures. The common orientation of the phenyl rings plays an important role in the formation of distinct hydrophobic layers in the crystals as described below.

An ORTEP III drawing of the molecular structure of (R)-ACL in the (R)-ACL:(S)-TPA salt together with the atomic numbering system is shown in Fig. 1b. Selected torsion angles for (S)-ACL and (R)-ACL in the (R)-ACL:(S)-TPA and (S)-ACL:(S)-TPA:W crystals are shown in Table 3. The absolute values of corresponding torsion angles in the two molecules are similar. This structural similarity indicates that the (S)-ACL and (R)-ACL molecules assume a relatively rigid chair conformation that is maintained in different environments.

Table 3 Selected geometrical parameters in the diastereomeric salts of the ACL:TPA system

	(R)-ACL:(S)-TPA	(S)-ACL:(S)-TPA:W
(i) Torsion angles (°) in TPA		
C111–C110–S101–N101	96.9(4)	112.2(7)
C110–S101–N101–C101	– 92.1(4)	– 71.2(6)
S101–N101–C101–C103	143.3(4)	146.3(5)
N101–C101–C103–C104	– 72.5(6)	– 73.4(8)
C101–C103–C104–C105	– 89.1(7)	– 58.2(11)
(ii) Dihedral angles between phenyl rings and the distances between the centroids of the rings		
A/B (°)	11.3(4)	2.6(5)
Cg(A)···Cg(B) (Å)[a]	4.099	4.305
(iii) Torsion angles (°) in ACL		
N1–C2–C3–C4	60.7(6)	– 70.9(9)
C2–C3–C4–C5	– 81.3(5)	82.0(7)
C3–C4–C5–C6	63.7(6)	– 59.3(8)
C4–C5–C6–C7	– 59.8(5)	61.3(6)
C5–C6–C7–N1	73.7(6)	– 79.2(5)
C6–C7–N1–C2	– 66.7(7)	64.0(10)
C7–N1–C2–C3	6.9(8)	3.6(12)
N1–C2–C3–N3	– 177.7(4)	165.1(6)
O2–C2–C3–N3	3.2(7)	– 16.9(10)
O2–C2–C3–C4	– 118.5(5)	107.2(8)

[a] Cg denotes the centroid of the phenyl ring

2.3
Crystal Structures

The crystal structures of the two diastereomeric salts drawn with the program WebLab Viewer Pro5 [10] are shown in Figs. 2 and 3. As shown in Fig. 2, four (S)-TPA molecules, labeled a, b, c and d, build a basic packing unit that is very similar in the two crystal structures.

The phenyl rings of (S)-TPA pack together and form hydrophobic layers. Only rings labeled B are involved in intermolecular contacts. Since the pat-

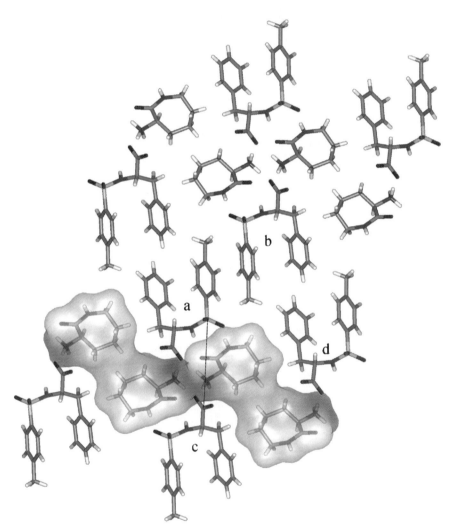

Fig. 2 Crystal structure of (R)-ACL : (S)-TPA; the solvent accessible surface of a column of (R)-ACL is shown

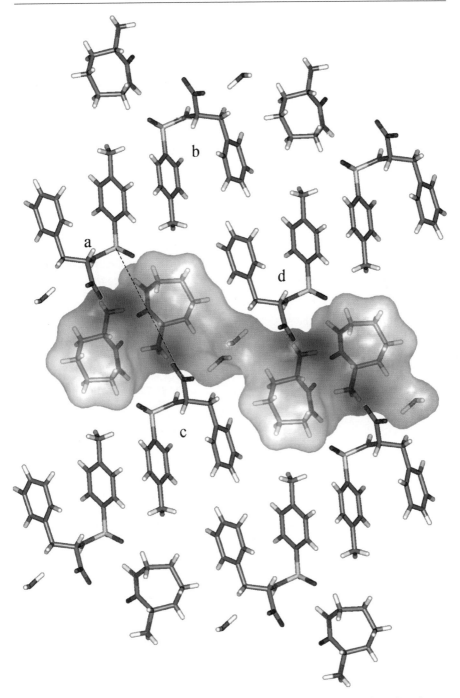

Fig. 3 Crystal structure of (S)-ACL : (S)-TPA : W; the solvent accessible surface of a column consisting of (S)-ACL and water molecules is shown

tern of hydrophobic packing in these crystals is essentially the same, it seems that the packing between the phenyl rings is crucial in building up these crystal structures. Between these hydrophobic layers the hydrophilic layers are formed. The molecular recognition between (S)-TPA and, (R)-ACL or (S)-ACL, essentially takes place in the hydrophilic layers. The intermolecular hydrogen bonds play decisive roles in determining the crystal packing in the hydrophilic layers. The geometrical parameters of the hydrogen bonds in these crystals are given in Table 4. Although the hydrogen bonding patterns observed between (S) or (R)-ACL and (S)-TPA molecules in the (S)-ACL:(S)-TPA:W and (R)-ACL:(S)-TPA crystals are essentially similar, more extended hydrogen bonds are formed in (S)-ACL:(S)-TPA:W.

As described above both (S)-ACL:(S)-TPA and (R)-ACL:(S)-TPA crystals have similar packing patterns. The major difference is observed in the hydrophilic layers. In (S)-ACL:(S)-TPA the hydrophilic layer is significantly expanded and the water molecules contribute to the expansion. The distance between the S101 in molecule **a** and the C102 atom in molecule **c** just across the hydrophilic layer seem to be a good measure to compare the width of the

Table 4 Hydrogen bonds in the ACL:TPA system

(S)-ACL:(S)-TPA

D–H $\cdots$ A[a]	D–H (Å)	H $\cdots$ A (Å)	D $\cdots$ A (Å)	∠ D–H $\cdots$ A (°)
N3–H $\cdots$ O2[i]	0.98	1.97	2.791(7)	139
N3–H $\cdots$ O102[i]	0.95	1.88	2.686(8)	141
N3–H $\cdots$ O101[ii]	0.99	2.19	2.813(7)	119
N3–H $\cdots$ O102[ii]	0.99	2.23	3.221(8)	175
O6–H $\cdots$ O6[b,iii]	0.92	2.06	2.868(8)	146
N101–H $\cdots$ O101[iv]	0.74	2.09	2.811(7)	165
N1–H $\cdots$ O103[v]	1.10	1.91	2.902(8)	148
O6–H $\cdots$ O101[vi]	1.11	1.75	2.777(8)	151

symmetry operation:
(i) $1-x, -1/2+y, -z$ (ii) $1-x, 1/2+y, -z$ (iii) $2-x, -1/2+y, -z$
(iv) $x, 1+y, z$ (v) x, y, z (vi) $1+x, y, z$

(R)-ACL:(S)-TPA:W

D–H $\cdots$ A	D–H (Å)	H $\cdots$ A (Å)	D $\cdots$ A (Å)	∠D–H $\cdots$ A (°)
N1–H $\cdots$ O104[i]	0.87(5)	2.14(5)	3.007(6)	177(4)
N3–H $\cdots$ O102[ii]	0.95	1.78	2.721(5)	168
N3–H $\cdots$ O102[iii]	0.95	1.78	2.725(5)	171
N3–H $\cdots$ O101[iv]	0.95	2.07	2.926(5)	149

symmetry operation:
(i) $x, 1+y, z$ (ii) $2-x, 1/2+y, 1-z$ (iii) x, y, z (iv) $2-x, -1/2+y, 1-z$

[a] A and D denote hydrogen bond acceptor and donor, respectively
[b] O6 designates the oxygen atom of the water molecule

hydrophilic layer. These distances shown by dotted lines are 9.42 and 7.89 Å, in (S)-ACL : (S)-TPA : W and (R)-ACL : (S)-TPA, respectively. The solvent accessible surfaces of the hydrophilic layers are also shown in Figs. 2 and 3.

2.4
Molecular Recognition

The crystal structures indicate that the phenyl groups of (S)-TPA tend to aggregate and form hydrophobic layers during crystallization. The polar groups aggregate to form hydrophilic layers between the hydrophobic layers. Very similar packing patterns are observed in the crystal structures of a chiral selector (+)-(1S)-1,1'-binaphthalene-2,2'-diyl phosphate and several amino acids [11]. In this case the hydrophilic layers are preferentially formed and concurrently hydrophobic layers are formed. The molecular recognition between the chiral selector and the selected molecules also takes place in the hydrophilic layers.

Sufficiently strong interaction between the substrate and the hydrophilic moiety of the chiral selector is required for chiral discrimination. If the dielectric constant is low, the polar groups in the hydrophilic layer can approach so close that the hydrophilic layer packs tightly. To estimate the stereochemical effect of the replacement of (R)-ACL by (S)-ACL in the crystal structure of (R)-ACL : (S)-TPA, the (S)-ACL molecule was superimposed to the (R)-ACL molecule in the crystal. Six atoms in the ring except the C3 atom are superimposed relatively well and the root mean square deviation (rmsd) is 0.429 Å. As shown in Fig. 4, the difference of the stereochemistry at C3 would greatly affect the intermolecular hydrogen bonds. If the (R)-ACL molecule is replaced by the (S)-ACL molecule, the hydrogen bond between the N3 atom and the O102 atom of the upper left (S)-TPA molecule in the crystal would be broken because the N3 atom no longer points towards the O102 atom of the molecule. The N3 atom instead now approaches the O102 atom of the lower left (S)-TPA molecule. The distance between these atoms would be too short (1.38 Å in this superposition) as a hydrogen bond and such a short contact causes a significant steric hindrance. As the conformational flexibility of molecule ACL is not that high as described in Sect. 2.2.1, the steric hindrance would not be relieved by changing its conformation. This indicates that the (S)-ACL molecule cannot be accommodated in this crystal structure and the lower left (S)-TPA molecule should shift away at least 1 Å to accommodate the (S)-ACL molecule. The shift implies an expansion of the hydrophilic layer. If the dielectric constant of the solvent is low, such an expansion is impossible. Therefore, under this situation, only the (R)-ACL isomer can be discriminated.

Under the condition of a medium dielectric constant, the interaction between the polar groups would be shielded to some extent and the hydrophilic layer can expand. Such expansion is realized in the crystal of

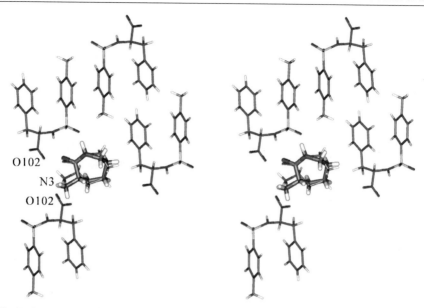

Fig. 4 Stereoscopic drawing of a part of the crystal structure of (R)-ACL:(S)-TPA. An (S)-ACL molecule is superimposed on the (R)-ACL molecule. The former molecule is drawn with *thick sticks*

(S)-ACL:(S)-TPA:W. Expansion of the hydrophilic layer, however, means that the crystal packing gets worse and the efficiency of the discrimination drops. In the (S)-ACL:(S)-TPA:W crystal, this problem is solved by a water molecule incorporated into the hydrophilic layer. The water molecule fills the void and realizes a relatively close packing. To understand the reason why the (R)-ACL molecule can not be discriminated in this case, the (R)-ACL molecule was superimposed on the (S)-ACL molecule in this crystal using a similar condition as described above. In this case rmsd is 0.430 Å. As shown in Fig. 5 the hydrogen bond between the N1 atom and the O103 atom of the upper left (S)-TPA molecule is maintained. The N3 atom in (R)-ACL, however, steps away from the O102 atom of the lower left (S)-TPA molecule and the hydrogen bond between these atoms in the (S)-ACL:(S)-TPA:W crystal is disrupted. The N3 atom now points toward the water molecule, but the distance (N···O = 3.68 Å) is too long for a hydrogen bond. The replaced (R)-ACL molecule would be only loosely recognized here at this recognition site. This implies that under the condition of the medium dielectric constant the (R)-molecule would not be fixed there and cannot be discriminated properly supposing the (S)-TPA molecule assumes this particular packing pattern.

In a solvent with a low dielectric constant, the electrostatic interactions between polar groups of the relevant molecules are strengthened and they can approach close enough. Under such a situation, (R)-ACL should be preferable to pack the hydrophilic layers efficiently. Because of this reason the

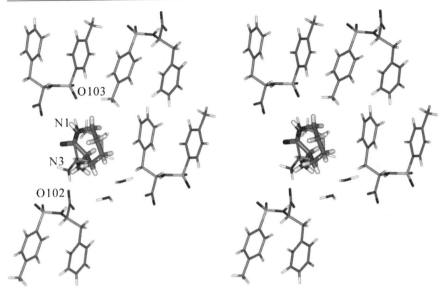

Fig. 5 Stereoscopic drawing of a part of the crystal structure of (S)-ACL:(S)-TPA:W. An (R)-ACL molecule is superimposed on the (S)-ACL molecule. The former molecule is drawn with *thick sticks*

(R)-ACL:(S)-TPA crystal was obtained from the solvent system with a ε of 5. On the other hand in the solvent system with medium dielectric constants, the polar groups should be separated. Under these circumstances a water molecule is necessary to maintain the close packing of the hydrophilic layer. The situation makes it possible that (S)-ACL is preferentially discriminated. Therefore, the (S)-ACL:(S)-TPA:W crystal was obtained from the solvent system with a ε of 38.

The crystal structures of (R)-ACL:(S)-TPA and (S)-ACL:(S)-TPA:W unequivocally show that the hydrophilic layers formed in these crystals can inherently accommodate both (R)-ACL and (S)-ACL isomers. Discrimination of one of the enantiomers may be a matter of selecting the suitable intermolecular interactions between the chiral selector and the enantiomer to be discriminated. The results strongly suggest that the modes of the intermolecular interactions can be controlled by adjusting the dielectric property of the employed solvent.

3
(RS)-2-Methylpyrrolidine: (R,R)-Tartaric Acid System

In the previous examples aromatic rings play important roles in molecular recognition. It is, therefore, particularly interesting whether aromatic rings are essential for DCR or not. We investigated the effect of dielectric properties of the solvent on resolution systems which contain no aromatic rings. Enantiomerically pure 2-methylpyrrolidine (MPRD) is an attractive key intermediate for pharmaceuticals such as ABT-239 that is a histamine H_3 receptor antagonist for Alzheimer disease [12]. In the course of finding the most suitable resolving agent for (RS)-MPRD, we found that (R,R)-tartaric acid (TA) is a suitable chiral selector as reported and this particular chiral resolution system clearly showed the DCR phenomenon [13].

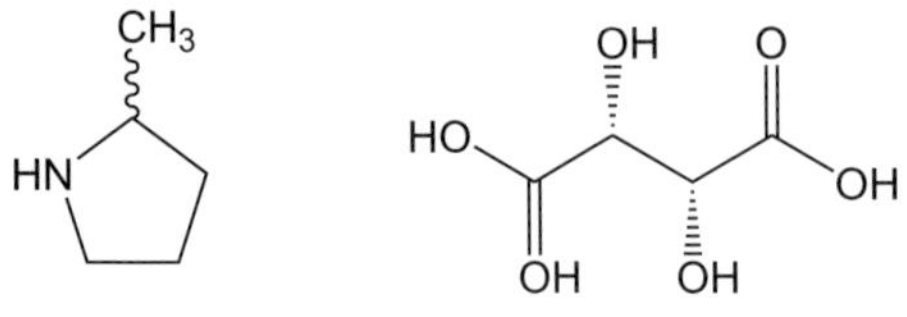

Scheme 2 Chemical structures of MPRD and TA

3.1
X-ray Analysis of the Diastereomeric Salts

The results of a series of resolution experiments obtained for the (RS)-MPRD:(R,R)-MA system in EtOH/water mixed solvents are given in Table 5. The configuration of MPRD enriched in the salt deposited by the resolution procedure is variable depending on the dielectric constant of the solvent used. The less-soluble diastereomeric salt of (R)-MPRD:(R,R)-TA was obtained from the mixed solvents with the dielectric constant $24 \leq \varepsilon \leq 29$. On the other hand (S)-MPRD:(R,R)-TA was obtained from the solvents with $30 \leq \varepsilon \leq 36$. Therefore, this resolution system clearly shows the DCR phenomenon.

The diastereomerically pure salt between (R)-MPRD and (R,R)-TA ((R)-MPRD:(R,R)-TA) was obtained from the mixed EtOH/water (23/1.2 w/w, ε 24.4) system and followed by recrystallization with the same solvent. The diastereomerically pure salt between (S)-MPRD and (R,R)-TA ((S)-MPRD:(R,R)-TA) was obtained from the mixed EtOH/water (12/1.6 w/w, ε 30) system and followed by recrystallization with the same solvent. Since the latter salt contains a water molecule of crystallization it will be designated as (S)-MPRD:(R,R)-TA:W hereafter. W denotes the water molecule.

Despite many attempts at crystallization, good single crystals of both salts could not be obtained. The crystals obtained are fragile and small; hence X-ray diffraction from both crystals was poor. X-ray analysis was undertaken with the same procedures described in Sect. 2.2. The crystallographic data and refinement results are summarized in Table 6. For (R)-MPRD:(R,R)-TA

Table 5 Resolution of (RS)-MPRD with (R,R)-TA from EtOH – H_2O solvent system

Solvent	w/w vs (R/S)-MPRD	Dielectric constant (ε)	De %[a]	R/S	Yield[b]	Resolution efficiency (E)[c]
100% EtOH	40	24	7	R	72	11
99% EtOH	35	24.4	37	R	64	47
97% EtOH	31	25	33	R	62	41
95% EtOH	24	26.5	16	R	64	20
94% EtOH	21	27	9	R	68	12
90% EtOH	17	29	5	R	42	4
89% EtOH	16	30	16	S	44	14
86% EtOH	12	32	19	S	33	12
83% EtOH	10	33	8	S	51	8
79% EtOH	8	35	12	S	54	13
74% EtOH	4	36	12	S	50	12

[a] diastereomeric excess = 100([A − B]/[A + B]), where A and B are the diastereomers
[b] Yield (%) is calculated based on (RS)-MPRD
[c] E (%) = yield (%) × de (%) × 2/100

Table 6 Crystallographic data of two diastereomeric salts of the MPRD: TA system

	(R)-MPRD : (R,R)-TA	(S)-MPRD : (R,R)-TA : W
formula	$C_5H_{11}N : C_4H_6O_6$	$C_5H_{11}N : C_4H_6O_6 : H_2O$
space group	$P2_1$, $Z = 2$	$P2_12_12_1$, $Z = 4$
crystal system	monoclinic	orthorhombic
formula weight	235.24	253.25
a (Å)	4.9095(9)	7.322(4)
b (Å)	17.009(3)	7.666(5)
c (Å)	6.698(1)	21.71(1)
β (°)	101.83(1)	–
V (Å^3)	547.4(2)	1218(1)
D calc (g cm^{-3})	1.427	1.380
θ (max)(°) with CuKα	68.23	68.22
observed reflections [$I > 2\sigma(I)$]	705	673
R[$I > 2\sigma(I)$]	0.048	0.058
final Δ max/σ $\Delta\rho$ (max; min) (eÅ^{-3})	0.00; 0.00	0.00; 0.00

all H atoms were located from difference Fourier synthesis. H atoms bonded to nitrogen and oxygen atoms were refined isotropically. Other H atoms were refined by the riding model and finally fixed. For (S)-MPRD : (R,R)-TA : W all H atoms were located from difference Fourier synthesis. H atoms bonded to nitrogen and oxygen atoms were refined isotropically, but other H atoms were refined by the riding model. In the final refinement cycle all H atoms were fixed. In the process of X-ray analysis a water molecule of crystallization was found in the crystal structure of (S)-MPRD:(R,R)-TA:W. The absolute configurations were fixed to the known ones in both crystal structures and not reconfirmed by the X-ray analyses independently.

3.2
Molecular Structures

The molecular structures of MPRD and TA in the (R)-MPRD:(R,R)-TA crystal depicted by ORTEPIII are shown in Fig. 6 with the atomic labeling system.

Selected torsion angles in both molecules in the two crystal structures are compared in Table 7. The two enantiomers of MPRD adopt significantly different conformations. (S)-MPRD takes an envelope conformation with the C2 atom being deviated significantly from the plane consisting of the other four atoms in the ring. (R)-MPRD takes, however, a twisted conformation with the C1 and C2 atoms being positioned at opposite sides with respect to the plane formed by the other three atoms in the ring. The absolute values of torsion angles in the rings indicate that (S)-MPRD takes a flatter conformation.

The molecules (R,R)-TA in both crystals take very similar extended conformations with the C6–C7–C8–C9 torsion angles being − 175.8(5) and − 176.6(7)° in (R)-MPRD:(R,R)-TA and (S)-MPRD:(R,R)-TA:W, respectively. In both molecules, there is an intramolecular hydrogen bond between a hydroxyl oxygen atom (O4) and a carboxyl oxygen atom (O5). Because of this intramolecular hydrogen bond the torsion angles of O4–C8–C9–O5 are

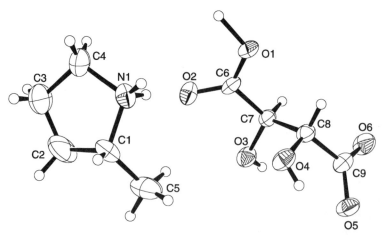

Fig. 6 Molecular structures of MPRD and TA in the (R)-MPRD:(R, R)-TA crystal. Atomic numbering is also shown

Table 7 Selected torsion angles (°) in the diastereomeric salts of the MPRD:TA system

Torsion Angles (°)	(R)-MPRD:(R,R)-TA	(S)-MPRD:(R,R)-TA:W
C(1)–N(1)–C(4)–C(3)	–11.9(10)	–0.2(8)
C(4)–N(1)–C(1)–C(2)	27.3(10)	–11.0(10)
C(4)–N(1)–C(1)–C(5)	150.8(8)	–136.7(7)
N(1)–C(1)–C(2)–C(3)	–32.9(11)	19.1(14)
C(5)–C(1)–C(2)–C(3)	–156.0(9)	142.1(11)
C(1)–C(2)–C(3)–C(4)	26.6(13)	–20.3(16)
C(2)–C(3)–C(4)–N(1)	–9.2(13)	12.5(14)
O(1)–C(6)–C(7)–O(3)	173.4(6)	–172.0(6)
O(1)–C(6)–C(7)–C(8)	48.8(8)	62.5(9)
O(3)–C(7)–C(8)–O(4)	–63.7(6)	–67.6(8)
O(3)–C(7)–C(8)–C(9)	60.2(7)	56.6(8)
C(6)–C(7)–C(8)–O(4)	60.0(6)	59.3(8)
C(6)–C(7)–C(8)–C(9)	–176.1(5)	–176.5(6)
O(4)–C(8)–C(9)–O(5)	4.0(9)	7.0(10)
C(7)–C(8)–C(9)–O(5)	–117.3(6)	–115.5(8)

almost similar in both crystals. The O2–C6–C7–O3 torsion angles in a similar chemical environment, however, take significantly different values in both molecules.

3.3
Crystal Structures

The crystal packing of both crystal structures drawn by the program Mercury [14] are shown in Fig. 7. The geometrical parameters of the hydrogen bonds are given in Table 8. The packing patterns of both structures are significantly different. In the (R)-MPRD:(R,R)-TA crystal, (R)-MPRD and (R,R)-TA form columns by crystallographic translation along the c axis as shown in Fig. 7a. Therefore, all (R)-MPRD molecules align with the nitrogen atoms being positioned on the same side of the column. In the crystal of (S)-MPRD:(R,R)-TA:W shown in Fig. 7b, however, (S)-MPRD and (R,R)-TA are arranged by two-fold screw symmetry. (S)-MPRD form columns with the nitrogen atoms being located on both sides of the columns. These columns are formed by hydrophobic interactions between aliphatic moieties of (S)-MPRD. Hydrophilic columns are formed by (R,R)-TA and water molecules. The nitrogen atoms of (S)-MPRD are in contact with the hydrophilic columns.

The common structural characteristic observed in both crystal structures is the strong intermolecular hydrogen bond between O1 and O5. The O1···O5 distances are 2.473(7) and 2.539(7) Å in (R)-MPRD:(R,R)-

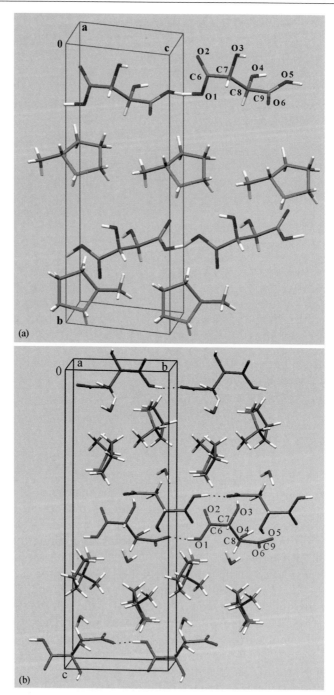

Fig. 7 Crystal structures of (R)-MPRD:(R,R)-TA (**a**) and (S)-MPRD:(R,R)-TA:W (**b**). Atomic numbering systems are also shown

Table 8 Hydrogen bonds in the MPRD:TA system

(R)-MPRD:(R,R)-TA

D–H $\cdots$ A[a]	D–H (Å)	H $\cdots$ A (Å)	D $\cdots$ A (Å)	∠ D–H $\cdots$ A (°)
N(1)–H $\cdots$ O(2)^{i}	0.82(8)	2.05(8)	2.861(9)	176(8)
N(1)–H $\cdots$ O(2)^{ii}	0.90(8)	2.02(8)	2.889(9)	160(7)
N(1)–H $\cdots$ O(2)^{ii}	0.90(8)	2.34(7)	2.954(8)	125(6)
O(3)–H $\cdots$ O(2)^{iii}	0.73(7)	2.10(8)	2.826(7)	172(7)
O(4)–H $\cdots$ O(5)^{i}	0.80(10)	2.13(10)	2.613(7)	120(9)
O(1)–H $\cdots$ O(5)^{iv}	1.25(10)	1.28(10)	2.473(7)	155(9)

symmetry operation:
(i) x, y, z (ii) $1 + x, y, z$ (iii) $-1 + x, y, z$ (iv) $1 + x, y, 1 + z$

(S)-MPRD:(R,R)-TA:W

D–H $\cdots$ A	D–H (Å)	H $\cdots$ A (Å)	D $\cdots$ A (Å)	∠ D–H $\cdots$ A (°)
N(1)–H $\cdots$ O(1)^{i}	0.97	2.39	3.057(8)	125
N(1)–H $\cdots$ O(2)^{ii}	0.97	2.11	2.800(8)	127
N(1)–H $\cdots$ O(7)^{i,b}	0.98	1.86	2.822(9)	166
O(3)–H $\cdots$ O(6)^{iii}	1.05	1.81	2.777(7)	150
O(4)–H $\cdots$ O(5)^{i}	0.93	2.21	2.606(7)	105
O(4)–H $\cdots$ O(3)^{iii}	0.93	2.01	2.866(7)	153
O(1)–H $\cdots$ O(5)^{iv}	1.07	1.50	2.539(7)	162
O(7)–H $\cdots$ O(6)^{i}	0.87	1.91	2.788(7)	176
O(7)–H $\cdots$ O(4)^{v}	0.98	2.34	2.853(7)	112

symmetry operation:
(i) x, y, z (ii) $-1/2 + x, 1/2 - y, 1 - z$ (iii) $1/2 + x, 3/2 - y, 1 - z$ (iv) $x, -1 + y, z$
(v) $-1 + x, y, z$

[a] A and D denote hydrogen bond acceptor and donor, respectively
[b] O(7) denotes the oxygen atom of the water molecule

TA and (S)-MPRD:(R,R)-TA:W, respectively. These strong hydrogen bonds are typical negative-charge-assisted hydrogen bonds ((–)CAHB) [15]. In (R)-MPRD:(R,R)-TA the hydrogen atom obtained from difference Fourier synthesis exists just between two oxygen atoms and represents well a proton-centered hydrogen bond. In (S)-MPRD:(R,R)-TA:W, however, the hydrogen atom is attached to the O1 atom. Nevertheless, the short O1$\cdots$O5 distance indicates that the hydrogen bond still belongs to the (–)CAHB class. By the strong hydrogen bonds, (R,R)-TA forms infinite chains perpendicular to the longest crystallographic axis in both crystal structures. The calculated density of the crystals (R)-MPRD:(R,R)-TA and (S)-MPRD:(R,R)-2:W are 1.427 and 1.380 g/cm^3, respectively. The density shows that the crystal (S)-MPRD:(R,R)-TA:W is more loosely packed than the crystal (R)-

MPRD:(R,R)-TA. Furthermore, if the water molecule of crystallization in the crystal (S)-MPRD:(R,R)-TA:W is removed from the lattice, the density significantly decreases to 1.282 g/cm^3. This indicates that the water of crystallization is essential to maintain the crystal structure of (S)-MPRD:(R,R)-TA:W.

3.4
Molecular Recognition

As described above, strong negative-charge-assisted hydrogen bonds govern the crystal packing of the two crystal structures. Recognition between substrate and selector molecules in both crystal structures is not in a one-to-one fashion. Instead multiple (R,R)-TA molecules connected by (−)CAHB form chiral spaces where the specific enantiomer of the molecule MPRD is recognized. In Fig. 8a, the recognition mode observed in the crystal (R)-MPRD:(R,R)-TA is shown. Multiple (R,R)-TA molecules connected by (−)CAHB form layers and (R)-MPRD is recognized at the chiral space created between the layers. In this figure (R,R)-MA molecules are depicted by solvent accessible surfaces to illustrate the landscape of the surface of the connected molecules. As shown in the figure the chiral space has an appropriate size to accommodate (R)-MPRD. In Fig. 8b a part of the crystal (S)-MPRD:(R,R)-TA:W is shown. The hydrogen-bonded (R,R)-TA molecules also form layers and are depicted by a solvent accessible surface. The chiral space created between the layers is significantly larger than that of the crystal (R)-MPRD:(R,R)-TA. Therefore, (S)-MPRD is too small in size to properly fill up this void. This means that the recognition is not perfect and the packing of molecules is not good enough to build up a rigid crystal. If the void is stuffed with water molecules, the packing will be materially improved as shown in Fig. 8c.

Only (R)-MPRD is accepted in the chiral space formed in the crystal (R)-MPRD:(R,R)-TA. To understand the reason for this preference we have calculated the difference of packing energy when (R)-MPRD is replaced by (S)-MPRD in the crystal structure. Firstly, the positions of hydrogen atoms of the structure shown in Fig. 8a were optimized and then the total energy of the structure was calculated by molecular mechanics. MMFF94x was used as the force field [16] and calculations were undertaken by the software MOE [17]. Secondly, one of the (R)-MPRDs was replaced by (S)-MPRD and the corresponding energy was calculated. The difference of energies is 29.2 kcal/mol in favor of (R)-MPRD. Similarly, the effect of replacing (S)-MPRD by (R)-MPRD in the crystal (S)-MPRD:(R,R)-TA:W was also evaluated. In this case (S)-MPRD is more favorable by 29.6 kcal/mol than (R)-MPRD. These results have demonstrated that the chiral spaces formed in both crystal structures preferentially accommodate the corresponding enantiomers.

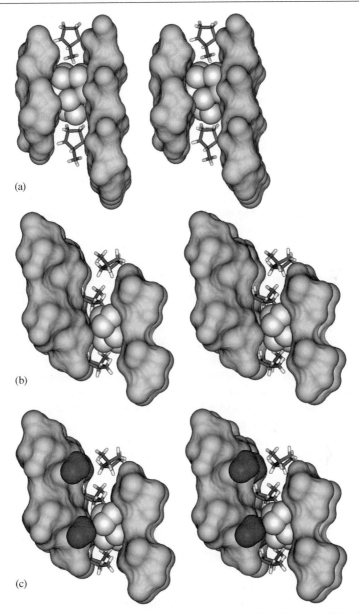

Fig. 8 Molecular recognition between MPRD and TA. **a** Molecular recognition between (R)-MPRD and (R,R)-TA. Columns of (R,R)-TA are represented by the solvent accessible surface. One molecule of (R)-MPRD is depicted by van der Waals *radii*. Other (R)-MPRDs are shown by *sticks*. **b** Molecular recognition between (S)-MPRD and (R,R)-TA : W. Columns of (R,R)-TA are represented by the solvent accessible surface. One molecule of (S)-MPRD is depicted by van der Waals *radii*. Other (R)-MPRDs are shown by *sticks*. Water molecules are omitted. **c** Molecular recognition between (S)-MPRD and (R,R)-TA : W. Water molecules are shown by the solvent accessible surface in *dark grey*

Since (R)-MPRD:(R,R)-TA and (S)-MPRD:(R,R)-TA:W are obtained in the solvent with dielectric constants of 24.4 and 30, respectively, the electrostatic interaction between molecules in the latter solution should be reduced at least by a factor of 24.4/30 compared to the former one. Actually in (S)-MPRD:(R,R)-TA:W, the O···O distance of (−)CAHB is significantly lengthened and the separation between the layers of hydrogen-bonded chains of (R,R)-TA is appreciably expanded. The chiral space created between the layers of (R,R)-TA plays a pivotal role in the process of resolution of MPRD and the dielectric property could affect the size and shape of the chiral space. The present results have illustrated that the chiral space formed by the layers of (R,R)-TA could accommodate both enantiomers supposing the separation between the layers can be varied as long as adequate packing to form crystals is maintained. As shown in Fig. 8c water molecules occupy depressions formed on the surface of the layers of (R,R)-TA. A water molecule has a suitable size and properties to fill the void to attain closer packing to form the crystal.

The present study has demonstrated that the DCR phenomenon can occur in a resolution system that does not contain aromatic rings. The chiral selector (R,R)-TA inherently forms chains by very strong (−)CAHBs. In the solution the chains should build up layers. Between the layers a specific chiral space would be created and one of the enantiomers of MPRD is recognized in this space. The shape and size of the chiral space will vary depending on the dielectric constant of the solvent used for the resolution process. If the size and shape of the chiral space are controlled, both enantiomers can be recognized. In the solvent with a larger dielectric constant, the hydrophilic layers tend to be separated and a larger chiral space will be available. With the assistance of the water molecule to attain suitable packing, the antipode can be recognized in the space. Molecular mechanics calculations showed that each enantiomer is preferentially accepted in the corresponding chiral space. The present study has demonstrated that DCR can be applied to flexible molecules.

4
(RS)-Phenyl-2-p-Tolyl Ethylamine: (S)-Mandelic Acid System

Phenyl-2-p-tolyl ethylamine (PTE) is used as an efficient chiral selector of chiral intermediates for insecticide such as chrysanthemic acid [18]. Mandelic acid (MA) can be used as a suitable resolving agent for enantiomer separation of PTE. Since mandelic acid is one of the most popular resolving agents, it is particularly interesting to know whether the DCR phenomenon can be observed in this resolution system. Moreover, it has been reported that solvents affect the resolution process in the (RS)-PTE:(S)-MA resolution system [19] and it was expected that this system should show the DCR phenomenon. Therefore, we have undertaken systematic experiments to investigate the ef-

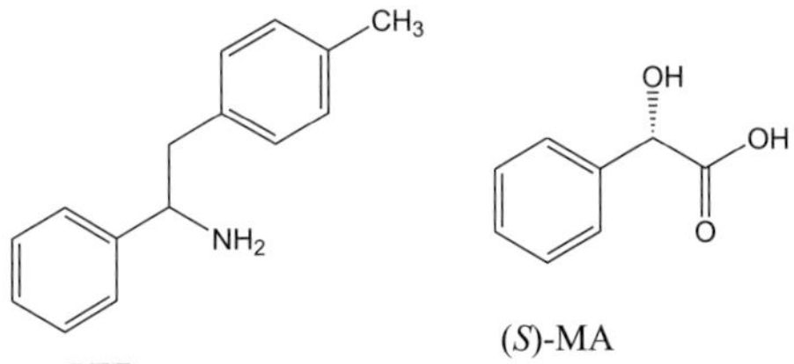

Scheme 3 Chemical structures of PTE and MA

fect of dielectric properties of solvents on the enantiomeric resolution. The salt between (R)-PTE and (S)-MA is designated as (R)-PTE:(S)-MA. Since the salt between (S)-PTE and (S)-MA contains a water molecule, it is designated as (S)-PTE:(S)-MA:W.

4.1
X-ray Analysis of the Diastereomeric Salts

Resolution experiments using ethanol and water as solvents are shown in Table 9. The results clearly show that the DCR phenomenon occurs as expected [20]. (R)- and (S)-PTE were obtained with highest efficiency when absolute ethanol ($\varepsilon = 24$) and 74% ethanol ($\varepsilon = 38$) were used, respectively. The crystals of (R)-PTE:(S)-MA and (S)-PTE:(S)-MA:W were further recrystallized using ethanol and water, respectively, to obtain single crystals for X-ray analysis.
X-ray
analysis was undertaken according to the procedures described in Sect. 2.2. The crystallographic data and refinement results are summarized in Table 10. All non-hydrogen atoms were refined using anisotropic displacement parameters. For both crystals all H atoms were located from difference Fourier synthesis and refined by the riding model. In the process of X-ray analysis a water molecule of crystallization was found in the crystal structure of (S)-PTE:(S)-MA:W. The hydrogen atoms of the water molecule could not be located from the difference Fourier synthesis. The absolute configurations were fixed to the known ones in both crystal structures and not reconfirmed by the X-ray analyses independently.

4.2
Molecular and Crystal Structures

The molecular structures of PTE and MA in the (R)-PTE:(S)-MA crystal depicted by ORTEPIII are shown in Fig. 9 with the atomic labeling system.
  The selected bond lengths, bond angles and torsion angles in (R)-PTE:(S)-MA and (S)-PTE:(S)-MA:W are compared in Table 11. In (R)-PTE:(S)-MA,

Table 9 Resolution of (RS)-PTE with (S)-MA in ethanol/water solvent system

Ethanol concentration %	ε	Solvent/(R/S)-PTE[a] w/w	De %[b]	R/S	Yield[c]	E[d]
100	24	12	96	R	16	31
95	27	10	28	S	38	21
90	29	7	35	S	47	33
85	32	8	68	S	41	56
74	38	10	99	S	34	67
60	46	11	66	S	44	58
50	51	16	48	S	54	52
30	62	46	36	S	58	42
16	69	120	98	S	34	67

* (S)-2/(RS)-1 = 1.0 molar ratio
[a] Solvent weight was determined by solubility of the solid substances at 50°
[b] diastereomeric excess (% de) = [A-B] × 100/(A+B), where A and B are both diastereomers
[c] Calculated based on (RS)-1
[d] Resolution efficiency (E, %) = yield (%) × diastereomeric excess (% de) × 2/100

Table 10 Crystallographic data of two diastereomeric salts of the PTE:MA system

	(R)-PTE:(S)-MA	(S)-PTE:(S)-MA:W
formula	$C_{15}H_{17}N : C_8H_8O_3$	$C_{15}H_{17}N : C_8H_8O_3 : H_2O$
space group	$P2_12_12_1, Z = 4$	$P2_12_12_1, Z = 4$
crystal system	orthorhombic	orthorhombic
formula weight	363.46	381.47
a (Å)	6.607(4)	6.135(1)
b (Å)	15.504(1)	11.901(2)
c (Å)	20.243(2)	27.907(5)
V (Å^3)	2073(1)	2037.5(6)
D calc(g cm^{-3})	1.164	1.243
θ_{max} (°) with CuKα	68.17	68.19
observed reflections [$I > 2\sigma(I)$]	1397	1564
$R(I > 2.00\sigma(I))$	0.065	0.043
final Δ max/σ	0.000	0.000
$\Delta\rho$ (max; min) (eÅ^{-3})	0.23; –0.18	0.13; –0.22

O1–C1 is significantly shorter than O2–C1. Whereas in (S)-PTE:(S)-MA:W, O1–C1 and O2–C1 have similar bond lengths. The conformations of PTE molecules are significantly different in both crystals. Torsion angles of N1–

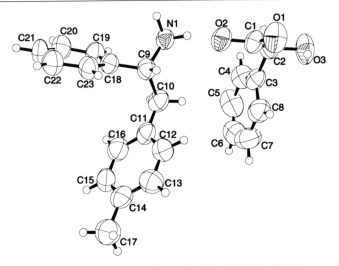

Fig. 9 Molecular structures of RPTE and MA in the (R)-PTE:(S)-MA crystal. Atomic numbering is also shown

Table 11 Selected geometrical parameters in the diastereomeric salts of the PTE:MA system

	(R)-PTE:(S)-MA	(S)-PTE:(S)-MA:W
O(1)–C(1)	1.232(9)	1.250(4)
O(2)–C(1)	1.270(7)	1.240(4)
O(3)–C(2)	1.414(8)	1.420(4)
N(1)–C(9)	1.495(8)	1.504(4)
N(1)–C(9)–C(10)	107.6(4)	106.9(2)
N(1)–C(9)–C(18)	111.6(5)	109.8(2)
O(1)–C(1)–C(2)	116.7(5)	115.5(3)
O(2)–C(1)–O(1)	123.0(5)	126.2(3)
O(2)–C(1)–C(2)	120.2(6)	118.3(3)
O(3)–C(2)–C(1)	111.8(5)	111.0(3)
O(3)–C(2)–C(3)	110.0(4)	110.7(2)
N(1)–C(9)–C(10)–C(11)	169.5(5)	−160.0(3)
N(1)–C(9)–C(18)–C(19)	51.5(8)	93.7(4)
N(1)–C(9)–C(18)–C(23)	−126.8(6)	−81.4(4)
O(1)–C(1)–C(2)–O(3)	1.9(8)	−14.7(4)
O(1)–C(1)–C(2)–C(3)	126.1(6)	107.6(3)
O(2)–C(1)–C(2)–O(3)	178.1(5)	166.9(2)
O(2)–C(1)–C(2)–C(3)	−57.7(8)	−70.8(4)
O(3)–C(2)–C(3)–C(4)	−109.5(6)	−134.4(3)
O(3)–C(2)–C(3)–C(8)	65.8(7)	47.6(4)

C9–C18–C19 and N1–C9–C18–C23 are markedly different in both crystals. The dihedral angles between two phenyl rings in PTE are 52.0(3)° and 86.3(2)° in (R)-PTE:(S)-MA and (S)-PTE:(S)-MA:W, respectively. The conformations of (S)-MA in both crystal structures are also significantly different but to a lesser extent than those of PTE. The torsion angles of C1–C2–C3–C8 are −59.3(8)° and −74.8(4)° in (R)-PTE:(S)-MA and (S)-PTE:(S)-MA:W, respectively. The geometrical differences indicate that chiral selectors and target molecules must undergo substantial conformational changes in order to form the crystals in which each enantiomer can be recognized by the same chiral selector.

Parts of the crystal structures of (R)-PTE:(S)-MA and (S)-PTE:(S)-MA:W are compared in Fig. 10. Both crystal structures are viewed along the c axis. In both crystal structures, the (S)-MA molecules similarly pack and form chiral space where PTE molecules are recognized. (S)-MA and PTE molecules are depicted by van der Waals spheres and sticks, respectively. The oxygen atom of the water molecule is colored in green.

In the crystal structure of (R)-PTE:(S)-MA, the width of chiral space is relatively narrow and only (R)-PTE can be accommodated in the chiral

Table 12 Hydrogen bonds in the PTE:MA system

(R)-PTE:(S)-MA D–H ⋯ A[a]	D–H (Å)	H ⋯ A (Å)	D ⋯ A (Å)	∠ D–H ⋯ A (°)
O(3)–H(22) ⋯ O(1)[i]	1.18	1.78	2.600(6)	121
N(1)–H(23) ⋯ O(2)[i]	0.97	1.82	2.730(7)	156
N(1)–H(24) ⋯ O(2)[ii]	1.02	1.89	2.857(5)	157
N(1)–H(25) ⋯ O(1)[iii]	1.08	1.73	2.776(7)	160

symmetry operation:
(i) x, y, z (ii) $1/2 + x, 1/2 - y, -z$ (iii) $1 + x, y, z$ (vi) $1 + x, y, z$

(S)-PTE:(S)-MA:W D–H ⋯ A	D–H (Å)	H ⋯ A (Å)	D ⋯ A (Å)	∠ D–H ⋯ A (°)
N(1)–H(22) ⋯ O(2)[i]	0.97	1.89	2.849(4)	173
N(1)–H(23) ⋯ O(3)[ii]	0.85	2.06	2.832(4)	150
N(1)–H(24) ⋯ O(1)[iii]	0.99	1.75	2.716(4)	165
O(3)–H(25) ⋯ O(1)[iii]	1.16	1.95	2.604(4)	111
O(3)–H(25) ⋯ O(4)[i]	1.16	1.82	2.776(5)	136
O(1) ⋯ O(4)[i]			3.039(6)	
O(2) ⋯ O(4)[ii]			2.757(6)	

symmetry operation:
(i) $1 - x, y, z$ (ii) $-x, -1/2 + y, 1/2 - z$ (iii) x, y, z

[a] A and D denote hydrogen bond acceptor and donor, respectively
[b] O4 designates the oxygen atom of the water molecule

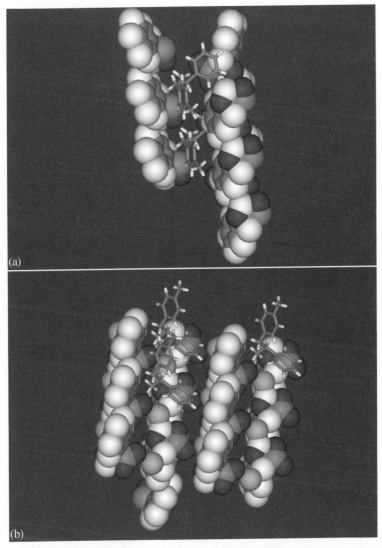

Fig. 10 Crystal structures of the PTE:MA system. **a** Crystal structure of (R)-PTE:(S)-MA. **b** Crystal structure of (S)-PTE:(S)-MA:W. MAs are drawn by van der Waals *radii* and PTE sticks. Oxygen atoms of water molecules are drawn by van der Waals *radii* in *light grey*

space. (S)-MA makes columns perpendicular to the b axis. Therefore, the width of the chiral space can be approximated by the half length of the b axis, i.e., roughly 7.8 Å. On the other hand in the crystal structure of (S)-PTE:(S)-MA:W, (S)-MA also makes columns perpendicular to the b axis. The neighboring columns to form the chiral space, however, are related by the translation along the b axis. Therefore, the width of the chiral space can be approximated by the length of the b axis, i.e., roughly 11.9 Å. The width of the chiral space is markedly wider than that of (R)-PTE:(S)-MA, and only (S)-PTE can be accommodated in this chiral space.

In Table 12, hydrogen bonds in the crystal structures are given. Since the hydrogen atoms of the water molecule could not be located from difference Fourier syntheses, only O···O distances are given for the hydrogen bonds in which O4 is supposed to act as hydrogen bond donor.

In the (R)-PTE:(S)-MA crystal, O2 is involved in two hydrogen bonds with the N atoms. O1, however, is involved in one hydrogen bond with the N atom. It is also involved in an intramolecular hydrogen bond with O3. Whereas in the (S)-PTE:(S)-MA:W crystal, O1, O2 and O3 act as hydrogen acceptors and form hydrogen bonds with the N atom. These oxygen atoms are also possibly involved in hydrogen bonds with the water molecules. O1 is also involved in an intramolecular hydrogen bond with O3 as in the (R)-PTE:(S)-MA crystal. It is noteworthy that O3 is involved in an intermolecular hydrogen bond in the (S)-PTE:(S)-MA:W crystal, but not in the (R)-PTE:(S)-MA crystal.

The markedly different hydrogen-bond schemes observed in the two crystal structures indicate that the dielectric property of the solvent influences the state of the carboxylate and affects molecular recognition. On the hydrophilic surface of (S)-MA columns formed in the (R)-PTE:(S)-MA crystal, (S)-PTE is preferentially recognized. On the other hand, (S)-PTE is preferentially accepted at the hydrophilic surfaces created by (S)-MA.

4.3
Molecular Recognition

The above results clearly provide us with insight into the molecular mechanism of the DCR phenomenon observed in the (RS)-PTE:(S)-MA resolution system. In the solution, (S)-MA should be highly inclined to aggregate to form columns. Between the particular columns, the chiral spaces are formed. In the solvent with the smaller dielectric constant, the columns come close together to form a relatively small chiral space. In addition, under the environment with the low dielectric constant, O2 forms hydrogen bonds with the ammonium ion. Therefore, only (R)-PTE can be recognized at the chiral space. In the solvent with the larger dielectric constant, however, the separation between the columns increases because of reduced electrostatic interactions between them. Water molecules must be incorporated between the columns to maintain the crystal packing. In media with a larger dielectric constant, O1, O2 and O3 can be equally involved in the intermolecular hydrogen bonds with the N atom. Accordingly, only (S)-PTE can be recognized in the larger chiral space. Both PTE and MA should undergo some degree of conformational changes to realize the optimum recognition.

5
(RS)-Cyclohexylethylamine: (S)-Mandelic Acid System

Enantiopure 1-cyclohexylethylamine(CHEA) is an important key compound for syntheses of several enzyme inhibitors such as an inosine monophosphate dehydrogenase inhibitor [21]. CHEA is also applied as a resolving agent for enantiomer separation of chiral acids [22]. Since CHEA has a flexible ring structure and an exocyclic asymmetric carbon, the structure is markedly different from the three target molecules described so far. Therefore, it is of interest to study whether the CHEA : MA system shows the DCR phenomenon or not, and we have undertaken systematic resolution experiments on the system [23, 24]. The salt between (R)-CHEA and (S)-MA is designated as (R)-CHEA : (S)-MA. The diastereomeric salt between (R)-CHEA and (S)-MA is designated as (R)-CHEA : (S)-MA : W because it is a hemi hydrate.

5.1
X-ray Analysis of the Diastereomeric Salts

Resolution experiments using various solvents with different dielectric constants are shown in Table 13. Although the diastereomeric salts were obtained only from three solvents, the DCR phenomenon was clearly observed. The diastereomeric salts of (S)-CHEA and (R)-CHEA are obtained from 2-propanol ($\varepsilon = 18$) and water ($\varepsilon = 78$), respectively. The single crystals of (S)-CHEA : (S)-MA and (R)-CHEA : (S)-MA were recrystallized from the solvents used for chiral separation.

Table 13 Resolution of (RS)-CHEA with (S)-MA in various solvent

Solvent	ε	Solvent volume vs (R/S)-CHEA (w/w)	Yield vs (R/S)-CHEA (%)	De %	Absolute configuration
chloroform	5	11.2	not crystallized		
99% chloroform	6	11.2	not crystallized		
2-PrOH	18	6.5	3	24	S
97% 2PrOH	20	5.0	20	16	S
MeOH	33	2.0	not crystallized		
81% EtOH	34	2.1	not crystallized		
88% MeOH	38	2.0	solidified		
		3.0	not crystallized		
DMSO	49	2.2	9	30	S
50% EtOH	51	2.5	not crystallized		
40% MeOH	60	1.0	solidified		
		2.0	not crystallized		
water	78	12.0	7	91	R

Table 14 Crystallographic data of two diastereomeric salts of the CHEA : MA system

	(S)-CHEA : (S)-MA	(R)-CHEA : (S)-MA : W
formula	$C_8H_{17}N : C_8H_8O_3$	$C_8H_{17}N : C_8H_8O_3 : 0.5H_2O$
space group	$P2_12_12$, $Z = 4$	$C2$, $Z = 4$
crystal system	orthorhombic	monoclinic
formula weight	279.38	288.39
a (Å)	5.62(1) Å	16.64(3) Å
b (Å)	15.45(3) Å	6.32(1) Å
c (Å)	19.26(4) Å	14.64(3) Å
β (°)	0	96.4(1)
V (Å^3)	1670(6) Å^3	1529(4) Å^3
D calc(g cm^{-3})	1.111 g/cm^3	1.252 g/cm^3
No. of observations ($I > 2.00\sigma(I)$)	535	1936
θ_{max} (°) with CuKα	68.14°	68.25°
R ($I > 2.00\sigma(I)$)	0.052	0.062
final Δ max/σ	0.001	0.000
$\Delta\rho$ (max; min) (eÅ^{-3})	0.18; –0.17	0.35; –0.25

X-ray analysis was undertaken according to the procedures described in Sect. 2.2. The crystallographic data and refinement results are summarized in Table 14. All non-hydrogen atoms were refined using anisotropic displacement parameters. For both crystals all H atoms were located from difference Fourier synthesis and refined by the riding model. The absolute configurations were fixed to the known ones in both crystal structures and not reconfirmed by the X-ray analyses independently.

5.2
Molecular and Crystal Structures

The molecular structures of CHEA and MA in the crystal (S)-CHEA : (S)-MA depicted by ORTEPIII are shown in Fig. 11 with the atomic labeling system.

The selected bond lengths, angles and torsion angles are given in Table 15. Two C – O bond lengths in MA are asymmetric in (S)-CHEA : (S)-MA. As observed in (R)-PTE : (S)-MA that was obtained from the solvent with a lower

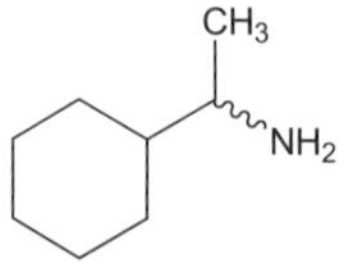

Scheme 4 Chemical structure of CHEA

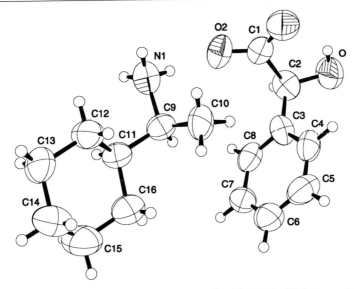

Fig. 11 Molecular structures of CHEA and MA in the (S)-CHEA:(S)-MA crystal. Atomic numbering is also shown

Table 15 Selected bond lengths (Å), bond angles (°) and torsion angles (°)

	(S)-CHEA:(S)-MA	(R)-CHEA:(S)-MA:W
O(1)–C(1)	1.21(2)	1.228(5)
O(2)–C(1)	1.26(2)	1.228(6)
O(3)–C(2)	1.42(1)	1.396(6)
N(1)–C(9)	1.54(1)	1.466(5)
O(1)–C(1)–O(2)	127(1)	124.0(4)
O(1)–C(1)–C(2)	113(1)	119.1(4)
O(2)–C(1)–C(2)	118(1)	117.0(3)
O(3)–C(2)–C(1)	111(1)	109.4(3)
O(3)–C(2)–C(3)	111(1)	110.8(4)
N(1)–C(9)–C(10)	108.7(9)	108.5(3)
N(1)–C(9)–C(11)	109.1(8)	112.3(3)
O(1)–C(1)–C(2)–O(3)	–18(1)	–17.3(5)
O(1)–C(1)–C(2)–C(3)	104(1)	105.9(5)
O(2)–C(1)–C(2)–O(3)	163(1)	161.8(4)
O(2)–C(1)–C(2)–C(3)	–73(1)	–75.0(5)
O(3)–C(2)–C(3)–C(4)	38(1)	7.2(5)
O(3)–C(2)–C(3)–C(8)	–144(1)	–170.8(4)
N(1)–C(9)–C(11)–C(12)	–61(1)	60.1(5)
N(1)–C(9)–C(11)–C(16)	173(1)	–66.8(5)

dielectric constant; the O2–C1 bond is significantly longer than O1–C1. Whereas they are symmetrical in (R)-CHEA : (S)-MA : W. O1–C1 is also symmetrical in (S)-PTE : (S)-MA : W, which was obtained from the solvent with a higher dielectric constant. The (S)-MA molecules in both crystal structures take relatively different conformations. The torsion angles of O3–C2–C3–C4 in (S)-CHEA : (S)-MA and (R)-CHEA : (S)-MA : W are 38 (1)° and 7.2(5)°, respectively. CHEA adopts a similar chair conformation in both crystal structures. The conformations of the ethylamine moieties, however, are significantly different. Although the torsion angles of N1–C9–C11–C12 agree within the experimental error in both crystal structures, the torsion angles of N1–C9–C11–C16 are significantly different with the values being 173(1) and −66.8(5) in (S)-CHEA : (S)-MA and (R)-CHEA : (S)-MA : W, respectively. These geometrical differences indicate that chiral selectors and target molecules should undergo structural change appreciably in order to realize the optimum molecular recognition. Only molecular recognition with sufficiently strong intermolecular interactions can produce crystals that can be purely separated.

Crystal structures of (S)-CHEA : (S)-MA and (R)-CHEA : (S)-MA : W are compared in Fig. 12. (S)-MA and CHE are depicted by van der Waals spheres and sticks, respectively. Water molecules are colored in green. In both crystal structures, the (S)-MA molecules similarly pack and build up chiral space where CHEA molecules are recognized.

In the crystal structure of (S)-CHEA : (S)-MA (Fig. 12a), the width of the chiral space is relatively wide but only (S)-CHEA can be accommodated in the chiral space. The width of the chiral space represented by the separation between the C1 atom of (S)-MA and the center of the phenyl group of another (S)-MA that is located at the opposite side of the chiral space is 9.33 Å. The width is shown by a yellow dotted line. In the crystal structure of (R)-CHEA : (S)-MA : W (Fig. 12b), (S)-MA forms columns perpendicular to the a axis. Therefore, the width of the chiral space is roughly a half length of the a axis, i.e., 8.32 Å.

The hydrophilic columns and layers in both crystal structures are highlighted in Fig. 13. Oxygen and nitrogen atoms are emphasized by van der Waals spheres. In Fig. 13a, the crystal structure of (R)-CHEA : (S)-MA : W is viewed along the b axis. The hydrophilic layers extend along the b axis and are perpendicular to the c axis. The separation between the hydrophilic layers is 14.64 Å. The crystal structure of (S)-CHEA : (S)-MA viewed along the a axis is shown in Fig. 13b. The hydrophilic columns extend along the a axis. They do not stack in parallel to each other. The spacing between the centers of the nearest-neighbor columns is 9.63 Å. Although the chiral space in (R)-CHEA : (S)-MA : W is narrower than that in (S)-CHEA : (S)-MA, the separation of the hydrophilic columns in the former crystal is significantly wider than that in the latter crystal.

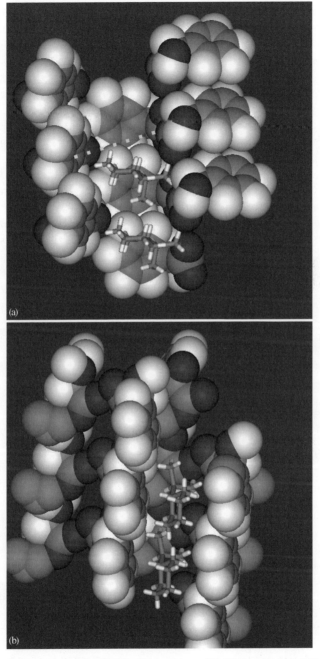

Fig. 12 Crystal structures of the CHEA : MA system. **a** (S)-CHEA : (S)-MA. MAs are drawn by van der Waals *radii* and PTE sticks. **b** (R)-CHEA : (S)-MA : W. MAs are drawn by van der Waals *radii* and PTE sticks. Oxygen and hydrogen atoms of water molecules are depicted by van der Waals *radii* in *light grey*

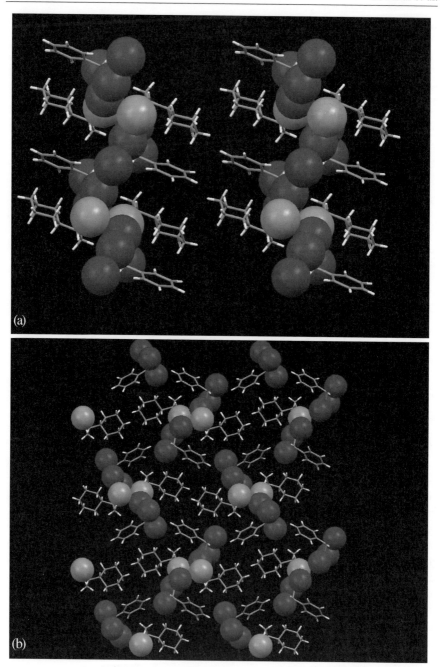

Fig. 13 Hydrophilic columns and layers. Oxygen and nitrogen atoms are depicted by van der Waals spheres in *red* and *blue*. Nitrogen atoms are depicted in *pale grey*. Other atoms are drawn by *sticks*. **a** the (R)-CHEA:(S)-MA:W crystal viewed along the *b* axis. **b** the (S)-CHEA:(S)-MA crystal viewed along the *a* axis

Table 16 Hydrogen bonds in the CHEA:MA system

(S)-CHEA:(S)-MA

D–H ··· A[a]	D–H (Å)	H ··· A (Å)	D ··· A (Å)	∠ D–H ··· A (°)
O(3)–H(7) ··· O(1)[i]	0.8989	1.9103	2.640(14)	136.95
N(1)–H(23) ··· O(1)[ii]	0.9063	2.0354	2.835(15)	146.46
N(1)–H(23) ··· O(2)[ii]	0.9063	2.5782	3.429(17)	156.56'
N(1)–H(24) ··· O(2)[iii]	0.9122	1.8786	2.777(16)	167.97
N(1)–H(25) ··· O(2)[i]	0.9000	1.9990	2.896(13)	174.41

symmetry operation:
(i) x, y, z (ii) $1/2 + x, 3/2 - y, 1 - z$ (iii) $- 1/2 + x, 3/2 - y, 1 - z$ (vi) $1 + x, y, z$

(R)-CHEA:(S)-MA:W

D–H ··· A	D–H (Å)	H ··· A (Å)	D ··· A (Å)	∠ D–H ··· A (°)
O(3)–H(7) ··· O(4)[b,i]	1.0596	1.8003	2.719(7)	142.55
N(1)–H(23) ··· O(1)[ii]	0.8078	2.1318	2.787(7)	138.25
N(1)–H(23) ··· O(3)[ii]	0.8078	2.4355	3.151(8)	148.17'
N(1)–H(24) ··· O(2)[iii]	0.9818	1.8420	2.783(7)	159.58
N(1)–H(25) ··· O(1)[iv]	0.9736	1.9335	2.852(7)	156.34
O(4)–H(26) ··· O(2)[v]	1.0713	1.6540	2.677(7)	157.91

symmetry operation:
(i) $x, 1 + y, z$ (ii) $x, - 1 + y, z$ (iii) x, y, z (iV) $- 5/2 - x, - 1/2 + y, - 1 - z$
(v) $- 2 - x, y, - 1 - z$

[a] A and D denote hydrogen bond acceptor and donor, respectively
[b] O4 designates the oxygen atom of the water molecule

In Table 16, hydrogen bonds in both crystal structures are given. In the crystal structure of (S)-CHEA:(S)-MA, O2 is involved in three hydrogen bonds with N1, whereas O1 is involved in one hydrogen bond with N1. In the crystal structure of (R)-CHEA:(S)-MA:W, however, O1 is involved in two hydrogen bonds with N1. O2 is involved in one hydrogen bond with N1 and one hydrogen bond with the water molecule. As observed in the PTE-MA system, two carboxyl oxygen atoms behave differently in both crystals. In the crystal obtained from the solvent with a smaller dielectric constant, O2 is highly inclined to forming a hydrogen bond with the ammonium ions but in the crystal obtained from the solvent with a larger dielectric constant O1 and O2 almost similarly make hydrogen bonds. It is interesting to note that O3 is involved in intermolecular hydrogen bonds in (R)-CHEA:(S)-MA:W, but not in (S)-CHEA:(S)-MA. A very similar hydrogen-bonding pattern is observed in the PTE-MA system.

The markedly different hydrogen-bond schemes observed in the two crystal structures indicate that the dielectric property of the solvent significantly influences the electronic states of the carboxylate group. The electronic states of the carboxylate group also should play an important role in DCR.

5.3
Molecular Recognition

The crystal structures described above illustrate a molecular mechanism of the DCR phenomenon observed in the (RS)-CHEA:(S)-MA resolution system. In the solution, (S)-MA is considered to be highly inclined to aggregate to form columns because (S)-MA pack in a similar fashion in both crystal structures. Between the columns, the chiral spaces are formed. In the solvent with a smaller dielectric constant, the columns come close together to form a chiral space where (S)-CHEA is better recognized than (R)-CHEA. In the solvent with a smaller dielectric constant, O2 of MA tends significantly towards forming hydrogen bonds with the ammonium ion. In the chiral space the structure of (S)-CHEA can fulfill this requirement and only (S)-CHEA can be recognized in the chiral space. In the solvent with a larger dielectric constant, however, the separation between the hydrophilic columns increases because of reduced electrostatic interactions between them. As water molecules must be incorporated between the columns to realize the close packing, hydrophilic layers are now formed instead of hydrophilic columns. The chiral space is formed between hydrophobic walls with a hydrophilic floor and ceiling as shown in Fig. 12b. The width of the chiral space is narrow in this case. In the solvent with the larger dielectric constant, O1, O2 and O3 can be equally involved in the intermolecular hydrogen bonds with the N atom. Accordingly, the structure of (R)-CHEA is more preferable to (S)-CHEA in this chiral space. In addition, both the CHEA and MA molecules can undergo conformational changes to obtain an optimum fit into the chiral space in order to form crystals to be precipitated.

6
Molecular Mechanism of DCR

We undertook systematic studies regarding the effect of dielectric properties of solvents on chiral resolution in four different resolution systems. Three popular chiral selectors were used in this study. The selectors are structurally variable. The chemical structures of the target molecules are also variable. These four resolution systems clearly showed that dielectric properties of the solvent employed significantly affect the chiral resolution, namely all of them showed the distinct DCR phenomena. Although certain conformational flexibility of chiral selectors and target molecules appears to be necessary, specific molecular prerequisites are not required for DCR. The results indicate that DCR is just dependent on the dielectric property of the solvent employed in the resolution process.

The crystal structures of these four resolution systems described in this paper clearly give us an idea of the common molecular mechanism of DCR. In

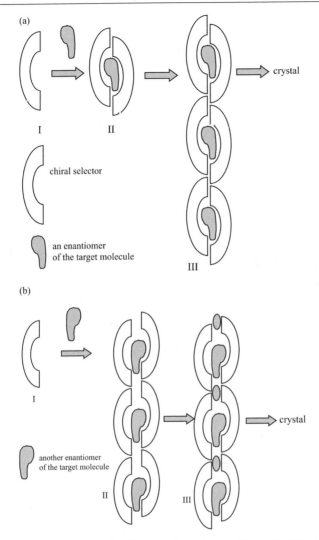

Fig. 14 Molecular mechanisms of DCR. **a** In the solvent with small dielectric constant. Chiral selectors (I) aggregate and an enantiomer of a target molecule is recognized in the chiral space formed by multiple chiral selectors (II). Basic units formed by chiral selectors and the target molecule further aggregate (III) to result in the crystal. **b** In the solvent with large dielectric constant. As hydrophilic moieties of chiral selectors (I) cannot approach, chiral space formed between chiral selectors are more voluminous (II). By filling small molecules up into the void (III), the aggregates finally build up the crystal

each resolution system, the chiral selectors pack very similarly in the crystals of both diastereomeric salts. The separations between the hydrophilic columns formed by chiral selectors, however, are different in both diastereomeric salt crystals. The separations in the crystals obtained from the solvent with the smaller dielectric constant are smaller than those in the crystals obtained from the solvent with the larger dielectric constant. Therefore, the chiral spaces formed by chiral selectors in both crystal structures are appreciably different and each enantiomer can be recognized in one of the chiral spaces by adopting the conformation that optimizes the molecular recognition. In the case where the separations between hydrophilic columns are large, small molecules such as water are required to fill the void to realize the close packing to be crystallized. In these four examples water is incorporated to realize a close packing.

The presumed molecular mechanism of DCR is summarized in Fig. 14. The concept of DCR is that a chiral selector can inherently recognize both enantiomers of a target molecule and the dielectric property of the employed solvent determine the enantiomer to be resolved. In the solvent with the smaller dielectric constant (Fig. 14a), the hydrophilic moieties of the optical selectors are close enough to form a chiral space where one of the enantiomers can be preferentially recognized (II). The close packing brings about the crystal (III). In the solvent with the larger dielectric constant (Fig. 14b), however, the hydrophilic moieties of the chiral selectors should be separated to generate a different chiral space that can be preferentially recognized by another enantiomer. In this case the packing between the chiral selectors is not close enough to enable a substantial void to be generated between the hydrophilic columns (II). If small molecules such as water are available to fill in the void and the close packing is realize, the diastereomeric salt can be obtained as stable crystal (III).

7
Conclusions and Scope

Chiral resolution has been regarded as a strict one-to-one molecular recognition between a specific enantiomer of a chiral selector and the corresponding specific enantiomer of a target molecule. Therefore, both of the enantiomers of chiral selectors have been considered to be absolutely necessary to resolve both enantiomers of the target molecules. The discovery and detailed study of molecular mechanisms of the DCR phenomenon, however, have clearly shown that we should completely change this traditional concept regarding chiral resolution. One chiral selector can inherently recognize both enantiomers of the target molecule. The series of studies described in this work demonstrates that DCR does not depend on specific chemical structures of chiral selectors and target molecules and shows that DCR can be observed in various resolution systems.

Chiral resolution is just a typical example of molecular recognition. The role of the solvent in the process of molecular recognition is not currently taken into account to consider the molecular mechanism. The discovery of

DCR, however, obviously teaches us the essential role of the solvent in various molecular recognition processes. Undoubtedly this discovery will enable new light to be shed on the molecular mechanisms of various important molecular recognition problems such as highly specific molecular recognition in biological systems.

Acknowledgements The authors thank Tomoko Kawamura and Rumiko Tanaka for their excellent technical assistance. One of the authors (N.H.) is grateful for financial support from the Research and Study Program of Tokai University Educational System General Research Organization.

References

1. Jacques J, Collet A, Wilen SH (1981) Enantiomers, Racemates and Resolutions. Wiley, New York, p 251
2. Sakai K, Sakurai R, Yuzawa A, Hirayama N (2003) Tetrahedron-Asymmetr 14:3713
3. Sakai K, Sakurai S, Hirayama N (2004) Tetrahedron-Asymmetr 15:1073
4. Jouyban A, Soltanpour S, Chan H-K (2004) Int J Pharm 269:353
5. Sakai K, Sakurai R, Akimoto T, Hirayama N (2005) Org Biomol Chem 3:360
6. Process-Auto (1998) Automatic Data Acquisition and Processing Package for Imaging Plate Diffractometer. Rigaku, Tokyo, Japan
7. Alltomare A, Cascarano G, Giacovazzo C, Guagliardi A, Burla M, Polidori G, Camalli M (1994) J Appl Cryst 27:435
8. CrystalStructure Ver.3.6.0 (2004) Crystal Structure Analysis Package, Rigaku and Rigaku/MSC
9. Burnett MN, Johnson CK (1996) ORTEP-III: Oak Ridge Thermal Ellipsoid Plot Program for Crystal Structure Illustration. Oak Ridge National Laboratory Report ORNL-6895
10. WebLab ViewerPro: Version 4.0 (2000) Molecular Simulations Inc., San Diego
11. Fujii I, Hirayama N (2002) Helv Chim Acta 85:2946
12. Gfesser GA, Faghih R, Bennani YL, Curtis MP, Esbenshade TA, Hancock AA, Cowart MD (2005) Bioorg Med Chem Lett 15:2559
13. Sakai K, Sakurai R, Hirayama N (2006) Cryst Growth Design 6:1606
14. Bruno IJ, Cole JC, Edgington PR, Kessler MK, Macrae CF, McCaben P, Pearson J, Taylor R (2002) Acta Crystallogr Sect B 58:389
15. Gilli P, Bertolasi V, Ferretti V, Gill G (1994) J Am Chem Soc 116:909
16. Halgren TA (1996) J Comp Chem 17:490
17. MOE (Molecular Operating Environment), version 2004.04 (2004) Chemical Computing Group Inc, Montreal, Canada
18. Saito K, Magari O (1998) Japan Kokai Tokkyo Koho JP 633035540 (Application JP 1986–181249)
19. Nohira H, Murata H, Asakura I, Terunuma D (1984) Japan Patent Kokai 59-110656
20. Sakai K, Sakurai R, Nohira H, Tanaka R, Hirayama N (2004) Tetrahedron-Asymmetr 15:3495
21. Iwanowicz EJ, Dhar M, Leftheris K, Liu C, Mitt T, Watterson SH, Barrish JC (2002) USP 6,420,403
22. Seki M, Yamada S, Kuroda T, Imashiro R, Shimizu T (2000) Synthesis, p 1677
23. Sakai K, Sakurai R, Hirayama N (2006) Tetrahedron-Asymmetr 17:1541
24. Sakai K, Sakurai R, Hirayama N (2006) Tetrahedron-Asymmetr 17:1812

From Racemates to Single Enantiomers – Chiral Synthetic Drugs over the last 20 Years

Hisamichi Murakami

Ex-Director of R&D Division, Yamakawa Chemical Industry Co., Ltd, Kitaibaraki, 319-1541 Ibaraki, Japan
HsMurakami@aol.com

1	Introduction	274
2	New Drugs Launched in 1985–2004 and their Chirality	275
3	Origins of New Drugs: Countries and Companies	277
4	Therapeutic Aspects	278
5	Structural Features of Single Enantiomers	281
5.1	Number of Chiral Centers	281
5.2	Molecular Weights	283
5.3	Hydrogen Bond Donors and Acceptors	284
6	Manufacturing Methods: Introduction of Chirality	285
6.1	Single Enantiomers with One Chiral Center	286
6.1.1	Chirality Pool	286
6.1.2	Optical Resolution	286
6.1.3	Asymmetric Syntheses	288
6.1.4	Racemic Switches	289
6.2	Single Enantiomers with Two or More Chiral Centers	289
6.2.1	Synthesis from Chiral Building Blocks	290
6.2.2	Resolution of Racemates	290
6.2.3	Separation of Diastereomer Mixtures	291
6.2.4	Crystallization-induced Asymmetric Transformations	291
6.2.5	Diastereoselective Reactions	292
7	Examples of Excellent Syntheses	292
7.1	Indinavir Sulfate (Crixivan)	292
7.2	Aprepitant (Emend)	293
7.3	Tadalafil (Cialis)	294
7.4	Sertralin (Zoloft)	295
8	Conclusions	296
References		298

Abstract Before 1985, most chiral drugs newly introduced annually to the market were racemates and *only three to five drugs* were launched yearly as single enantiomers. Since then, the situation has changed dramatically, with racemates being introduced only rarely to the market. In 2004, all the new synthetic drugs (13 in the world) were launched as single enantiomers.

This article summarizes the change of chiral drugs in the last two decades from racemates to single enantiomers and outlines the structural features of these new drugs and the manufacturing technologies applied to introduce chirality to them – chirality pool, various types of resolutions and asymmetric syntheses, of which the new technologies developed recently are highlighted.

Keywords Asymmetric synthesis · Chirality pool · Resolution · Single enantiomer · Synthetic drug

Abbreviations
ACE	Angiotensin converting enzyme
CMV	Cytomegalovirus
CNS	Central nervous system
CSA	10-Camphorsulfonic acid
DBTA	O,O'-Dibenzoyl-tartaric acid
DTTA	O,O'-Ditoluoyl-tartaric acid
HMG-CoA	3-Hydroxy-3-methylglutaryl coenzyme A
IPA	Isopropyl alcohol
MA	Mandelic acid
MBA	α-Methylbenzylamine
Mw	Molecular weight
NCE	New chemical entitiy
NCS	N-Chlorosuccinimide
NK1	Neurokinin 1
SMB	Simulated moving bed
TA	Tartaric acid
p-TSA	p-Toluenesulfonic acid

1
Introduction

In 1984, E.J. Ariens presented a paper, in which he declared the therapeutically non-active isomer in a racemate should be regarded as an impurity (50% or more). He also pointed out that in clinical pharmacology, and particularly in pharmacokinetics, neglect of stereoselectivity in action leads to the performance of expensive "highly sophisticated scientific nonsense" [1]. At that time, most of chiral synthetic drugs on the market were used as racemates and the proportion of single enantiomers was only 12% of chiral drugs [2, 3]. After two decades, most of newly launched chiral synthetic drugs are single enantiomers, and only one or two racemates or diastereomers are introduced in a year. For the first time in 2004, all the approved chiral synthetic drugs went to market as single enantiomers. In this article, the course of these changes proceeded in these two decades, that is "from racemates to single enantiomers", will be confirmed and the technologies supporting this revolution will be reviewed.

2
New Drugs Launched in 1985–2004 and their Chirality

In the 20 years between 1985 and 2004, 754 new drugs (new chemical and biological entities) were launched in the world. These drugs were classified according to their source and chirality, and the number of drugs appeared in each four-year period are illustrated in Table 1. These data are mainly obtained from the chapters "To Market, to Market" of Annual Reports in Medicinal Chemistry, published annually under the sponsorship of the Division of Medicinal Chemistry of the American Chemical Society [4].

As shown in Table 1, among the total 754 entities, synthetic drugs constitute the largest group of 550 (73%), followed by semisynthetics (128) and biologicals (76). The number of newly launched drugs decreases through this period constantly, from 204 in 1985–1988 to 113 in 2001–2004.

With synthetic drugs, the proportion of achiral drugs decreased gradually from 48% (1985–1988) to 35% (2001–2004). The most striking observations are the increase of single enantiomers in chiral drugs and the decrease of racemates, which dominated two-thirds of chirals in 1985–1988. Afterwards they gradually declined and then dramatically dropped to only 3 at the opening of the 21st century. In the year 2004, among the 16 newly approved

Table 1 New drugs launched in 1985–2004

Year	1985–88	1989–92	1993–96	1997–00	2001–04	Total
Grand total	204	141	160	136	113	754
Synthetics total	146	104	116	109	75	550
Achiral	70	46	49	46	26	237
Chiral total	76	58	67	63	49	313
Racemates	50	32	30	22	3	137
Diastereomers	2	5	0	0	2	9
Single enantiomers (SE)	24	21	37	41	44	167
SE[a]/Chiral (%)	31.6	36.2	55.2	65.6	89.8	53.5
Achirals/Synthetics (%)	47.9	44.2	42.2	41.8	34.7	43
Semisynthetics total	41	24	28	20	15	128
Biologics total	17	13	16	7	23	76
Natural	13	7	1	1	4	26
Recombinant	4	6	15	6	19	50

[a] Single enantiomers

synthetic drugs, 13 chirals were all single enantiomers and the rest, only three, were achiral. These changes to synthetic drugs with chirality are illustrated in Fig. 1.

The move from racemates to single enantiomers has been driven by the guidelines issued by regulatory authorities in the late 1980s, which require the applicants of racemic drug candidates to submit scientific evidences to use racemates not as single enantiomers [5, 6]. As a consequence, pharmaceutical companies had to investigate the stereochemical features of their drug candidates in the early stage of development and, in most cases, they selected single enantiomers for further development. Racemates were chosen only exceptionally, when the single enantiomer racemizes easily in vitro and/or in vivo, individual enantiomers have similar pharmacological and toxicological profiles, or the use of racemates results in some synergic effects to exhibit better pharmacological or toxicological outcome, etc.

The number of achiral drugs also decreased significantly but the proportion of 35–40% has been kept in synthetic drugs. Newer therapeutic groups, such as angiotensin II receptor antagonists, tyrosine kinase inhibitors all consist of achiral compounds.

The new launches of semisynthetic drugs have nearly halved in the last ten years, from 84 new entities in 1985–1994 to 44 in 1995–2004. This decrease of new launches is mainly ascribed to the cease of development of cephalosporin antibiotics (from 19 new launches in the former ten years to only three in the latter). A slow-down of the appearance of steroids and macrolide antibiotics is also observed, from 18 and 15 to 11 and 8.

Although the biological drugs including recombinant products were launched at a nearly constant pace to the total of 76 entities in these 20 years, natural products extracted from plants, animals or fermentation broths decreased significantly at the end of 20th century. On the contrary, approved

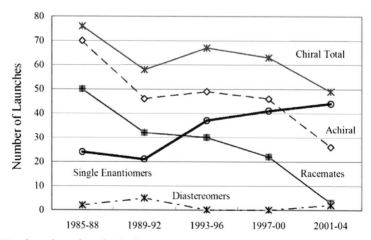

Fig. 1 New launches of synthetic drugs: 1985–2004

numbers of recombinant products, first launched at the end of 1980s, is increasing constantly and in 2004 reached the total of 50.

3
Origins of New Drugs: Countries and Companies

In Table 2, the numbers of new synthetic drugs launched in each four-year period are listed according to their chirality and countries of origin. The largest number of chiral and single enantiomeric drugs have their origins in

Table 2 Origins of chiral synthetic drugs: 1985–2004

Originated in	Type[a]	1985–88	1989–92	1993–96	1997–00	2001–04	Total
USA	Synth.	33	19	26	37	31	146
	Chiral	22	11	14	19	25	91
	SE	9	7	11	14	25	66
Japan	Synth.	35	29	37	20	17	138
	Chiral	15	18	25	14	11	83
	SE	7	5	12	5	9	38
UK	Synth.	11	17	13	13	7	61
	Chiral	4	10	6	5	4	29
	SE	1	2	4	5	4	16
Switzerland	Synth.	6	9	10	5	5	35
	Chiral	4	4	7	3	2	20
	SE	2	2	3	3	1	11
France	Synth.	16	2	7	5	5	35
	Chiral	8	1	4	3	3	19
	SE	1	0	2	2	3	8
Germany	Synth.	14	8	9	13	3	47
	Chiral	6	4	3	8	1	22
	SE	0	1	0	6	0	7
Italy	Synth.	17	4	4	3	0	28
	Chiral	9	2	1	2	0	14
	SE	3	2	1	1	0	7
Sweden	Synth.	3	3	1	2	2	11
	Chiral	2	2	1	2	1	8
	SE	0	1	1	1	1	4
Other countries	Synth.	11	13	9	11	5	49
	Chiral	6	6	6	7	2	27
	SE	1	1	3	4	1	10
Total	Synth.	146	104	116	109	75	550
	Chiral	76	58	67	63	49	313
	SE	24	21	37	41	44	167

[a] Synth. = synthetics, Chiral = racemates + diastereomers + single enantiomers (SE)

the USA, followed by Japan and the UK. In these three countries, 204 chiral drugs (65.0% of total 314) were discovered, of which 121 entities are single enantiomers (72.0% of total 167). In Switzerland, France and Germany, a significant number of single enantiomers (26, 15.5%) are also originated. Chiral synthetic drugs discovered in these six countries account for 84.4% of chiral drugs (265 entities) and 87.5% of single enantiomers (147).

The change to single enantiomers proceeded most rapidly in the USA and the UK as early in the 1993–1996 period, when the proportion of single enantiomers in chiral drugs has reached to 78% in the USA and 67% in the UK. After 1997, all the 10 new chiral drugs originated in the UK went to market as single enantiomers, and later in 2001–2004, 25 new chiral drugs discovered in the USA were also launched entirely as single enantiomers.

As to the originators, Merck leads the discovery of single enantiomers with 10 NCE's in the past 20 years, followed by AstraZeneca (8), Hoffmann-La Roche (7), Novartis (7), and Glaxo SmithKline (6). Pfizer itself discovered 4 NCE's in this period, but as a result of mergers with Warner-Lambert and Pharmacia, 11 NCE's have their origins in the Pfizer group.

4
Therapeutic Aspects

The aspects in the therapeutic areas are indicated in Table 3, where single enantiomeric new drugs were actively developed.

1) Cardiovascular Agents: 32 NCE's were launched in this area, of which angiotensin converting enzyme (ACE) inhibitors are ranked as the top of all therapeutic groups with 15 new drugs. The first ACE inhibitor, captopril, was introduced to the market in 1981, followed by enalapril in 1984, then 15 new entities of this type were launched from 1987 to the end of the 20th century. Angiotensin II receptor antagonists, recently developed potent and selective antihypertensives led by losartan (1994), are all achiral compounds except for valsartan (1996), which is an optically active L-valine derivative, and candesartan silexetil (1997), a racemic prodrug.
The second active area of the cardiovascular agents is antihyperlipoproteinemic agents with five synthetic HMG-CoA reductase inhibitors (statins) and a new type of small molecule cholesterol absorption inhibitor (ezetimibe, 2002). Lovastatin and two other statins introduced in 1980's are semisynthetic compounds derived from fermentation products and not counted in Table 3. Six antithrombotics have been introduced to the market in this period, including entities with various chemical structures from prostaglandin to oligopeptides.
2) The second largest therapeutic group is antivirals, including anti-HIV agents. Seven reverse transcriptase inhibitors and eight protease inhibitors

Table 3 Single enantiomers by therapeutic categories

Therapeutic categories	1985–88	1989–92	1993–96	1997–00	2001–04	Total
Cardiovascular agents						
Antihypertensives	4	7	6	2	1	20
ACE inhibitors	(4)	(5)	(5)	(1)	(0)	(15)
Others	(0)	(2)	(1)	(1)	(1)	(5)
Antihyperlipoproteinemics	0	0	1	2	3[a]	6
Antithrombotics	1	0	0	4	1	6
Antivirals						
HIV reverse transcriptase inhibitors	0	1	2	2[b]	2	7
HIV protease inhibitors	0	0	3	3	2	8
Others	0	0	2	0	1	3
Antineoplastics						
Antimetabolites	1	1	4	1	3	10
Pt complexes	0	0	1	1	0	2
Peptides and others	1	0	2	0	3[c]	6
Antibacterials						
Antibiotics (carbapenem, etc.) (two adjuvants included)	3[d]	2[e]	2	1	2	10
Quinolones and others	0	0	1	2[f]	1	4
Antiglaucomas	0	0	3	1	2	6
Antidepressants	0	2	0	0	3	5
Hormonals (peptides, etc.)	0	0	1	3	0	4
Antidiabetics	0	0	0	2	1	3
Antiemetics	0	0	1	0	2[g]	3
Antimigraines	0	0	0	1	2	3
Total	10	13	29	25	29	106
Grand total[h]	24	21	37	41	44	167

[a] ezetimibe
[b] efavirenz
[c] bortezomib
[d] cilastatin
[e] tazobactam
[f] linezolid
[g] aprepitant
[h] of all single enantiomers in all therapeutic categories

were introduced to the market to overcome HIV infections. As to reverse transcriptase inhibitors, semisynthetic products, zidovudine (1987) and didanosine (1991) were launched at first, then six nucleosides, e.g., zalcitabine (1992) and stavudine (1994) followed. To manufacture these compounds, several processes have been developed and some of them are derived from natural or fermented nucleosides, such as adenosine, guanosine or uridine.

Other than nucleosides, efavirenz was introduced to the market in 1998 as a unique non-nucleoside HIV reverse transcriptase inhibitor.

The first HIV protease inhibitor, saquinavir, was launched in 1995, followed by seven related compounds and used in combination with reverse transcriptase inhibitors. These are peptidomimetic compounds and have relatively complex structures with large molecular weights and many chiral centers.

Apart from anti-HIV drugs, three antiviral compounds for herpes, and AIDS-related CMV infections have been introduced. Two antiviral compounds for influenza, launched in 1999, were not included in this survey, because these are semisynthetic compounds derived from natural resources.

3) Antineoplastics, with 18 new entities, constitute the third group, of which ten antimetabolites, two platinum complexes, four hormonal peptides and two others were launched. As antimetabolites, two antifolates, raltitrexed (1996) and pemetrexed (2004), have been introduced, besides eight nucleosides. Bortezomib (2003) is the first proteasome inhibitor for the treatment of multiple myeloma, and chemically, it is the first boronic acid derivative as a pharmaceutical. Amrubicin (2002) is also the first completely synthetic anthracycline antibiotic.

4) In the field of antibacterials, eight β-lactam antibiotics, two β-lactamase inhibitors as the adjuvants to β-lactam antibiotics, three quinolone derivatives and a new class of antibacterial agent, linezolid (2000), have entered to market. Imipenem was launched in 1985 as the first carbapenem antibiotic, followed by panipenem (1994) and three 1β-methyl-carbapenems, meropenem (1994), biapenem and ertapenem (2002).

5) Six antiglaucomas were launched in these 20 years, of which two carbonic anhydrase inhibitors, dorzolamide (1995) and brinzolamide (1998), have appeared in addition to four prostaglandins, represented by unoprostone (1994).

6) Five antidepressants were introduced to the market, of which sertralin (1990) and paroxetine (1991) are now blockbusters. Escitalopram (2002) is a racemic switch, and atomoxetine (2003) the first non-stimulant drug for attention deficit hyperactivity disorder (ADHD). Duloxetine (2004) is approved for the treatment of stress urinary incontinence and diabetic peripheral neuropathic pain, in addition to the treatment of major depression.

7) In the fields of antiemetics and antimigraines, achiral or racemic compounds have been used initially, but single enantiomers were recently introduced to the market; ramosetron (1996), palonosetron (2003) and aprepitant (2003) as antiemetics, zolmitriptan (1997), eletriptan (2001) and frovatriptan (2002) as antimigraines. Among them, aprepitant is the first NK1 receptor antagonist, while the other antiemetics are 5HT3 receptor antagonists.

5
Structural Features of Single Enantiomers

A number of large and complicated molecules are observed in recent new drugs especially in single enantiomeric synthetics. Number of chiral centers, molecular weights and number of hydrogen bond donors in the drug molecules were selected as the measures of structural complexity and their change in the recent twenty years were surveyed.

5.1
Number of Chiral Centers

Table 4 shows the numbers of single enantiomers and racemates by the number of chiral centers and the launched years.

1) The numbers of chiral centers of single enantiomers distribute from one to over ten, but the entities with one or two chiral centers account for more than half of all the single enantiomers. Among 17 entities with seven or more chiral centers, 14 are peptides consisting of 7 to 38 amino acid units, two are oligonucleotides (fomvirsen (1998) and pegaptanib (2004)), and one polymeric Zn complex of amino acid derivative (polaprezinc (1994)).
2) The number of single enantiomers having one chiral center increased significantly after 1995. While only 14 entities with one chiral center were launched in ten years from 1985 to 1994, 44 have entered the market in the period from 1995 to 2004. On the other hand, launches of racemates

Table 4 New chiral synthetic drugs: The number of chiral centers

	No. of Chiral centers	Launched in 1985–88	1989–92	1993–96	1997–00	2001–04	Total	Ratio (%)
Single Enantiomers	1	6	4	11	21	16	58	34.7
	2	4	9	5	6	11	35	20.9
	3	5	5	6	3	3	22	13.2
	4	2	1	6	4	5	18	10.8
	5–6	4	1	5	2	5	17	10.2
	≥ 7	3	1	4	5	4	17	10.2
	Total	24	21	37	41	44	167	100.0
Racemates	1	44	28	27	14	3	116	84.7
	2	5	3	1	4	0	13	9.5
	≥ 3	1	1	2	4	0	8	5.8
	Total	50	32	30	22	3	137	100.0

with one chiral center had dramatically decreased from 89 in 1985–1994 to 27 in 1995–2004, clearly reflecting the switch from racemates to single enantiomers.

3) Single enantiomers with two chiral centers had also increased moderately from 15 in 1985–1994 to 20 in 1995–2004.

4) Fifty-seven new single enantiomeric drugs having three to six chiral centers were launched in the past 20 years, of which 26 appeared in 1985 to 1994, and 31 in 1995 to 2004. The largest group in this category is ACE inhibitors with 13 entities, of which ten were launched in the first ten years. Tied for second with eight each are the antineoplastics with nucleoside moieties and HIV protease inhibitors; the latter were all introduced after 1995. Seven each of β-lactam antibiotics and prostaglandins, of which four entities are antiglaucomas, have also entered the market. These five groups of new drugs constitute about three quarters of the total 57 new entries (Table 5).

5) Contrary to single enantiomers, most of the racemates (116 of total 137) have only one chiral center and the racemates with two chiral centers are merely 13, of which most compounds are the mixture of two antipodes with the same relative configuration such as (R^*,R^*) or (R^*,S^*). Only three compounds are the mixture of four isomers and the two of them (loxoprofen (1986) and troglitazone (1997)) have labile chiral centers easily converting in the body. Introduced in 1985 to 1997, five racemates with three or four chiral centers, all consist of two enantiomers.

As the increasing number of isomers causes a sharp decrease in the proportion of the active isomers, inventors should inevitably select structures with the definite relative configuration devising any stereoselective synthetic schemes to reduce the number of isomers.

Table 5 Major drug groups: The number of chiral centers (1985–2004)

Number of chiral centers	1	2	3	4	5	6	$\geqq 7$	Total
ACE inhibitors	0	2[a]	10[b]	0	3	0	0	15
HIV protease inhibitors	0	0	2	3	3	0	0	8
Nucleosides								
Antineoplastics	0	0	2	6	0	0	0	8
Antivirals	2	5	0	0	0	0	0	7
β-Lactam antibiotics	0	1	2	3	0	2	0	8
Prostaglandins	0	0	0	1	6	0	0	7[c]
Peptides	0	0	0	0	2	0	13	15

[a] Besides these, captopril was launched in 1981
[b] Besides these, appeared enalapril (1984), and fosinopril (1991, diastereomer)
[c] Besides these, four diastereomers and one racemate have been launched

5.2
Molecular Weights

As a measure of the size of drug molecules, molecular weights (Mws) of new drugs were surveyed and listed in Table 6. In case a drug is sold as its salt, the molecular weight of the free form is usually adopted. The average Mws of the entities launched in each four-year period were calculated, excluding the ones with larger Mws than 800, which are peptides or oligonucleotides.

1) As shown in Table 6, the Mws of most single enantiomers reside in the range of 200 and 500 (76%), and only 17 large molecules with Mws 500–800 (10%) are used, except for 17 peptides and 2 oligonucleotides (Mw > 800). The average Mws in 1985–1992 (about 340) have increased sharply to 408 in 1993–1996 and the level of 370 is kept thereafter. This is explained by the increase in the large molecules with Mws over 500 after 1995, represented by HIV protease inhibitors.
2) Of the 17 drugs with Mws 500–800, 12 are administered orally. Besides eight HIV protease inhibitors with Mws 505.6–721.0, tenofovir disoproxil fumarate (antineoplastic, 519.4), aprepitant (antiemetic, 534.4), atorvastatin (antihyperlipoproteinemic, 558.6) and montelukast (antiallergic, 586.2) are sold as tablets or capsules.
3) In comparisons with single enantiomers, achiral drugs are smaller and racemates are about the same in their Mws. Particularly to be commented,

Table 6 Molecular weights of synthetic drugs

	Molecular Weight	Launched in 1985–88	1989–92	1993–96	1997–00	2001–04	Total
Single Enantiomers	< 200	0	0	0	1	3	4
	200–300	10	7	8	11	8	44
	300–400	5	6	8	8	14	41
	400–500	5	7	11	10	9	42
	500–600	0	0	1	4	5	10
	600–800	1	0	4	1	1	7
	> 800	3	1	5	6	4	19
	Total	24	21	37	41	44	167
	Average Mw[a]	341.5	336.6	408.2	373.7	373.5	367.0
Racemates	Number of drugs	50	32	30	22	3	137
	Average Mw[a]	349.7	359.6	377.0	373.8	437.0	364.3
Achirals	Number of drugs	70	46	49	46	26	237
	Average Mw[a]	318.4	314.6	328.5	339.8	389.5	331.4

[a] Entities with Mws > 800 are excluded

the average Mws of these two categories are constantly increasing year to year.
4) As a whole, the average size of drug molecules has gone up in the last 20 years: the ratio of average Mws of (2001–2004)/(1985–1992) for single enantiomers is 1.09, for racemates 1.25 and 1.22 for achirals. In 2001–2004, the average Mws of single enantiomers is smaller than those of the other groups.

5.3
Hydrogen Bond Donors and Acceptors

C.A. Lipinski defined the number of OH and NH groups in a drug molecule as "hydrogen bond donors" and the number of oxygen and nitrogen atoms as "hydrogen bond acceptors", recognizing them as measures of drug permeability and absorption in the body, along with Mws and log P's (P: partition coefficient of a drug compound in 1-octanol-water). He proposed the "rule of 5" upon the investigation of a large number of marketed and investigational drugs [7]. The rule predicts a poor absorption or permeation in the body for a drug compound, which meets one or two of the four criteria shown below:

1) There are more than 5 hydrogen bond donors.
2) The molecular weight is over 500.
3) The log P is over 5.
4) There are more than 10 hydrogen bond acceptors.

Here the numbers of hydrogen bond donors of the new synthetic drugs launched after 1985 are listed in Table 7.

As shown in Table 7, about 90% of single enantiomers have 1 to 5 hydrogen bond donors, and only 6 chemical entities (4.0%) have 6 or 7 donors. Among these six drugs, only droxidopa (Mw 213.2) and saquinavir (670.8) are administered orally and other four (Mw 285–483) by injection.

The average numbers of hydrogen bond donors calculated for the entities having less than 10 donors are as follows: single enantiomers 2.37, racemates 1.41 and achirals 1.80. The implication of these differences in the number of hydrogen bond donors is not certain.

As to the single enantiomers and racemates, their average numbers of chiral centers are 2.43 and 1.20, respectively, and closely resemble the number of hydrogen bond donors. But the numbers of hydrogen bond donors scatter in wide range from zero to over five for all single enantiomeric and racemic drugs with one to five chiral centers. No apparent correlations are observed between these two numbers.

Hydrogen bond acceptors are also checked and 12 single enantiomers are found to have 11 to 15 acceptors, excluding peptides and oligonucleotides. Of these 12 drugs, seven have large Mws over 500, of which four are HIV protease inhibitors and administered orally. Among racemates and achiral drugs, enti-

Table 7 Hydrogen bond donors: Synthetic drugs 1985–2004

	H-bond donors	Launched in 1985–88	1989–92	1993–96	1997–00	2001–04	Total
Single Enantiomers	0	0	0	1	2	4	7
	1	5	7	5	10	9	36
	2	7	7	10	9	5	38
	3	4	2	6	6	8	26
	4	1	1	7	7	7	23
	5	2	1	2	1	5	11
	6–9	1	2	1	0	2	6
	≥ 10	(4)	(1)	(5)	(6)	(4)	(20)
	Total	20 + (4)	20 + (1)	32 + (5)	35 + (6)	40 + (4)	147 + (20)
	Average[a]	2.58	2.40	2.71	2.19	2.72	2.37
Racemates	No. of drugs	50	32	29 + (1)	20 + (2)	3	134 + (3)
	Average[a]	1.60	1.34	1.21	1.52	2.33	1.41
Achirals	No. of drugs	70	46	48 + (1)	46	26	236 + (1)
	Average[a]	1.90	1.65	2.02	1.76	1.85	1.80

[a] Drugs with more than 10 hydrogen bond donors were omitted for calculation of the average

ties having more than ten hydrogen bond acceptors are rarer than in single enantiomeric drugs (only three in racemates and five in achirals).

6
Manufacturing Methods: Introduction of Chirality

How to introduce the chirality to the target drug molecule is the inevitable subject in the syntheses of single enantiomeric drugs and innumerable procedures have been developed and applied in the pharmaceutical industry.

Single enantiomeric drugs launched between 1985 and 2004 belong to diverse structure types and have one to more than ten chiral centers. Here, for convenience's sake, these drugs were divided into the entities having only one chiral center and those with two or more chiral centers, to inspect and classify the methods of introducing the chirality.

Usually, several processes are developed to manufacture a single target drug molecule, and it is not certain which process is actually adopted in commercial production. Moreover, process changes are often reported even after the launch of new drugs. The manufacturing processes referred to the individual drug compounds in this chapter are estimated according to the author's best knowledge.

6.1
Single Enantiomers with One Chiral Center

As shown in Table 4, 58 entities with one chiral center were launched in 1985–2004. Their manufacturing methods are classified conventionally, as follows:

1) Chirality pool: syntheses from optically active intermediates or building blocks
2) Optical resolution of the appropriate racemates, in the course of any stage in the manufacturing processes
3) Asymmetric synthesis, including enzymatic or biological procedures.

Chirality pool originally meant the relatively inexpensive natural or fermented products, such as carbohydrates, amino acids, oxy-acids, alkaloids, terpenes, etc. Recently, a variety of optically active compounds are produced on a large scale by various methods and available easily from a number of suppliers. Therefore it is relevant to include these products as important constituents of chirality pool.

The chirality of single enantiomeric drugs with one chiral center launched in 1985–2004 are introduced by the three procedures, according to the author's estimation or supposition, as follows:

Chirality pool: 26
Optical resolution: 27
Asymmetric synthesis: 5

6.1.1
Chirality Pool

Nearly a half of single enantiomeric drugs with one chiral center are produced by introducing their chirality, using optically active compounds or chirality pool as the starting material or intermediates and combining with achiral compounds at an appropriate stage of the synthetic sequence.

Natural amino acids are the most frequently used chirality pool (of 14 drugs), followed by unnatural amino acids and amino acid analogs (6), optically active glycidol derivatives (6). Most frequently used amino acids are L-Cys (bucillamine (1987), telmesteine (1992) and fudosteine (2001)), L-Glu (raltitrexed (1996), carglumic acid (2003) and pemetrexed (2004)) and L-Val (valaciclovir (1995) and valsartan (1996)).

6.1.2
Optical Resolution

Resolution is the most frequently utilized technology to produce single enantiomeric new drugs. As shown in Table 8, 27 new single enantiomeric drugs launched in 1985–2004 are estimated to be manufactured by using any kind

Table 8 Optical resolution: Single enantiomers with one chiral center

Drug Name (Launched in)	Resolving Agents	Substrates
Resolution of starting materials		
Flunoxaprofen (1987)	L-ephedrine	4-(α-cyanoethyl)phenol
Dexrazoxane (1992)[a]	not disclosed	1,2-diaminopropane
Tamsulosin (1993)	(R) MBA[b]	substd. phenylacetone
Tiagabine (1996)	L-TA	Et 3-piperidinecarboxylate
Clopidogrel (1998)	L-TA	2-Cl-phenylglycine Me ester
Repaglinide (1998)	N-Ac-L-TA	1-Ar-isopentylamine
Ramatroban (2000)	(S)-MBA[b]	3-oxo-tetrahydrocarbazole
Levocetirizine (2001)	L-TA	4-Cl-benzhydrylamine
Cinacalcet (2004)	not disclosed	1-α-naphthylethylamine
Resolution of intermediates		
Ropivacaine (1996)	L-DBTA	pipecolic anilide
Pramipexole (1997)	L-TA	2-aminothiazole deriv.
Rivastigmine (1997)	L-DTTA	N,N-diMe-MBA deriv.
Bepotastine (2000)	N-Ac-L-Phe	4-diarylmethoxy-piperidine
Levosimendan (2000)	L-TA	4-Me-pyridazinone deriv.
Escitalopram (2002)[c]	L-TA	dimethylaminobutanol deriv.
Duloxetine (2004)	(S)-MA	thiophene deriv.
Mitiglinide (2004)[d]	(R)-MBA	2-benzylsuccinic acid
Resolution at the final step		
Denopamine (1988)	D-Ac-Phe	racemate
Dexibuprofen (1994)[a]	(S)-Me-Ph-butylamine	racemate (ibuprofen)
Ramostron (1996)	D-DBTA	racemate
Dexfenfluramine (1997)[a]	(+)-camphoric acid	racemate (fenfluramine)
Levalbuterol (1999)[a,d]	D-DTTA	penultimate intermediate
Dexmedetomidine (2000)[a]	L-TA	racemate (medetomidine)
Frovatriptan (2002)	(+)-CSA	racemate
Atomoxetine (2003)[d]	(S)-MA	racemate
Eszopiclone (2004)[a,e]	D-DBTA	racemate (zopiclone)
Pregabalin (2004)[d,e]	(S)-MA	racemate

[a] Racemic switch
[b] Reductive amination with a prochiral ketone
[c] Process change to SMB separation is reported
[d] Asymmetric synthesis developed as the alternative process
[e] Enzymatic kinetic resolutions are developed

of resolution, of which the majority of drugs are produced via diastereomer crystallization using resolving agents, except for one by continuous chromatographic (SMB) resolution (escitalopram, 2002) [8].

Diastereomer crystallization is well known as a classical resolution and apt to be thought of as out-of-date technology, but as a sharp increase of single enantiomeric drugs with one chiral center after 1995, the number of new drugs manufactured by using this technology also increased dramatically: from only 5 entities in 1985–1994 to 22 in 1995–2004.

Most frequently used resolving agents are natural L-tartaric acid and its bis-aroylated derivatives (D- or L-dibenzoyl- and ditoluoyl-), followed by (R)- or (S)-mandelic acid.

Basic resolving agents, represented by α-methylbenzylamine are used in five resolutions. In the two cases, the amine is condensed with prochiral ketones and reduced to diastereomeric secondary amine mixtures, followed by separation by crystallization and hydrogenolysis, to give optically active primary amines (tamsulosin (1993) and ramatroban (2000)). Benzylsuccinic acid is efficiently resolved by the salt formation with (R)-α-methylbenzylamine for the production of mitiglinide (2004). Alternative preparation via asymmetric synthesis is also developed for this entity.

Eszopiclone (2004), the racemic switch of a hypnotic, zopiclone, can be manufactured by classical resolution of the racemate or by enzymatic resolution of precursors [9]. For pregabalin (2004), a classical resolution process using (S)-mandelic acid as the resolving agent was established at a relatively early stage of its development [10], but a more efficient manufacturing process by asymmetric synthesis [11] or enzymatic procedure is reportedly implemented in the commercial production. The newer processes by asymmetric reduction of precursor ketones are also developed for levalbuterol (1999) and atomoxetine (2003) (see next section).

In general, resolutions at an early step of the synthesis are recommended as the golden rule from the economical and environmental points of view. Actually in this survey, only 7 resolutions are carried out at the starting step, 11 at the intermediate steps, and 9 at the final step. Four of the nine resolved at the last stage are so-called racemic switches of generic drugs, and it may be reasonable to resolve the relatively cheap racemates (Sect. 6.1.4).

6.1.3
Asymmetric Syntheses

Five new chemical entities are manufactured by using asymmetric syntheses. Early in 1993, the commercial production of (S)-1,2-propanediol, the key intermediate of levofloxacin, was commenced by a catalytic asymmetric hydrogenation of hydroxyacetone. An asymmetric alkylation process was first applied in the production of efavirenz (1998), a unique non-nucleoside HIV reverse transcriptase inhibitor. The key step, addition of cyclopropylacetylene to a prochiral ketone in the presence of chiral aminoalchohol and BuLi, is achieved with an excellent enantioselectivity of 98% [12].

For the stereoselective oxidation of the thioether group of omeprazole precursor to its (S)-sulfoxide, esomeprazole (2000), two processes, Sharpless oxidation using a Ti tartrate catalyst and a microbial oxidation, have been developed, of which the former is estimated to be implemented in the commercial production [13, 14].

Asymmetric reduction of the keto group with chiral borane reagents is utilized in the manufacture of brinzolamide and montelukast (both in 1998). This procedure is also reported to be adopted in the improved syntheses of levalbuterol (1999) [15] and atomoxetine (2003) [16] instead of classical resolutions.

6.1.4
Racemic Switches

From the mid-1980s, the single enantiomeric version of racemic drugs in the market have been intensively developed as "racemic switches", with the expectation that they would have a higher efficacy, better pharmacokinetic profiles or reduced side effects and so on, in comparisons with the parent racemates. Some of them have been launched and, today, 13 racemic switches are identified on the market, of which AstraZeneca's antiulcerative esomeprazole is the most successful one [13]. All these drugs have one chiral center with the exception of dexmethylphenidate, which has two chiral centers.

These switched single enantiomers are manufactured by any of the methods depicted in this chapter. Seven drugs are produced by using classical resolution, of which, four are resolved at the final step of the syntheses as referred in Sect. 6.1.2. Dexmethylphenidate (2002) is also probably made by the resolution of its *threo*-racemate, methylphenidate. Asymmeric syntheses are applied to the commercial syntheses of levofloxacin (1993), levalbuterol (1999) and esomeprazole (2000). A new technology of SMB chromatographic separation is reportedly utilized in the production of escitalopram (2002). Levobupivacaine(2000) is produced by using Cbz-L-Lys as the starting material. As to the latest switch, eszopiclone (2004), several methods are developed for its manufacture, including classical resolution or chromatographic separation of the racemate and enzymatic kinetic resolution of the intermediate ester by using an immobilized lipase.

6.2
Single Enantiomers with Two or More Chiral Centers

Over 100 new drugs with more than one chiral centers were launched in these 20 years. But, as shown in Tables 4 and 5, the entities with seven or more chiral centers are almost peptides, and the majority of drugs with four to six chiral centers belong to specific drug groups with their characteristic structures and synthetic methodologies, exemplified by prostaglandins or β-lactam antibiotics, etc. In this section, syntheses of single enantiomers having two or three chiral centers are scrutinized and the characteristic methodologies different from those for the entities with only one chiral center will be highlighted.

There are five fundamental methods to construct single enantiomers having two or more chiral centers.
1) Synthesis from chiral building blocks
2) Resolution of racemates having definite relative configurations
3) Separation of the diastereomer mixture derived from a single enantiomeric precursor
4) Crystallization-induced asymmetric transformation
5) Diastereoselective reaction

These methods will be outlined with some typical examples.

6.2.1
Synthesis from Chiral Building Blocks

Two or more optically active building blocks including chirality pool are combined successively. By repeating this procedure, large compounds having multiple chiral centers such as peptides or oligonucleotides can be constructed. ACE inhibitors are, in principle, produced by this method, but in many cases, stereoselective reactions are introduced to reduce the number of necessary chiral intermediates (Sect. 6.2.3).

As simpler examples, pidotimod (1993, immunostimulant) is synthesized by condensing L-pyroglutamic acid with (R)-thiazolidine-4-carboxylate; taltirelin (2000, CNS stimulant) by condensing L-His with L-ProNH$_2$, and then with L-dihydroorotic acid derivative.

6.2.2
Resolution of Racemates

The racemic compounds having two or three chiral centers and the definite relative configurations such as (R^*,R^*) and (R^*,S^*) or *cis* and *trans* can be synthesized by stereoselective reactions and resolved to single enantiomers by a classical resolution using appropriate resolving agents (Table 9). Sertralin (1990, antidepressant) and voriconazole (2002, antifungal) are the typical examples. Their racemates with *cis*- or *trans*-configurations are resolved to single enantiomers at the final step of the syntheses. But, regarding sertralin, the commercial process was recently reformed to an extremely rationalized scheme, starting the synthesis with continuous chromatographic resolution of the racemic starting material, a tetralone derivative, to its single enantiomers (see Sect. 7.4) [17].

From Racemates to Single Enantiomers 291

Table 9 Optical resolution: Single enantiomers with two chiral centers

Drug name (Launched in)	Resolving agents	Substrates
Resolution of starting materials or Intermediates		
Paroxetine (1991)	L-TA	penultimate intermediate
Moxifloxacin (1999)	L- and D-TA	N-Bn-diazabicyclononane
Emtricitabine (2003)	enzyme	penultimate intermediate
Resolution of racemates		
Sertraline (1990)[a]	(R)-MA	cis-racemate
Barnidipine (1992)	L-malic acid	diastereomers
Dexmethylphenidate (2002)[b]	BNDHP[c]	threo-racemate
Voriconazole (2002)	(R)-(+)-CSA	(R^*,S^*)-racemate

[a] The process change to SMB completed
[b] Racemic switch
[c] (R)-(–)-binaphthyl-2,2'-diyl hydrogen phosphate

6.2.3
Separation of Diastereomer Mixtures

Starting with chiral building blocks, condensation reactions with appropriate prochiral compounds generate the second chiral centers in the products to afford diastereomer mixtures. Separation of the diastereomers can be achieved by the physical methods as crystallization or chromatography. In many cases, condensation reactions proceed in some diastereoselective fashion to give one isomer in more or less larger proportions.

For example, enalapril maleate (1984) is synthesized by condensing a dipeptide, L-Ala-L-Pro with 4-phenyl-2-oxobutanoate and the resulted imine is hydrogenated over Pd catalyst to give a 87 : 13 mixture of (S,S,S)- and (R,S,S)-isomers, which is crystallized as the maleate salt to give the pure (S,S,S)-maleate. Most of ACE inhibitors are produced by using the similar procedures [18].

Palonosetron (2003, antiemetic) is synthesized by utilizing a chiral building block, (S)-3-aminoquinuclidine, and an intermediate having a C = C double bond is hydrogenated over Pd catalyst to give a 7 : 3 diastereomer mixture, which is separated by one crystallization as hydrochloride to afford the pure (S,S)-isomer [19].

6.2.4
Crystallization-induced Asymmetric Transformations

When the second chiral center, generated as described above, is apt to isomerize under mild conditions and the desired diastereomer crystallizes more easily than the undesired isomer, the desired isomer may be obtained in a nearly quantitative yield by adjusting the reaction conditions. This technology is utilized in the manufacturing processes of two new pharmaceuticals, aprepitant and tadalafil, both launched in 2003 (Sect. 7).

6.2.5
Diastereoselective Reactions

By utilizing the preexisting chiral center as an intramolecular chiral auxiliary, in some favorable cases, the second chiral center may be generated in a very high stereoselectivity. Typical examples of this category are found in the syntheses of indinavir (1996, diastereoselective alkylation and epoxidation), aprepitant (2003, hydrogenation over Pd catalyst), and rosuvastatin (2003, chelation-controlled borane reduction of 1,3-hydroxyketone).

7
Examples of Excellent Syntheses

In this chapter, the recently developed manufacturing processes of four complicated single enantiomeric drugs, having two to five chiral centers, are presented to illustrate how various technologies are combined to elaborate simplified and rationalized processes.

7.1
Indinavir Sulfate (Crixivan)

This is an HIV protease inhibitor having five chiral centers, launched by Merck in 1996. The commercialized process starts with (1S,2R)-1-amino-2-indanol (**1**), which is prepared by a classical resolution of the racemate or by an asymmetric epoxidation of indene followed by Ritter reaction. **1** is acylated with phenylpropionyl chloride and derived to the acetonide **2**, which is treated with allyl bromide to give (S)-**3** in 97 : 3 stereoselectivity and 95% yield. Without purification, **3** is reacted with NCS in the presence of NaI in a two phase mixture of isopropyl acetate and aqueous sodium bicarbonate, followed by the treatment of obtained iodohydrin with a base to give the epoxide **4** in a quantitative yield (diastreomer ratio 97 : 3). After one recystallization, purified **4** is condensed with (S)-**5** and the protective group on the piperazine ring is replaced to give indinavir, which is purified as sulfate [20–22] (Scheme 1).

The two of five chiral centers in the drug molecule are effectively generated in two successive highly efficient diastereoselective reactions. The other three chiral centers come from **1** and **5**; the latter is also supplied through a resolution by diastereomer salt formation of the racemate.

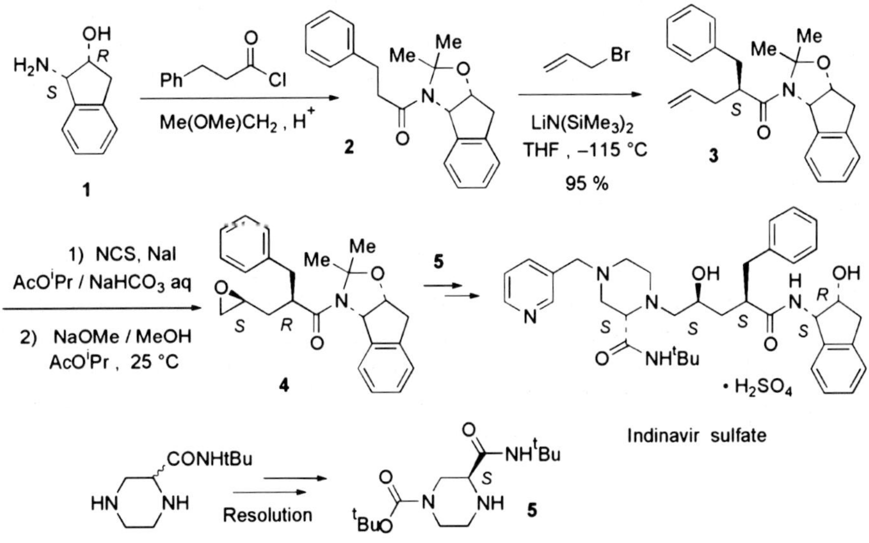

Scheme 1 Synthesis of Indinavir sulfate

7.2
Aprepitant (Emend)

This is a new antiemetic to control chemotherapy-induced nausea and vomiting, also developed by Merck and launched in 2003. In addition to its new mechanism of action, neurokinin 1 receptor antagonist, ingenious methodologies devised for its synthesis are noteworthy. The structural features of this entity are cis-substituted morpholine acetal skeleton and three neighboring chiral centers. Several synthetic routes were intensively explored to elaborate a highly refined process. Starting with an optically active alcohol obtained by an asymmetric reduction of a ketone, the second chiral center is constructed by crystallization-induced asymmetric transformation on the morpholine ring, then the adjacent third chiral center is introduced by a stereoselective hydrogenation of an imino group [23] (Scheme 2).

(R)-1-Phenylethanol 7 produced by an asymmetric reduction of the corresponding ketone is condensed with lactam-lactol 6 to give lactam-acetal 8 as a 55 : 45 diastereomer mixture in nearly quantitative yield. The necessary isomer (R,R)-8 was found to be an easily crystallizing compound, contrary to the lower melting (S,R)-isomer. To accelerate the equilibration between the isomers, solvents and bases were thoroughly investigated. The crude mixture of 8 is dissolved in heptane and the potassium salt of 3,7-dimethyl-3-octanol and seed crystals are added. Kept at a low temperature for 5 hrs, the isomerization proceeds to a diastereomer ratio 96 : 4. By separating the crystals, (R,R)-8 of optical purity > 99% is obtained in an overall yield of 83–85%

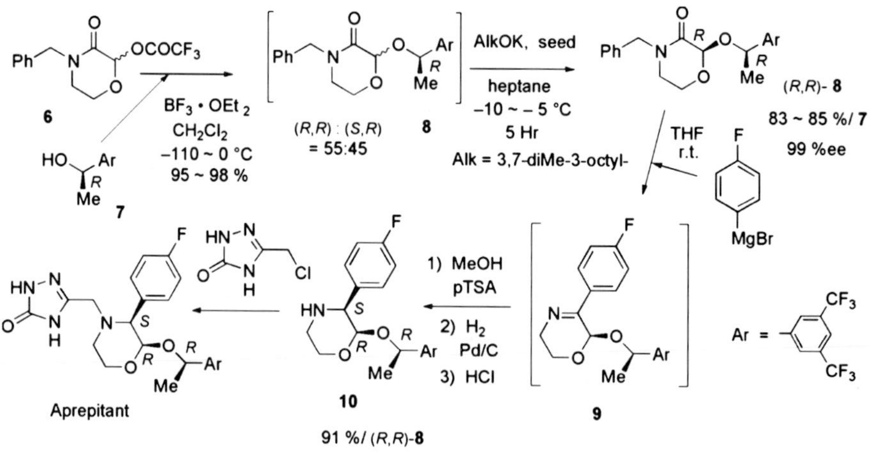

Scheme 2 Synthesis of Aprepitant

based on **7**. Furthermore, even by starting with **7** of lower optical purity (91% ee), optically pure (R,R)-**8** is obtained in a slightly lower yield.

In the next step, (R,R)-**8** is reacted with a Grignard reagent to give a cyclic imine **9**, by going through some complicated steps. The reaction mixture of **9** is quenched with methanol and two equivalents of p-toluenesulfonic acid, then it is hydrogenated in the presence of 5% Pd/C to give (R,R,S)-**10**, isolated as hydrochloride in the yield of 91% from (R,R)-**8**. At the final step, 3-oxo-1,2-triazole ring is introduced on the nitrogen atom of morpholine moiety to give the end product. This excellent process was awarded the Green Chemisty Challenge Award in 2005 [24].

7.3
Tadalafil (Cialis)

After the launch of sildenafil (Pfizer 1998), two PDE5 receptor antagonists, vardenafil (Bayer) and tadalafil (Lilly ICOS) followed it in 2003. Of these three drugs for the treatment of erectile dysfunction (ED), tadalafil is a single enantiomer with two chiral centers, whereas both sildenafil and vardenafil are achiral compounds with very similar structures. Tadalafil was originally invented at a French laboratory of Glaxo in collaboration with ICOS. After Glaxo discontinued the project in 1997, ICOS succeeded to acquire the approval from FDA in cooperation with Lilly.

The synthesis route of tadalafil fundamentally follows the scheme invented at Glaxo [25], but entirely improved reaction conditions, devised by researchers at Lilly ICOS, changed it to a nearly quantitative and highly efficient process [26, 27]. They found that the key intermediate **13**, synthesized by Pictet-Spengler reaction of D-Trp methyl ester hydrochloride (**11**) with

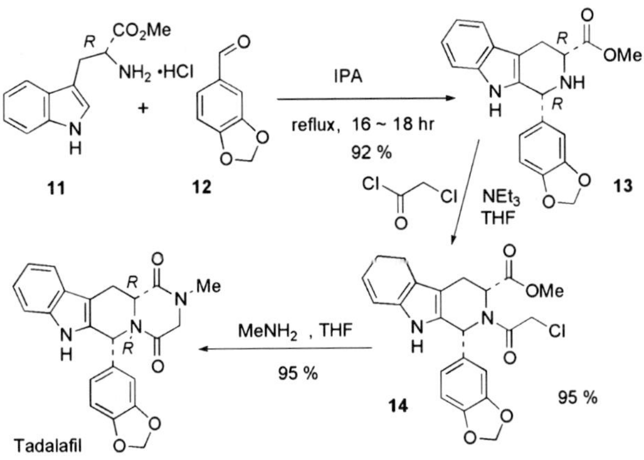

From Racemates to Single Enantiomers

Scheme 3 Synthesis of Tadalafil

piperonal (**12**), easily epimerizes and the necessary (R,R)-isomer is sparingly soluble in isopropanol (IPA). As illustrated in Scheme 3, **13** obtained as a *cis/trans* mixture (6 : 4) is refluxed in IPA for 15 hrs to give the crystalline (R,R)- or *cis*-isomer in the yield of 92%. The key to success was the selection of IPA as the reaction solvent to realize an efficient crystallization-induced asymmetric transformation.

7.4
Sertralin (Zoloft)

Approved in 1990, (Pfizer), Sertralin (Zoloft) is one of the most prescribed antidepressant, and manufactured through an efficient resolution by salt formation with (R)-mandelic acid [28]. Because the resolution was carried out at the final step of the synthesis and a large amount of hazardous waste was produced in the process, extensive efforts to improve the process were made at the Pfizer laboratories. After all the possibilities for the efficient preparation of (4S)-tetralone were evaluated, continuous chromatographic separation of the racemic tetralone **16** was adopted as the commercial process [17]. Extremely pure (S)-**16** is continuously separated on an amylose-based chiral stationary phase in 98% yield and (4R)-isomer is also recovered quantitatively, which can be racemized with base.

Prior to the commercialization of chromatographic separation, the synthesis of sertraline had been highly streamlined by the proper selection of reaction solvents and the catalyst to a cleaner and more environmentally benign process [29]. As depicted in Scheme 4, in the condensation of **15** with methylamine, the change of the solvent to ethanol, resulted in a 95% conversion without using any dehydrating agent, on account of the low-solubility of imine **16** in ethanol. By using Pd/CaCO$_3$ as the catalyst in the hydrogena-

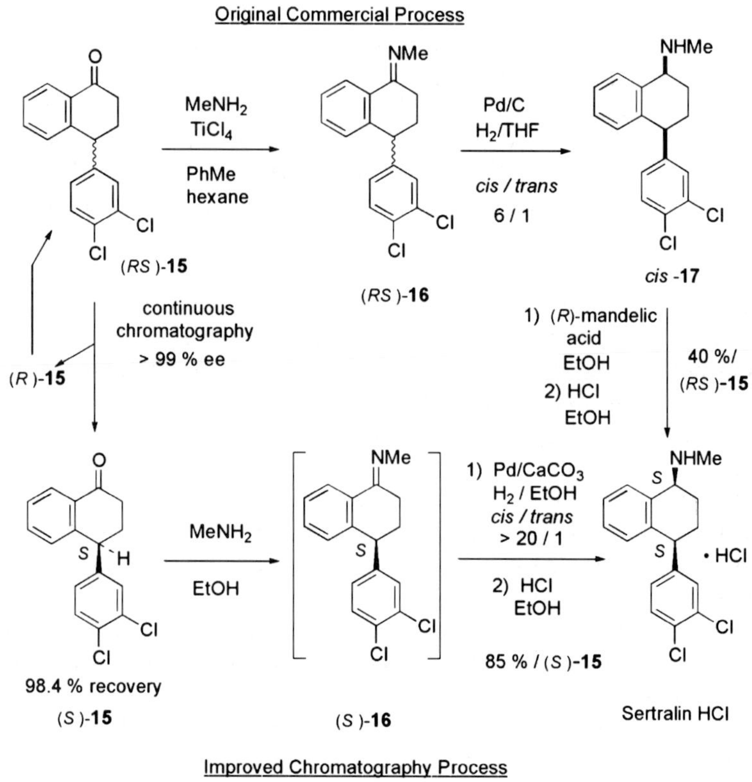

Scheme 4 Synthesis of Sertralin hydrochloride

tion, the *cis/trans* ratio was improved to 20/1 from 6/1 with Pd/C catalyst. Moreover, as this catalyst was found to perform better in ethanol, the hydrogenation reaction without isolating **16** turned out to be feasible. A large solvent requirement in number and quantity was reduced dramatically and the use of hazardous reagents was eliminated. For these improvements, the Green Chemistry Challenge Award in 2002 was given to Pfizer [30].

The combination of the continuous chromatographic separation of racemic tetralone **15** to the improved reactions depicted above realized a resolution at the starting step of the sertralin synthesis and doubled the plant capacity of the later steps.

8
Conclusions

As illustrated in Fig. 1, the switch from racemic drugs to single enantiomers has steadily advanced through the 1990s, and now at the beginning of the

21st century, nearly all the chiral synthetic drugs are launched as single enantiomers. In the two decades, from 1985 to 2004, 551 new synthetic drugs were launched, of which 168 single enantiomers, 137 racemates and 9 diastereomers are counted. According to E.J. Ariens, among 1200 synthetic drugs used in 1982, only 58 were sold as single enantiomers [2]. Seven more single enantiomers were launched in 1983 and 1984 by the author's search. Thus, among the total 233 single enantiomeric synthetic drugs now in use, nearly three quarters were introduced to the market after 1985.

About one-third (58 entities) of 168 single enantiomers launched after 1985 have one chiral center, the next one third (56) two or three chiral centers, and the rest (54) four or more chiral centers. Nearly half of the entities with one chiral center (27 drugs) are produced by resolution of the racemates using resolving agents, except for one by SMB chromatographic separation (Table 8). It is noteworthy that the majority of these drugs produced by using resolution were launched in the last ten years (22 among total 27). The next half (26 drugs) are constructed by using chirality pool as building blocks, which are natural or unnatural amino acids and optically active glycidol derivatives. The rest, the last five entities, are manufactured employing various types of asymmetric syntheses.

In the manufacture of single enantiomers with two or more chiral centers, a number of technologies are often utilized in combination. Stereoselective reactions mediated by preexisting chirality in the intermediates are efficient means for constructing complicated molecules with multiple chiral centers and successfully applied in the production of some new drugs. Crystallization-induced asymmetric transformation of diastereomer mixtures is the most effective means if applicable.

Chromatographic separation of enantiomer mixtures, especially SMB technology, has recently been introduced to the production of drugs or their intermediates, giving impacts to the industry with its high separation ability and productivity [16]. This technology enables the revolutionary process improvement which cannot be achieved using existing methodologies, such as asymmetric reactions or classical resolutions, represented by the case of sertralin (Sect. 7.4).

Notwithstanding heavy competition from newer technologies, classical resolution is the method of choice in many drug syntheses, because of its ease and simplicity in application and robustness of the process. Its usefulness is obvious from the fact that the number of new drugs produced by using this technology in commercial syntheses has significantly increased in these ten years as shown in Tables 8 and 9.

On the other hand, the manufacturing processes of several drugs that relied on the classical resolution have already changed or are on the way to newer processes, such as chromatographic separation, enzymatic resolutions or asymmetric syntheses. This is an inevitable progress and the author expects the improvements and innovations of optical resolution will be stimulated in these environments, supported by the better understanding of the phenomena working in the process of resolution, both in solution and crystal phases, as discussed in the preceding chapters of this volume. Anyway, an efficient manufacture of drug molecules can not be achieved by relying on any single technology, but an relevant combination of several means are vital. The author believes that optical resolution is one of those indispensable technologies.

References

1. Ariens EJ (1984) Eur J Clin Pharmacol 26:663
2. Ariens EJ, Wuis EW, Veringa EJ (1988) Biochem Pharmacol 37:9
3. Millership JS, Fitzpatrick A (1993) Chirality 5:573
4. Annual Reports in Medicinal Chemistry, vol 21(1986)–37(2002) Academic Press, San Diego; vol 38(2003)–40(2005) Elsevier Inc, San Diego
5. DeCamp WH (1989) Chirality 1:2
6. Sheldon RA (1993) Chirotechnology – Industrial synthesis of optically active compounds. Marcel Dekker Inc, New York, p 69
7. Lipinski CA, Lombardo F, Dominy BW, Feeney PJ (1997) Advanced Drug Delivery Reviews 23:3
8. Rouhi AM (2004) Chem & Eng News 82: June 14, p 47, 52
9. Palomo JM, Mateo C, Fernandez-Lorente G, Solares LF, Diaz M, Sanchez VM, Bayod M, Gotor V, Guisan JM, Fernandez-Lafuente R (2003) Tetrahedron Asymmetry 14:429
10. Hoekstra MS, Sobieray DM, Schwindt MA, Mulhern TA, Grote TM, Huckabee BK, Hendrickson VS, Franklin LC, Granger EJ, Karrick GL (1997) Org Process Res Dev 1:26
11. Hoge G (2003) J Am Chem Soc 125:10219, cf. [4] vol 40, p 464
12. Pierce ME (1998) J Org Chem 63:8536
13. Rouhi AM (2003) Chem & Eng News 81, May 5, p 45, 52, 56
14. Federsel H-J (2003) Chirality 15:S128
15. Gao Y, Hong Y, Zepp CM (to Sepracor, Inc) (1995) US 5 442 118
16. Li J, Liu KK-C, Sakya S (2004) Mini-Reviews Med Chem 4:1105–1109
17. Quallich GJ (2005) Chirality 17:S120
18. Sheldon RA (1993) Chirotechnology – Industrial synthesis of optically active compounds. Marcel Dekker Inc, New York, p 367
19. Kowalczyk BA, Dvorak CA (1996) Synthesis 1996:816
20. Askin D, Eng KK, Rossen K, Purick RM, Wells KM, Volante RP, Reider PJ (1994) Tetrahedron Lett 35:673
21. Maligres PE, Upadhyay V, Rossen K, Cianciosi SJ, Purick RM, Eng KK, Reamer RA, Askin D, Volante RP, Reider PJ (1995) Tetrahedron Lett 36:2195
22. Maligres PE, Weissman SA, Upadhyay V, Cianciosi SJ, Reamer RA, Purick RM, Sager J, Rossen K, Eng KK, Askin D, Volante RP, Reider PJ (1996) Tetrahedron 52:3327
23. Brands KMJ, Payack JF, Rosen JD, Nelson TD, Candelario A, Huffman MA, Zhao MM, Li J, Craig B, Song ZJ, Tschaen DM, Hansen K, Devine PN, Pye PJ, Rossen K, Dormer PG, Reamer RA, Welch CJ, Mathre DJ, Tsou NN, McNamara JM, Reider PJ (2003) J Am Chem Soc 125:2129
24. Ritter SK (2005) Chem & Eng News 83: Jun 27, p 40

25. Daugan A, Grondin P, Ruault C, le Monnier de Gouville A-N, Coste H, Linget JM, Kirilovsky J, Hyafil F, Labaudiniere R (2003) J Med Chem 46:4525–4533
26. Orme MW, Martinelli MJ, Doecke CW, Pawlak JM, Chelius EC (to Lilly ICOS) (2004) WO 2004/011 463
27. Dunn PJ (2005) Org Process Res Dev 9:88
28. Williams M, Quallich GJ (1990) Chem & Ind (London) May 21, p 315
29. Taber GP, Pfisterer DM, Colberg JC (2004) Org Process Res Dev 8:385
30. Rouhi AM (2002) Chem & Eng News 80: Apr 22, p 30

Author Index Volumes 251–269

Author Index Vols. 26–50 see Vol. 50
Author Index Vols. 51–100 see Vol. 100
Author Index Vols. 101–150 see Vol. 150
Author Index Vols. 151–200 see Vol. 200
Author Index Vols. 201–250 see Vol. 250

The volume numbers are printed in italics

Ajayaghosh A, George SJ, Schenning APHJ (2005) Hydrogen-Bonded Assemblies of Dyes and Extended π-Conjugated Systems. 258: 83–118
Albert M, Fensterbank L, Lacôte E, Malacria M (2006) Tandem Radical Reactions. 264: 1–62
Alberto R (2005) New Organometallic Technetium Complexes for Radiopharmaceutical Imaging. 252: 1–44
Alegret S, see Pividori MI (2005) 260: 1–36
Amabilino DB, Veciana J (2006) Supramolecular Chiral Functional Materials. 265: 253–302
Anderson CJ, see Li WP (2005) 252: 179–192
Anslyn EV, see Houk RJT (2005) 255: 199–229
Appukkuttan P, Van der Eycken E (2006) Microwave-Assisted Natural Product Chemistry. 266: 1–47
Araki K, Yoshikawa I (2005) Nucleobase-Containing Gelators. 256: 133–165
Armitage BA (2005) Cyanine Dye–DNA Interactions: Intercalation, Groove Binding and Aggregation. 253: 55–76
Arya DP (2005) Aminoglycoside–Nucleic Acid Interactions: The Case for Neomycin. 253: 149–178

Bailly C, see Dias N (2005) 253: 89–108
Balaban TS, Tamiaki H, Holzwarth AR (2005) Chlorins Programmed for Self-Assembly. 258: 1–38
Balzani V, Credi A, Ferrer B, Silvi S, Venturi M (2005) Artificial Molecular Motors and Machines: Design Principles and Prototype Systems. 262: 1–27
Barbieri CM, see Pilch DS (2005) 253: 179–204
Barchuk A, see Daasbjerg K (2006) 263: 39–70
Bayly SR, see Beer PD (2005) 255: 125–162
Beer PD, Bayly SR (2005) Anion Sensing by Metal-Based Receptors. 255: 125–162
Bertini L, Bruschi M, de Gioia L, Fantucci P, Greco C, Zampella G (2007) Quantum Chemical Investigations of Reaction Paths of Metalloenzymes and Biomimetic Models – The Hydrogenase Example. 268: 1–46
Bier FF, see Heise C (2005) 261: 1–25
Blum LJ, see Marquette CA (2005) 261: 113–129
Boiteau L, see Pascal R (2005) 259: 69–122
Bolhuis PG, see Dellago C (2007) 268: 291–317
Borovkov VV, Inoue Y (2006) Supramolecular Chirogenesis in Host–Guest Systems Containing Porphyrinoids. 265: 89–146

Boschi A, Duatti A, Uccelli L (2005) Development of Technetium-99m and Rhenium-188 Radiopharmaceuticals Containing a Terminal Metal–Nitrido Multiple Bond for Diagnosis and Therapy. 252: 85–115

Braga D, D'Addario D, Giaffreda SL, Maini L, Polito M, Grepioni F (2005) Intra-Solid and Inter-Solid Reactions of Molecular Crystals: a Green Route to Crystal Engineering. 254: 71–94

Brebion F, see Crich D (2006) 263: 1–38

Brizard A, Oda R, Huc I (2005) Chirality Effects in Self-assembled Fibrillar Networks. 256: 167–218

Bruce IJ, see del Campo A (2005) 260: 77–111

Bruschi M, see Bertini L (2007) 268: 1–46

del Campo A, Bruce IJ (2005) Substrate Patterning and Activation Strategies for DNA Chip Fabrication. 260: 77–111

Chaires JB (2005) Structural Selectivity of Drug-Nucleic Acid Interactions Probed by Competition Dialysis. 253: 33–53

Chiorboli C, Indelli MT, Scandola F (2005) Photoinduced Electron/Energy Transfer Across Molecular Bridges in Binuclear Metal Complexes. 257: 63–102

Collin J-P, Heitz V, Sauvage J-P (2005) Transition-Metal-Complexed Catenanes and Rotaxanes in Motion: Towards Molecular Machines. 262: 29–62

Collyer SD, see Davis F (2005) 255: 97–124

Commeyras A, see Pascal R (2005) 259: 69–122

Coquerel G (2007) Preferential Crystallization. 269: 1–51

Correia JDG, see Santos I (2005) 252: 45–84

Costanzo G, see Saladino R (2005) 259: 29–68

Credi A, see Balzani V (2005) 262: 1–27

Crestini C, see Saladino R (2005) 259: 29–68

Crich D, Brebion F, Suk D-H (2006) Generation of Alkene Radical Cations by Heterolysis of β-Substituted Radicals: Mechanism, Stereochemistry, and Applications in Synthesis. 263: 1–38

Cuerva JM, Justicia J, Oller-López JL, Oltra JE (2006) Cp_2TiCl in Natural Product Synthesis. 264: 63–92

Daasbjerg K, Svith H, Grimme S, Gerenkamp M, Mück-Lichtenfeld C, Gansäuer A, Barchuk A (2006) The Mechanism of Epoxide Opening through Electron Transfer: Experiment and Theory in Concert. 263: 39–70

D'Addario D, see Braga D (2005) 254: 71–94

Danishefsky SJ, see Warren JD (2007) 267

Darmency V, Renaud P (2006) Tin-Free Radical Reactions Mediated by Organoboron Compounds. 263: 71–106

Davis F, Collyer SD, Higson SPJ (2005) The Construction and Operation of Anion Sensors: Current Status and Future Perspectives. 255: 97–124

Deamer DW, Dworkin JP (2005) Chemistry and Physics of Primitive Membranes. 259: 1–27

Dellago C, Bolhuis PG (2007) Transition Path Sampling Simulations of Biological Systems. 268: 291–317

Deng J-Y, see Zhang X-E (2005) 261: 169–190

Dervan PB, Poulin-Kerstien AT, Fechter EJ, Edelson BS (2005) Regulation of Gene Expression by Synthetic DNA-Binding Ligands. 253: 1–31

Dias N, Vezin H, Lansiaux A, Bailly C (2005) Topoisomerase Inhibitors of Marine Origin and Their Potential Use as Anticancer Agents. 253: 89–108

DiMauro E, see Saladino R (2005) *259*: 29–68
Dittrich M, Yu J, Schulten K (2007) PcrA Helicase, a Molecular Motor Studied from the Electronic to the Functional Level. *268*: 319–347
Dobrawa R, see You C-C (2005) *258*: 39–82
Du Q, Larsson O, Swerdlow H, Liang Z (2005) DNA Immobilization: Silanized Nucleic Acids and Nanoprinting. *261*: 45–61
Duatti A, see Boschi A (2005) *252*: 85–115
Dworkin JP, see Deamer DW (2005) *259*: 1–27

Edelson BS, see Dervan PB (2005) *253*: 1–31
Edwards DS, see Liu S (2005) *252*: 193–216
Ernst K-H (2006) Supramolecular Surface Chirality. *265*: 209–252
Ersmark K, see Wannberg J (2006) *266*: 167–197
Escudé C, Sun J-S (2005) DNA Major Groove Binders: Triple Helix-Forming Oligonucleotides, Triple Helix-Specific DNA Ligands and Cleaving Agents. *253*: 109–148
Van der Eycken E, see Appukkuttan P (2006) *266*: 1–47

Fages F, Vögtle F, Žinić M (2005) Systematic Design of Amide- and Urea-Type Gelators with Tailored Properties. *256*: 77–131
Fages F, see Žinić M (2005) *256*: 39–76
Faigl F, Schindler J, Fogassy E (2007) Advantages of Structural Similarities of the Reactants in Optical Resolution Processes. *269*: 133–157
Fantucci P, see Bertini L (2007) *268*: 1–46
Fechter EJ, see Dervan PB (2005) *253*: 1–31
Fensterbank L, see Albert M (2006) *264*: 1–62
Fernández JM, see Moonen NNP (2005) *262*: 99–132
Fernando C, see Szathmáry E (2005) *259*: 167–211
Ferrer B, see Balzani V (2005) *262*: 1–27
De Feyter S, De Schryver F (2005) Two-Dimensional Dye Assemblies on Surfaces Studied by Scanning Tunneling Microscopy. *258*: 205–255
Flood AH, see Moonen NNP (2005) *262*: 99–132
Fogassy E, see Faigl F (2007) *269*: 133–157
Fujimoto D, see Tamura R (2007) *269*: 53–82
Fujiwara S-i, Kambe N (2005) Thio-, Seleno-, and Telluro-Carboxylic Acid Esters. *251*: 87–140

Gansäuer A, see Daasbjerg K (2006) *263*: 39–70
Garcia-Garibay MA, see Karlen SD (2005) *262*: 179–227
Gelinck GH, see Grozema FC (2005) *257*: 135–164
Geng X, see Warren JD (2007) *267*
George SJ, see Ajayaghosh A (2005) *258*: 83–118
Gerenkamp M, see Daasbjerg K (2006) *263*: 39–70
Giaffreda SL, see Braga D (2005) *254*: 71–94
de Gioia L, see Bertini L (2007) *268*: 1–46
Greco C, see Bertini L (2007) *268*: 1–46
Grepioni F, see Braga D (2005) *254*: 71–94
Grimme S, see Daasbjerg K (2006) *263*: 39–70
Grozema FC, Siebbeles LDA, Gelinck GH, Warman JM (2005) The Opto-Electronic Properties of Isolated Phenylenevinylene Molecular Wires. *257*: 135–164

Guiseppi-Elie A, Lingerfelt L (2005) Impedimetric Detection of DNA Hybridization: Towards Near-Patient DNA Diagnostics. *260*: 161–186
Di Giusto DA, King GC (2005) Special-Purpose Modifications and Immobilized Functional Nucleic Acids for Biomolecular Interactions. *261*: 131–168

Haase C, Seitz O (2007) Chemical Synthesis of Glycopeptides. *267*
Hansen SG, Skrydstrup T (2006) Modification of Amino Acids, Peptides, and Carbohydrates through Radical Chemistry. *264*: 135–162
Heise C, Bier FF (2005) Immobilization of DNA on Microarrays. *261*: 1–25
Heitz V, see Collin J-P (2005) *262*: 29–62
Herrmann C, Reiher M (2007) First-Principles Approach to Vibrational Spectroscopy of Biomolecules. *268*: 85–132
Higson SPJ, see Davis F (2005) *255*: 97–124
Hirayama N, see Sakai K (2007) *269*: 233–271
Hirst AR, Smith DK (2005) Dendritic Gelators. *256*: 237–273
Holzwarth AR, see Balaban TS (2005) *258*: 1–38
Houk RJT, Tobey SL, Anslyn EV (2005) Abiotic Guanidinium Receptors for Anion Molecular Recognition and Sensing. *255*: 199–229
Huc I, see Brizard A (2005) *256*: 167–218

Ihmels H, Otto D (2005) Intercalation of Organic Dye Molecules into Double-Stranded DNA – General Principles and Recent Developments. *258*: 161–204
Indelli MT, see Chiorboli C (2005) *257*: 63–102
Inoue Y, see Borovkov VV (2006) *265*: 89–146
Ishii A, Nakayama J (2005) Carbodithioic Acid Esters. *251*: 181–225
Ishii A, Nakayama J (2005) Carboselenothioic and Carbodiselenoic Acid Derivatives and Related Compounds. *251*: 227–246
Ishi-i T, Shinkai S (2005) Dye-Based Organogels: Stimuli-Responsive Soft Materials Based on One-Dimensional Self-Assembling Aromatic Dyes. *258*: 119–160

James DK, Tour JM (2005) Molecular Wires. *257*: 33–62
Jones W, see Trask AV (2005) *254*: 41–70
Justicia J, see Cuerva JM (2006) *264*: 63–92

Kambe N, see Fujiwara S-i (2005) *251*: 87–140
Kano N, Kawashima T (2005) Dithiocarboxylic Acid Salts of Group 1–17 Elements (Except for Carbon). *251*: 141–180
Kappe CO, see Kremsner JM (2006) *266*: 233–278
Kaptein B, see Kellogg RM (2007) *269*: 159–197
Karlen SD, Garcia-Garibay MA (2005) Amphidynamic Crystals: Structural Blueprints for Molecular Machines. *262*: 179–227
Kato S, Niyomura O (2005) Group 1–17 Element (Except Carbon) Derivatives of Thio-, Seleno- and Telluro-Carboxylic Acids. *251*: 19–85
Kato S, see Niyomura O (2005) *251*: 1–12
Kato T, Mizoshita N, Moriyama M, Kitamura T (2005) Gelation of Liquid Crystals with Self-Assembled Fibers. *256*: 219–236
Kaul M, see Pilch DS (2005) *253*: 179–204
Kaupp G (2005) Organic Solid-State Reactions with 100% Yield. *254*: 95–183
Kawasaki T, see Okahata Y (2005) *260*: 57–75
Kawashima T, see Kano N (2005) *251*: 141–180

Kay ER, Leigh DA (2005) Hydrogen Bond-Assembled Synthetic Molecular Motors and Machines. 262: 133–177
Kellogg RM, Kaptein B, Vries TR (2007) Dutch Resolution of Racemates and the Roles of Solid Solution Formation and Nucleation Inhibition. 269: 159–197
King GC, see Di Giusto DA (2005) 261: 131–168
Kirchner B, see Thar J (2007) 268: 133–171
Kitamura T, see Kato T (2005) 256: 219–236
Komatsu K (2005) The Mechanochemical Solid-State Reaction of Fullerenes. 254: 185–206
Kremsner JM, Stadler A, Kappe CO (2006) The Scale-Up of Microwave-Assisted Organic Synthesis. 266: 233–278
Kriegisch V, Lambert C (2005) Self-Assembled Monolayers of Chromophores on Gold Surfaces. 258: 257–313

Lacôte E, see Albert M (2006) 264: 1–62
Lahav M, see Weissbuch I (2005) 259: 123–165
Lambert C, see Kriegisch V (2005) 258: 257–313
Lansiaux A, see Dias N (2005) 253: 89–108
Larhed M, see Nilsson P (2006) 266: 103–144
Larhed M, see Wannberg J (2006) 266: 167–197
Larsson O, see Du Q (2005) 261: 45–61
Leigh DA, Pérez EM (2006) Dynamic Chirality: Molecular Shuttles and Motors. 265: 185–208
Leigh DA, see Kay ER (2005) 262: 133–177
Leiserowitz L, see Weissbuch I (2005) 259: 123–165
Lhoták P (2005) Anion Receptors Based on Calixarenes. 255: 65–95
Li WP, Meyer LA, Anderson CJ (2005) Radiopharmaceuticals for Positron Emission Tomography Imaging of Somatostatin Receptor Positive Tumors. 252: 179–192
Liang Z, see Du Q (2005) 261: 45–61
Lingerfelt L, see Guiseppi-Elie A (2005) 260: 161–186
Liu S (2005) 6-Hydrazinonicotinamide Derivatives as Bifunctional Coupling Agents for ^{99m}Tc-Labeling of Small Biomolecules. 252: 117–153
Liu S, Robinson SP, Edwards DS (2005) Radiolabeled Integrin $\alpha_v\beta_3$ Antagonists as Radiopharmaceuticals for Tumor Radiotherapy. 252: 193–216
Liu XY (2005) Gelation with Small Molecules: from Formation Mechanism to Nanostructure Architecture. 256: 1–37
Luderer F, Walschus U (2005) Immobilization of Oligonucleotides for Biochemical Sensing by Self-Assembled Monolayers: Thiol-Organic Bonding on Gold and Silanization on Silica Surfaces. 260: 37–56

Maeda K, Yashima E (2006) Dynamic Helical Structures: Detection and Amplification of Chirality. 265: 47–88
Magnera TF, Michl J (2005) Altitudinal Surface-Mounted Molecular Rotors. 262: 63–97
Maini L, see Braga D (2005) 254: 71–94
Malacria M, see Albert M (2006) 264: 1–62
Marquette CA, Blum LJ (2005) Beads Arraying and Beads Used in DNA Chips. 261: 113–129
Mascini M, see Palchetti I (2005) 261: 27–43
Matsumoto A (2005) Reactions of 1,3-Diene Compounds in the Crystalline State. 254: 263–305
McGhee AM, Procter DJ (2006) Radical Chemistry on Solid Support. 264: 93–134
Meyer B, Möller H (2007) Conformation of Glycopeptides and Glycoproteins. 267
Meyer LA, see Li WP (2005) 252: 179–192

Michl J, see Magnera TF (2005) *262*: 63–97
Milea JS, see Smith CL (2005) *261*: 63–90
Mizoshita N, see Kato T (2005) *256*: 219–236
Möller H, see Meyer B (2007) *267*
Moonen NNP, Flood AH, Fernández JM, Stoddart JF (2005) Towards a Rational Design of Molecular Switches and Sensors from their Basic Building Blocks. *262*: 99–132
Moriyama M, see Kato T (2005) *256*: 219–236
Murai T (2005) Thio-, Seleno-, Telluro-Amides. *251*: 247–272
Murakami H (2007) From Racemates to Single Enantiomers – Chiral Synthetic Drugs over the last 20 Years. *269*: 273–299
Mutule I, see Suna E (2006) *266*: 49–101

Nakayama J, see Ishii A (2005) *251*: 181–225
Nakayama J, see Ishii A (2005) *251*: 227–246
Neese F, see Sinnecker S (2007) *268*: 47–83
Nguyen GH, see Smith CL (2005) *261*: 63–90
Nicolau DV, Sawant PD (2005) Scanning Probe Microscopy Studies of Surface-Immobilised DNA/Oligonucleotide Molecules. *260*: 113–160
Nilsson P, Olofsson K, Larhed M (2006) Microwave-Assisted and Metal-Catalyzed Coupling Reactions. *266*: 103–144
Niyomura O, Kato S (2005) Chalcogenocarboxylic Acids. *251*: 1–12
Niyomura O, see Kato S (2005) *251*: 19–85
Nohira H, see Sakai K (2007) *269*: 199–231

Oda R, see Brizard A (2005) *256*: 167–218
Okahata Y, Kawasaki T (2005) Preparation and Electron Conductivity of DNA-Aligned Cast and LB Films from DNA-Lipid Complexes. *260*: 57–75
Oller-López JL, see Cuerva JM (2006) *264*: 63–92
Olofsson K, see Nilsson P (2006) *266*: 103–144
Oltra JE, see Cuerva JM (2006) *264*: 63–92
Otto D, see Ihmels H (2005) *258*: 161–204

Palchetti I, Mascini M (2005) Electrochemical Adsorption Technique for Immobilization of Single-Stranded Oligonucleotides onto Carbon Screen-Printed Electrodes. *261*: 27–43
Pascal R, Boiteau L, Commeyras A (2005) From the Prebiotic Synthesis of α-Amino Acids Towards a Primitive Translation Apparatus for the Synthesis of Peptides. *259*: 69–122
Paulo A, see Santos I (2005) *252*: 45–84
Pérez EM, see Leigh DA (2006) *265*: 185–208
Pilch DS, Kaul M, Barbieri CM (2005) Ribosomal RNA Recognition by Aminoglycoside Antibiotics. *253*: 179–204
Pividori MI, Alegret S (2005) DNA Adsorption on Carbonaceous Materials. *260*: 1–36
Piwnica-Worms D, see Sharma V (2005) *252*: 155–178
Polito M, see Braga D (2005) *254*: 71–94
Poulin-Kerstien AT, see Dervan PB (2005) *253*: 1–31
Procter DJ, see McGhee AM (2006) *264*: 93–134

Quiclet-Sire B, Zard SZ (2006) The Degenerative Radical Transfer of Xanthates and Related Derivatives: An Unusually Powerful Tool for the Creation of Carbon–Carbon Bonds. *264*: 201–236

Ratner MA, see Weiss EA (2005) *257*: 103–133
Raymond KN, see Seeber G (2006) *265*: 147–184
Rebek Jr J, see Scarso A (2006) *265*: 1–46
Reckien W, see Thar J (2007) *268*: 133–171
Reiher M, see Herrmann C (2007) *268*: 85–132
Renaud P, see Darmency V (2006) *263*: 71–106
Robinson SP, see Liu S (2005) *252*: 193–216

Saha-Möller CR, see You C-C (2005) *258*: 39–82
Sakai K, Sakurai R, Hirayama N (2007) Molecular Mechanisms of Dielectrically Controlled Resolution (DCR). *269*: 233–271
Sakai K, Sakurai R, Nohira H (2007) New Resolution Technologies Controlled by Chiral Discrimination Mechanisms. *269*: 199–231
Sakamoto M (2005) Photochemical Aspects of Thiocarbonyl Compounds in the Solid-State. *254*: 207–232
Sakurai R, see Sakai K (2007) *269*: 199–231
Sakurai R, see Sakai K (2007) *269*: 233–271
Saladino R, Crestini C, Costanzo G, DiMauro E (2005) On the Prebiotic Synthesis of Nucleobases, Nucleotides, Oligonucleotides, Pre-RNA and Pre-DNA Molecules. *259*: 29–68
Santos I, Paulo A, Correia JDG (2005) Rhenium and Technetium Complexes Anchored by Phosphines and Scorpionates for Radiopharmaceutical Applications. *252*: 45–84
Santos M, see Szathmáry E (2005) *259*: 167–211
Sauvage J-P, see Collin J-P (2005) *262*: 29–62
Sawant PD, see Nicolau DV (2005) *260*: 113–160
Scandola F, see Chiorboli C (2005) *257*: 63–102
Scarso A, Rebek Jr J (2006) Chiral Spaces in Supramolecular Assemblies. *265*: 1–46
Scheffer JR, Xia W (2005) Asymmetric Induction in Organic Photochemistry via the Solid-State Ionic Chiral Auxiliary Approach. *254*: 233–262
Schenning APHJ, see Ajayaghosh A (2005) *258*: 83–118
Schmidtchen FP (2005) Artificial Host Molecules for the Sensing of Anions. *255*: 1–29 Author Index Volumes 251–255
Schindler J, see Faigl F (2007) *269*: 133–157
Schoof S, see Wolter F (2007) *267*
De Schryver F, see De Feyter S (2005) *258*: 205–255
Schulten K, see Dittrich M (2007) *268*: 319–347
Seeber G, Tiedemann BEF, Raymond KN (2006) Supramolecular Chirality in Coordination Chemistry. *265*: 147–184
Seitz O, see Haase C (2007) *267*
Senn HM, Thiel W (2007) QM/MM Methods for Biological Systems. *268*: 173–289
Sharma V, Piwnica-Worms D (2005) Monitoring Multidrug Resistance P-Glycoprotein Drug Transport Activity with Single-Photon-Emission Computed Tomography and Positron Emission Tomography Radiopharmaceuticals. *252*: 155–178
Shinkai S, see Ishi-i T (2005) *258*: 119–160
Sibi MP, see Zimmerman J (2006) *263*: 107–162
Siebbeles LDA, see Grozema FC (2005) *257*: 135–164
Silvi S, see Balzani V (2005) *262*: 1–27
Sinnecker S, Neese F (2007) Theoretical Bioinorganic Spectroscopy. *268*: 47–83
Skrydstrup T, see Hansen SG (2006) *264*: 135–162
Smith CL, Milea JS, Nguyen GH (2005) Immobilization of Nucleic Acids Using Biotin-Strept(avidin) Systems. *261*: 63–90

Smith DK, see Hirst AR (2005) *256*: 237–273
Specker D, Wittmann V (2007) Synthesis and Application of Glycopeptide and Glycoprotein Mimetics. *267*
Stadler A, see Kremsner JM (2006) *266*: 233–278
Stibor I, Zlatušková P (2005) Chiral Recognition of Anions. *255*: 31–63
Stoddart JF, see Moonen NNP (2005) *262*: 99–132
Strauss CR, Varma RS (2006) Microwaves in Green and Sustainable Chemistry. *266*: 199–231
Suk D-H, see Crich D (2006) *263*: 1–38
Suksai C, Tuntulani T (2005) Chromogenetic Anion Sensors. *255*: 163–198
Sun J-S, see Escudé C (2005) *253*: 109–148
Suna E, Mutule I (2006) Microwave-assisted Heterocyclic Chemistry. *266*: 49–101
Süssmuth RD, see Wolter F (2007) *267*
Svith H, see Daasbjerg K (2006) *263*: 39–70
Swerdlow H, see Du Q (2005) *261*: 45–61
Szathmáry E, Santos M, Fernando C (2005) Evolutionary Potential and Requirements for Minimal Protocells. *259*: 167–211

Taira S, see Yokoyama K (2005) *261*: 91–112
Takahashi H, see Tamura R (2007) *269*: 53–82
Tamiaki H, see Balaban TS (2005) *258*: 1–38
Tamura R, Takahashi H, Fujimoto D, Ushio T (2007) Mechanism and Scope of Preferential Enrichment, a Symmetry-Breaking Enantiomeric Resolution Phenomenon. *269*: 53–82
Thar J, Reckien W, Kirchner B (2007) Car–Parrinello Molecular Dynamics Simulations and Biological Systems. *268*: 133–171
Thayer DA, Wong C-H (2007) Enzymatic Synthesis of Glycopeptides and Glycoproteins. *267*
Thiel W, see Senn HM (2007) *268*: 173–289
Tiedemann BEF, see Seeber G (2006) *265*: 147–184
Tobey SL, see Houk RJT (2005) *255*: 199–229
Toda F (2005) Thermal and Photochemical Reactions in the Solid-State. *254*: 1–40
Tour JM, see James DK (2005) *257*: 33–62
Trask AV, Jones W (2005) Crystal Engineering of Organic Cocrystals by the Solid-State Grinding Approach. *254*: 41–70
Tuntulani T, see Suksai C (2005) *255*: 163–198

Uccelli L, see Boschi A (2005) *252*: 85–115
Ushio T, see Tamura R (2007) *269*: 53–82

Varma RS, see Strauss CR (2006) *266*: 199–231
Veciana J, see Amabilino DB (2006) *265*: 253–302
Venturi M, see Balzani V (2005) *262*: 1–27
Vezin H, see Dias N (2005) *253*: 89–108
Vögtle F, see Fages F (2005) *256*: 77–131
Vögtle M, see Žinić M (2005) *256*: 39–76
Vries TR, see Kellogg RM (2007) *269*: 159–197

Walschus U, see Luderer F (2005) *260*: 37–56
Walton JC (2006) Unusual Radical Cyclisations. *264*: 163–200
Wannberg J, Ersmark K, Larhed M (2006) Microwave-Accelerated Synthesis of Protease Inhibitors. *266*: 167–197
Warman JM, see Grozema FC (2005) *257*: 135–164

Warren JD, Geng X, Danishefsky SJ (2007) Synthetic Glycopeptide-Based Vaccines. *267*
Wasielewski MR, see Weiss EA (2005) *257*: 103–133
Weiss EA, Wasielewski MR, Ratner MA (2005) Molecules as Wires: Molecule-Assisted Movement of Charge and Energy. *257*: 103–133
Weissbuch I, Leiserowitz L, Lahav M (2005) Stochastic "Mirror Symmetry Breaking" via Self-Assembly, Reactivity and Amplification of Chirality: Relevance to Abiotic Conditions. *259*: 123–165
Williams LD (2005) Between Objectivity and Whim: Nucleic Acid Structural Biology. *253*: 77–88
Wittmann V, see Specker D (2007) *267*
Wolter F, Schoof S, Süssmuth RD (2007) Synopsis of Structural, Biosynthetic, and Chemical Aspects of Glycopeptide Antibiotics. *267*
Wong C-H, see Thayer DA (2007) *267*
Wong KM-C, see Yam VW-W (2005) *257*: 1–32
Würthner F, see You C-C (2005) *258*: 39–82

Xia W, see Scheffer JR (2005) *254*: 233–262

Yam VW-W, Wong KM-C (2005) Luminescent Molecular Rods – Transition-Metal Alkynyl Complexes. *257*: 1–32
Yashima E, see Maeda K (2006) *265*: 47–88
Yokoyama K, Taira S (2005) Self-Assembly DNA-Conjugated Polymer for DNA Immobilization on Chip. *261*: 91–112
Yoshikawa I, see Araki K (2005) *256*: 133–165
Yoshioka R (2007) Racemization, Optical Resolution and Crystallization-Induced Asymmetric Transformation of Amino Acids and Pharmaceutical Intermediates. *269*: 83–132
You C-C, Dobrawa R, Saha-Möller CR, Würthner F (2005) Metallosupramolecular Dye Assemblies. *258*: 39–82
Yu J, see Dittrich M (2007) *268*: 319–347

Zampella G, see Bertini L (2007) *268*: 1–46
Zard SZ, see Quiclet-Sire B (2006) *264*: 201–236
Zhang W (2006) Microwave-Enhanced High-Speed Fluorous Synthesis. *266*: 145–166
Zhang X-E, Deng J-Y (2005) Detection of Mutations in Rifampin-Resistant *Mycobacterium Tuberculosis* by Short Oligonucleotide Ligation Assay on DNA Chips (SOLAC). *261*: 169–190
Zimmerman J, Sibi MP (2006) Enantioselective Radical Reactions. *263*: 107–162
Žinić M, see Fages F (2005) *256*: 77–131
Žinić M, Vögtle F, Fages F (2005) Cholesterol-Based Gelators. *256*: 39–76
Zipse H (2006) Radical Stability—A Theoretical Perspective. *263*: 163–190
Zlatušková P, see Stibor I (2005) *255*: 31–63

Subject Index

ACL–TPA 217, 227
Acyl-DL-amino acids, asymmetric transformation 100
Additives, tailor-made 168
Aggregation 159
–, supersaturated solution 164
Alanine benzenesulfonate 95
Alaninol 181
Aldehydes, racemization 88
Aliphatic acids, racemization 88
Alkylamines, 1-substituted 137
Amines, chiral 16
–, achiral 153
Amino acids 83
–, racemization 86
–, salts, asymmetric transformation 102
–, –, asymmetric transformation, chiral reagents 108
–, –, racemization 94
1,2-Aminoalcohol 178
2-Amino-1-butanol 180
Amino-ε-caprolactam 235
Anti-conglomerate 16
Aprepitant (Emend) 293
1-Arylethylamines 144
AS3PC 22, 30
Asp(OMe) 112
Aspartic acid 87
Asymmetric syntheses 288
Asymmetric transformation 83
Auto-seeded polythermic programmed preferential crystallization (AS3PC) 22, 30

Bromocamphor-8-sulfonic acid 179
Brucine 162
Butanetetrol 10

Camphor sulfonic acid 162, 179

Chalcones 178
CHEA/(S)-MA 226
Chiral additive 203
Chiral centers 281
Chiral compound 83
Chiral resolution 233
Chirality 159, 285
Chirality pool 286
Conglomerates 5, 6, 16
Cooling program 36
Crystal growth 168
Crystal habit modification 199
–, tailored inhibitor 201
Crystal size, controlled 48
Crystal structure 62
Crystallization, complexity system 55
–, enantiomeric resolution 55
Crystallization-induced asymmetric ransformations 85, 291
Cyclicity 30
Cyclohexylethylamine 261
Cysteine 171

DCR 199, 217, 233
–, molecular mechanism 268
Derivative resolving agent 133
Diastereomer mixtures, separation 291
Diastereomeric resolution, asymmetric transformation 107
Diastereomeric salts 133
–, formation 199
–, X-ray analysis 236, 246
Diastereomers, asymmetric transformation 106
Diastereoselective reactions 292
Dielectrically controlled resolution (DCR) 199, 217, 233
Duloxetine 201
Dutch resolution 163, 175

Enantiomeric enrichment   159
Enantiomeric purification   16
Enantiomeric resolution   53, 98
Enantiomers   159, 281
–, asymmetric transformation   98
–, distillation   140
–, in solution, assembly model   61
Enantiomorphs   169
Entrainment   1, 16, 28
–, solution   28
Ephedrine   182
Epitaxial transition   53

4-Fluorophenyl glycines   179

Glutamic acid   87

Hemihedrism   10
Heterochiral assemblies   133
Heterosolvates   5
Homochiral assemblies   133
Homochirality, biomolecular   54
HPG·oTS, asymmetric transformation   102
Hydrogen bond donors/acceptors   284
Hydroxyphenylglycine o-toluenesulfonate   96

Indinavir   201, 292

Lysine   90
Lysine p-aminobenzenesulfonate   94

Mandelic acid   181, 254, 261
–, enantiopure   168, 202
Methionine   90, 171
Methionine p-chlorobenzenesulfonate   96
Methionine hydrochloride   95
Methylpyrrolidine   245
MMT, resolution   213
Molecular recognition   233, 243, 252, 260, 268
Monoester, achiral, dicarboxylic acid monoester   152
MPRD/(R,R)-TA   224

Nucleation   18, 168, 172
–, inhibition/inhibitors   159, 174, 185
–, Dutch resolution   184

Optical resolution   83, 133, 199, 286
Orlistat   201

PC (preferential crystallization), binary systems   19
–, detectable metastable racemic compound   43
–, polymorphism   46
PEA:(R)-MA salt   206
Pharmaceutical intermediates   114
Phase diagrams   1
Phencyphos   183
Phenylalanine   90
Phenylbutylamine   188
Phenylethyl alcohols, 1-(substituted), dicarboxylic acid monoester   150
Phenylethylamine (PEA)   168, 202
Phenylglycine   179
Phenyl-2-p-tolyl ethylamine   254
Phosphoric acids, cyclic   175, 181
Polyethyleneglycol   172
Polyethylenimine   172
Polymorphic transition   53, 61, 67
Polymorphism   15
Preferential crystallization (PC)   1, 19, 43
–, in situ racemization   48
Preferential enrichment   53, 56, 72
PTE–MA resolution system   219, 228

Quasi-racemate-type resolving agent mixture   155
Quinazolinones, atropisomeric   139
Quinine/quinidine   162

R–S epitaxy   45
Racemates, resolution   290
Racemic compounds   93
Racemic mixture (conglomerate)   88, 93
Racemic switches   289
Racemization   83
– transformation   85
Replacing crystallization   18
Resolving agents   159
Rule of reversal   169

Salicylaldehyde   87
Second harmonic generation (SHG)   14
Seed crystals   76
Seeded isothermal preferential crystallization (SIPC)   22, 28

Subject Index

Seeded polythermic preferential crystallization (S3PC)   22, 29
Seeding   1, 34
Seeds, surfaces   46
Serine   87
Sertralin (Zoloft)   295
Sertraline   201
Shape recognition   17
Sodium ammonium tartaric salt, tetrahydrated   17
Sodium bisulfite   178
Solid solution   180
Solubility   159
Solvates   15
–, ternary system, preferential crystallization   32
Solvent effect   233
Space filler   199, 208
Stirring mode/rate   37
Structural similarity   133
Sulfonic acids   178
Supersaturated solutions, aggregation   163
Supersaturation   159
Symmetry-breaking complexity system   53
Synthesis-resolution   84

Tadalafil (Cialis)   294
(R,R)-Tartaric acid system   245
Ternary system   24
Thiamphenicol   175
Threonine   19, 169
N-Tosyl-(S)-phenylalanine system   235
Tryptophan   87
Tyrosine   87

Valine   171

Werner's complexes   4

X-ray diffraction/crystallography   12, 233
XRPD   12, 181

Printing: Krips bv, Meppel
Binding: Stürtz, Würzburg